Newnes Building Services Pocket Book

Newnes

Building Services Pocket Book

Edited by
John Knight and
Peter Jones

Newnes
An imprint of Butterworth-Heinemann
Linacre House, Jordan Hill, Oxford OX2 8DP
225 Wildwood Avenue, Woburn, MA 01801-2041
A division of Reed Educational and Professional Publishing Ltd

A member of the Reed Elsevier plc group

OXFORD AUCKLAND BOSTON
JOHANNESBURG MELBOURNE NEW DELHI

First Published 1995
Reprinted 1999, 2000

British Library Cataloguing in Publication Data
Newnes Building Services (Mechanical)
Pocket Book
I. Jones, Peter II. Knight, John
696

ISBN 0 7506 2148 6

Typeset by Tek-Art, Croydon, Surrey
Printed in Great Britain

Contents

Introduction

This book is intended as a guide to students and as a manual for mid-career engineers engaged in the practical design of mechanical services in buildings. Environmental systems are covered, together with the design of plumbing and drainage installations. In particular, fire protection is dealt with, including smoke ventilation and a design procedure for sprinklers. Limitations on the space available have prevented the inclusion of steam and compressed air systems, which occur occasionally in commercial applications but are more usually associated with industrial processes.

All aspects of design are covered, from the initial strategic decisions, through detailed design, down to commissioning. It is assumed that the reader has some knowledge of the physical processes that occur in building services and has access to the CIBSE Guide and other relevant design information. The book is intended to supplement the information and procedures obtainable from the current CIBSE Guide and from related papers and design notes published by BSRIA, showing how these can be applied to the practical, detailed design of systems.

The first chapter deals with efficiency in the use of energy and with the environmental and hygienic considerations that are of increasing social importance. The performance characteristics of most common types of environmental system are discussed, together with procedures for determining design loads, component sizes and methods of control.

It is often wrongly assumed that the design processes associated with building services are simple and straightforward, compared with those of other types of engineering. This book addresses such a misconception and explains some of the complexities that arise when the continuously varying thermal loads in a building are assessed.

Many existing heating and air conditioning systems have inefficient and oversized central plant, mainly because of unrealistic performance specifications, arising from a lack of feedback from operating experience. These problems are also considered and appropriate solutions offered. The practical considerations and design techniques to rationalize the choice of components, aid prefabrication and reduce the time and cost of commissioning are dealt with in some detail.

Contributors

Bruce Boulton MIFireE, MSFPE, MIFS
Has over 40 years' experience in the fire protection industry, currently Chief Fire Protection Engineer with Haden Young.

Mike James
Served an apprenticeship with Stitson White followed by a career spanning over 30 years with G. N. Haden, Donald Rudd and Partners, Haden Young and currently with J. E. Greatorex as Senior Public Health Engineer.

Peter Jones MSc, CEng, FInstE, MASHRAE, FCIBSE
Currently in practice as an independent consulting engineer after extensive experience for most of his career as Air Conditioning Consultant with Haden Young Ltd. Author of two well-established text books on air conditioning.

John Knight FCIBSE
Currently practising as a consultant and lecturer after a career with Haden Young Ltd as Chief Engineer.

Ted Prentice CEng, MIEE, MCIBSE
Served an electrical engineering apprenticeship and specialized in control system design for road and rail vehicles before a career with Haden Young as Manager of the Controls and Instrumentation Department until retirement.

1 Energy and Environmental Considerations

1.1 INTRODUCTION

Up to now the building industry in this country has been influenced by short-term economic strategies and high interest rates in such a way that design decisions have been dominated by the need to reduce capital cost and space requirement to a minimum.

However, public opinion, long-term commercial viability and higher standards must eventually influence design to give greater consideration to such factors as:

(i) low energy use;
(ii) low maintenance requirement;
(iii) more flexibility of building use;
(iv) better system hygiene.

This section is devoted to design techniques which address these factors.

1.2 TYPICAL BUILDING ENERGY USE PIE CHART

Since lower energy use is likely to be the subject of increasing public attention it is useful to have an awareness of how various components of building services use energy, and where significant savings can be made. The pie charts in Figure 1.1 give a broad indication of the energy intake proportions of various elements of building services. This is based on a typical UK commercial office building air conditioned with a four-pipe dry fan coil system without heat recovery or free cooling features and 55 hour/week occupied period. The top chart is based upon the annual energy intake metered at the building. The proportions of energy usage by various components would vary for different types of building, usage and air conditioning system.

The lower chart is for the same building, but in this case the electrical power usage is based on the fossil fuel energy consumed at the power stations to produce the electrical energy required. The chart assumes that electrical energy delivered to the building requires three times as much fossil fuel energy per useful kWh as natural gas. This multiplier is affected by the average efficiency of electricity generating stations which is

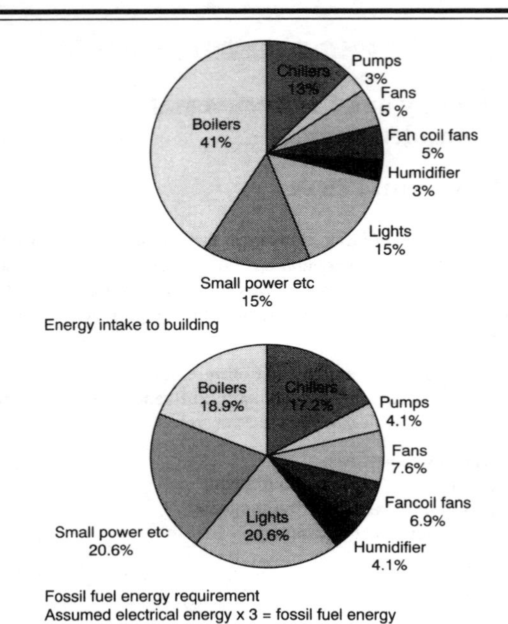

Figure 1.1 Building annual energy intake

changing due to the increasing proportion of gas-fired dual cycle stations. These must be considered as a rough guide and the multiplier should be based upon the latest information from UK energy statistics. From an environmental point of view the CO_2 emission (kg per kWh delivered) for electricity is also approximately three times that of natural gas based on the Building Research Establishment Environmental Assessment Method Report 1993 [1].

1.3 THE EFFECT OF DUCT SPACE ON THE ENERGY USED FOR AIR DISTRIBUTION

From the pie chart it can be seen that fan energy is a significant proportion of the building energy intake and a major part of this is expended in overcoming the resistance of ducts and duct fittings. For a given air flow this can be reduced in a number of ways, but unfortunately most of these have an effect on the planning and space allocation for services. Consideration should be given to the following factors when planning:

(a) Increasing duct diameter by 5 per cent reduces the energy in overcoming duct resistance by approximately 18 per cent.
(b) Feeding air into the middle of a duct instead of from one end, reduces the resistance of that duct (and consequently the energy required to overcome that resistance) to 12.5 per cent (result of half the run and half the velocity).
(c) In ceiling voids where space is restricted, splitting the distribution into several ducts instead of using one main duct, enables the air velocity and consequently the resistance to be reduced. (Two ducts of a given size have 25 per cent of the resistance of one duct of the same size, when conveying the same airflow.)
(d) Locating the air handling unit in the centre of the area to be served instead of at one end of the area reduces the horizontal duct resistance by up to 87 per cent for similar sized ducts. (The effect of half the travel and half the velocity.)
(e) The major part of duct resistance is usually in the fittings. Duct work should have a high priority when co-ordinating multiple services, to enable duct runs to be straight with the minimum number of directional changes. Sets in electrical cables incur no energy loss. Air duct sets incur a high energy loss.

1.4 FAN AND PUMP EFFICIENCIES

Pump efficiencies can vary considerably, dependent on the pump design and more importantly, where the operating point falls on the characteristic curve of the pump.

When selecting a pump, part of the system curve should be plotted to intersect the pump characteristic curve and so establish the probable operating point.

When selecting the duty of a pump it is normal to add margins to the design flow rate and system pressure drop to allow for balancing inaccuracies and system changes. These margins usually result in a shift in the operating point which, to some extent can be anticipated. It is worth checking the operating point for a number of different pumps to see which one gives the lowest power consumption. Figure 10.23 shows a typical pump characteristic curve, with design system curve and probable operating point.

Similar selection procedures apply to fans and these are dealt with in Chapter 6. There are, however, some important differences between selecting fans and pumps.

With pumps the outlet velocity has little significance on the power requirement or the flow generated noise. With fans the outlet velocity will have a considerable effect on both, requiring careful consideration when selecting a fan and designing the fan outlet connection. This is dealt with in in Chapters 6 and 7 but is mentioned here as it can have a marked effect on fan power and energy.

High efficiency electric motors are now readily available as an alternative to standard motors and will give a significant reduction in electrical consumption, as will avoiding oversized motors.

1.5 THE EFFECT OF VARIABLE FLOW AIR AND WATER DISTRIBUTION SYSTEMS ON ENERGY CONSUMPTION

Secondary chilled and heating water distribution systems are mostly designed with a constant flow rate and a constant pump

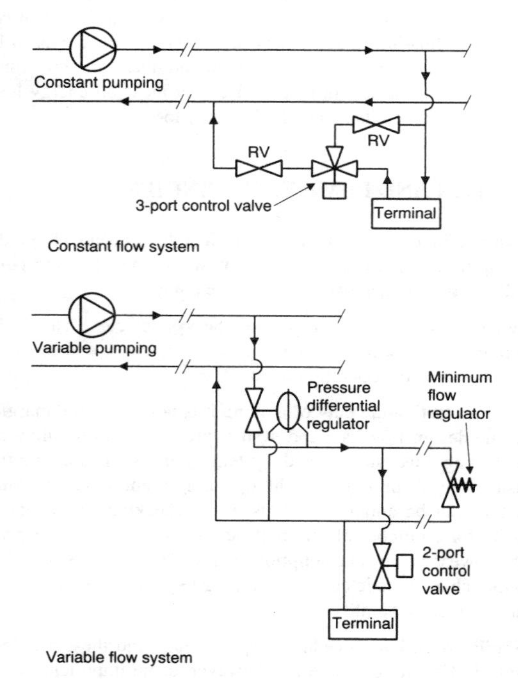

Figure 1.2 Constant and variable flow systems

head. Local unit control under this system is achieved by changing the flow rate through the terminal and by passing the balance of the flow directly back to the return main. This means that the pumping energy remains constant at any thermal load on the system and the pump flow rate must be the sum of all the terminal peak flow rates including calculation and selection margins plus a further margin for inaccurate balancing. In contrast, if the secondary distribution is designed as a variable flow system, where the whole system flow rate is the minimum required by the terminals (none being bypassed to the return main), then the pump duty is less and the power requirements reduces with the thermal load.

If a variable speed pump is used with this system the distribution pump energy is likely to be less than 15 per cent of the pump energy associated with the equivalent constant flow system. The hydraulic design of variable flow systems is a little more complex because the pressures vary with the load. This is dealt with in Chapter 10.

The distribution energy savings associated with variable air volume (VAV) systems, although significant, are not as dramatic as those associated with variable flow water systems. This is mainly due to the need to maintain high minimum air flows. VAV system design is dealt with in Chapter 3.

1.6 FREE COOLING AND THE ENERGY USED FOR REFRIGERATION

Free cooling is a term used for the reduction or elimination of mechanical refrigeration load by using outside air instead of recirculated air, or by the operation of the refrigeration plant as a thermosiphon, the compressor then being off. When used in an all-air system of air conditioning, such as most VAV plants, it means that if the outside air enthalpy is lower than the enthalpy of the return air there is a reduction of refrigeration load by using all outside air. When the outside air temperature is at or below that required of the cooling coil, the refrigeration load is zero and the refrigeration plant can be shut down. This temperature is typically about 10°C and by using the BSRIA data [2] on the occurrence of wet and dry bulb temperatures it can be seen that the dry-bulb temperature is at or below 10°C for about 59 per cent of the year.

With a dry fan coil system it is possible to use the primary air cooling coil to cool the secondary water to the fan coils (without using the refrigeration plant) when the outside air temperature is below about 5°C (approximately 23 per cent of the year) This technique also saves boiler energy since the cooling coil is then preheating the air. The arrangement and control of dry fan coil systems are described in Figure 3.37.

1.7 NIGHT OPERATION OF FANS TO PRECOOL BUILDINGS

Modern, tight, well-insulated buildings lose heat very slowly after the air conditioning system shuts down. Air plants are usually arranged so that their air intake and exhaust dampers shut when the air handling plant stops, and with the building entrance doors closed there is very little infiltration or heat loss.

After sunset there is generally a marked drop in outside air temperature giving scope for precooling the building structure and contents by operating the fans at low speed with the dampers on full fresh air. This operation requires monitoring controls which permit the fans to operate only when:

(a) The building is likely to require cooling the following day.
(b) The building is not cooled sufficiently to require heating the following day.
(c) The temperature rise (exhaust air – outside air) is high enough to justify operating the fans.

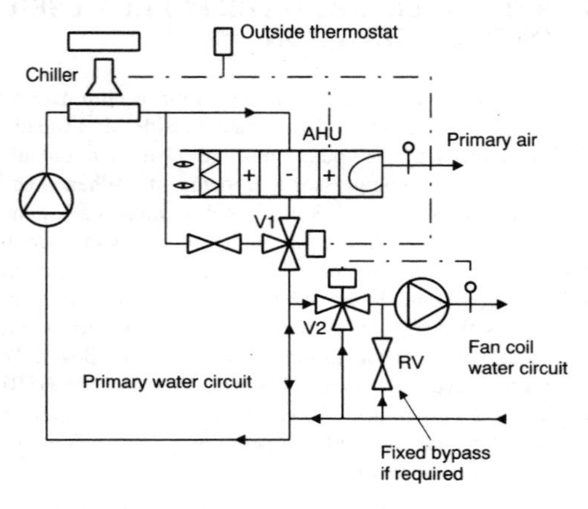

Figure 1.3 Dry fan coil system free cool cycle

EXAMPLE 1.1

Examine the economics of night operation of the primary air system of an office fan coil air conditioning system to reduce the refrigeration energy. Assume that by operating the extract fan without the supply fan running, but with the air handling unit mixing dampers set on full fresh air it is possible to ventilate the space at the rate of ll s^{-1} m^{-2} (treated floor area) and when operating in this mode the extract fan operates against a total pressure of 0.5 kPa and at an efficiency of 60 per cent.

Assume that the mechanical cooling plant requires an electrical power input of 1 W to provide 2 W of cooling when operating during the occupied period

1. Electrical power to operate fan

   $$= \frac{1.0 \times 0.5}{0.6} = 0.83 \text{ W m}^{-2}$$

2. Heat removed from structure by night fan operation per degree air temperature rise is given by

   $$\frac{1 \text{ l s}^{-1} \times 1°\text{C} \times 358}{273 + 23°\text{C}} = 1.22 \text{ W m}^{-2}\ °\text{K}^{-1} \text{ temperature rise}$$

3. Electrical power to operate cooling plant to remove 1.22 W m^{-2} (heat) = 0.61 W m^{-2} minimum.

4. Air temperature rise (outside air temperature to common extract temperature before fan) required to ensure viable operation of extract fan for night structure precooling

   $$= \frac{0.83}{0.61} = 1.36°\text{K (allow 1.5°K)}$$

In this example if we assume that air is likely to be extracted from the building at a temperature of 20°C then it would be worth operating the extract fan when the external air temperature dropped below 18.5°C. It is likely that outside air temperature will fall at a higher rate than the internal MRT of the building so that economic operation of the extract fan will be sustained through the night until the temperature rise dropped below 1.5°K or normal day operation of the air conditioning plant commenced. Note that calculation of the air temperature rise likely to be achieved is very complex and unreliable because of assumptions necessary to deal with the distribution of an air flow over the internal building surfaces. It is a simple matter to measure and monitor the performance actually achieved. The actual setting of the controls for the night cooling regime should be based on this measured performance of the building and system.

1.8 HEAT RECOVERY FROM EXHAUST AIR AND THE ENERGY REJECTED FROM REFRIGERATION SYSTEMS

Heat recovery from refrigeration

Where heat rejection from the air conditioning refrigeration plant is via a closed water condenser circuit this can be used to provide low grade heat for air preheating or reheating. In the winter when the heat rejected from the refrigeration plant may not be sufficient to provide heat required, it may be increased as required by removing heat from the exhaust air with a cooling coil (Figure 1.4).

Any system of heat pumping requires electrical energy to operate the refrigeration compressor; e.g. refrigeration with a cooling coefficient of performance (COP) of 3 will require 1 kW of electrical power to produce 4 kW of heating, quite apart from the extra fan power needed to overcome the resistance of the additional heat exchangers (such as the cooler coil mentioned above) necessary to effect the heat transfer process.

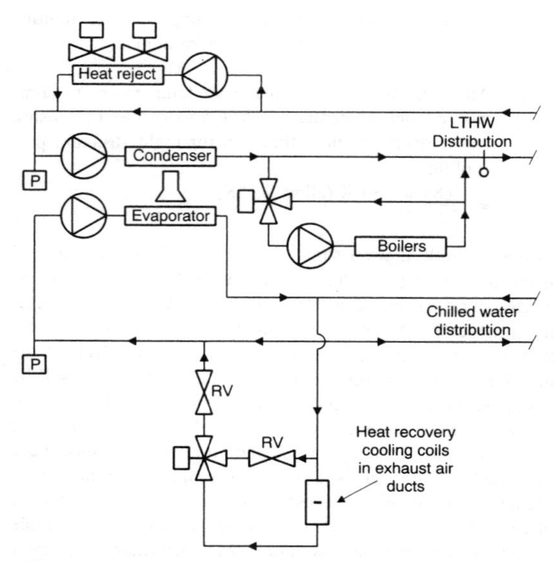

Typical arrangement of heat recovery from closed condenser water circuit with sequential control of heat rejection, exhaust air heat recovery, and boiler heat make-up.

Figure 1.4 Heat recovery from condenser water

1.8.1 Heat recovery from exhaust air

The typical energy intake pie chart, Figure 1.1, indicates that, for a normal building occupied for 10 hours per day, boiler energy is the largest single item. Such energy is used in the following ways:

(1) Offsetting the fabric and infiltration heat losses which occur 24 hours/day, although during the day these may be largely offset by casual heat gains.
(2) Hotwater supply to taps, showers and kitchen equipment, including standing losses. However, this load is often dealt with independently by point-of-use electric or gas water heaters and not from the main space heating boiler plant.
(3) Heating the outside air introduced by the air conditioning or mechanical ventilation system. This occurs only during the period of plant operation. It is a high load, usually more than the fabric and infiltration load. It can be reduced significantly by transferring heat from the exhaust air to preheat the outside air through a system of heat exchangers.

$$\text{Efficiency of heat transfer (sensible)} = \frac{t_l - t_{oa}}{t_{ea} - t_{oa}} \qquad (1.1)$$

where:

t_l is temperature leaving outside air heat exchanger

t_{oa} is outside air temperature

t_{ea} is temperature of exhaust air entering exhaust air heat exchanger.

Run-around coil systems comprise a preheat coil in the outside air intake protected by a prefilter and usually controlled by a three-port control valve to prevent overheating (Figure 1.5). In some applications this control is omitted and the pump switched off above limiting outside air temperatures. This coil is connected via a pumped, water/glycol circulation piping system to a cooling coil in the exhaust air.

Prefilter protection is important for the exhaust air coil which will run wet with condensate in cold weather and therefore be more subject to fouling. Typical sensible heat transfer efficiencies for run-around coil systems range from about 0.35 to 0.6, dependent on the coil design.

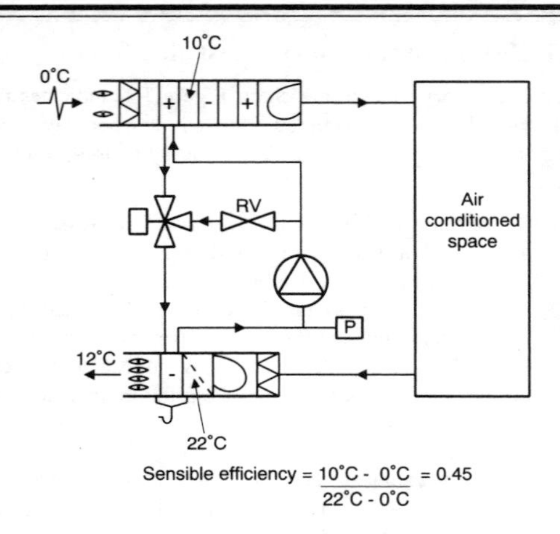

Figure 1.5 Run-around coil system

1.8.2 Economics

Similar to other heat recovery systems, run-around coil systems require additional energy and maintenance which must be taken into account:

Extra fan energy is needed to overcome the resistance of the coils and their prefilters, and this occurs for the operating period of the air plants. Extra fan power (kW)

$$= \frac{\text{extra air total resistance (kPa)} \times \text{air flow (m}^3\,\text{s}^{-1})}{\text{fan efficiency} \times \text{motor efficiency} \times \text{drive efficiency}} \qquad (1.2)$$

The circulating pump power is relatively small and occurs only when heat is being recovered.

The boiler power should be reduced by the heat transferred through the run-around system at design conditions. The run-around system can also be used to transfer cooling capacity from the exhaust air in hot weather when the outside temperature is more than 2°C above the exhaust air temperature. This gives a small reduction in the required mechanical refrigeration capacity.

Unlike other direct heat recovery arrangements, run-around coils can be used with a number of separate recovery and preheat coils on a common system and the recovery coils can be remote from the preheat coils.

1.8.3 Heat wheels

Heat wheels use a large rotating wheel of extended surface material through which the incoming outside air and the exhaust air pass in adjacent ducts. Various types of extended surface are used, some of which are hygroscopic and transfer moisture as well as heat from the exhaust air to the intake air. Heat transfer capacity is controlled by varying the speed of rotation. Cross-contamination will occur to some extent and good prefiltration of both air streams is essential. Even so, the extended surface requires periodic cleaning (Figure 1.6).

Heat wheels can attain higher heat transfer efficiencies than run-around coils but require more space. Note that high efficiency is sometimes not necessary, e.g. primary air for a four-pipe fan coil system is rarely required above 10°C and run-around coil efficiency is entirely adequate for this application.

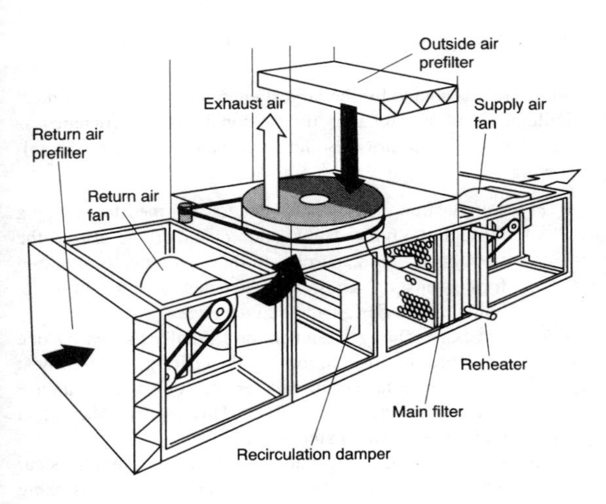

Figure 1.6 Heat exchange wheel

1.8.4 Plate heat exchangers

These comprise plates of non-corrosive material closely spaced in such a way that the spaces are alternately connected to incoming outside air and outgoing exhaust air, on a counter-flow basis. Plate heat exchangers are limited in the sizes available. They have a higher efficiency than run-around coils, need filter protection, cleaning facilities, a condensation drain and

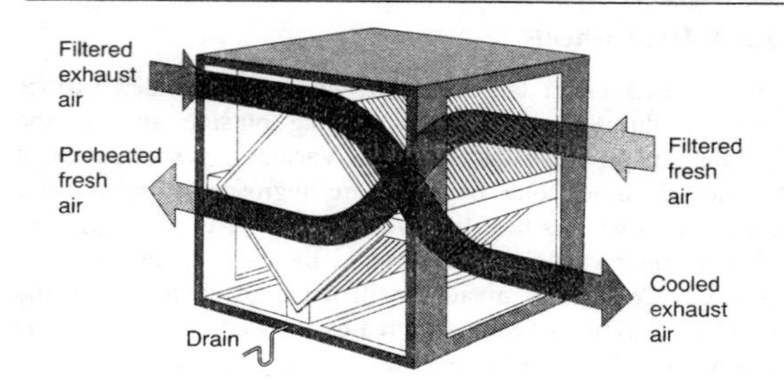

Figure 1.7 Plate-type heat exchanger

outside air ducts must be adjacent to exhaust air ducts (Figure 1.7).

1.9 BOILER SEASONAL EFFICIENCIES

Boiler fuel accounts for the largest energy requirement of most buildings and maximizing the seasonal boiler efficiency is therefore a high priority. Boiler inefficiency is due to the following:

(1) Combustion inefficiencies due to unburned fuel passing through the boiler combustion chamber and also to the flow of air that is surplus to that theoretically required for combustion. These inefficiencies are inherent in the boiler burner design and maintenance.
(2) Excessive flue gas temperature, usually caused by one or more of the following: overfiring, fouled heat exchange surface or insufficient velocity and turbulence at the heat exchange surface. This is also inherent in boiler design and maintenance.
(3) Boiler casing, convection and radiation losses occur when the boiler contains hot water, whether it is being fired or not. They can be minimized by selecting fewer, larger units as these present less surface area for the loss of heat than do a greater number of small boilers. Better insulation also reduces this loss.

Control system and boiler selection should ensure that the minimum potential boiler output is always selected to deal with the load as this reduces the losses and gives more stable control of flow temperatures. The control system should also ensure that the water circulation of unfired units is stopped after a short interval, to dissipate the residual heat. As a boiler cools down its temperature and its heat loss reduce.

(4) Boiler casing convection losses. When a boiler stops firing and still contains hot water, heat is lost from the heat exchange surface by the air drawn through the boiler by the chimney draught and, in the case of forced draught burners, by the prepurge operation of the fan. These standing losses may be minimized by the measures described for reducing boiler casing convection losses and also by the boiler burner design. Features such as air inlet and flue outlet shut off dampers and bottom flue outlet connections are effective in reducing standing losses. Reducing the water and thermal capacity of the boiler heat exchanger also reduces the standing loss that occurs at each shutdown.

(5) System operating temperatures. Boiler efficiencies are increased by reducing the water temperatures passing through them. The increased temperature difference between combustion gas and water enables more heat to be extracted from the combustion gas, particularly at the back end of the boiler where the gas is at a lower temperature. Condensing boilers are the ultimate example of exploiting this fact and their application is described in Chapter 2.

(6) Boiler efficiency. The efficiency quoted in manufacturers' catalogues only applies to a new, clean boiler unit operating continuously at full load. In practice, boilers operate for most of the time in a partly fouled condition at part load and often intermittently. Hence, the factors described above should always be considered since they may reduce the manufacturers' quoted efficiency of say 80 per cent to a seasonal operating efficiency of less than 60 per cent.

1.10 REFRIGERATION SEASONAL EFFICIENCIES

The design and operation of refrigeration plant is dealt with in Chapter 8. However, a few notes on the factors relating to seasonal efficiency are relevant here.

Chillers, like boilers, are often judged on their catalogue efficiency, usually quoted as the coefficient of performance (COP).

$$\text{Theoretical COP} = \frac{\text{refrigerating effect (kW)}}{\text{rate of work done in compression}} \tag{1.3}$$

$$\text{Practical COP} = \frac{\text{refrigeration capacity (kW)}}{\text{input power (kW)}} \tag{1.4}$$

The COP quoted will be for the unit operating continuously at full load with clean heat exchangers, for defined chilled water temperatures and flow rates and defined condenser cooling water temperatures and flow rates. In the case of air-cooled machines, entering air temperatures and flow rates are relevant. However, the COP of a water chiller is significantly affected by changes in these conditions and by part load operation. The effect of these changes are:

(1) Chilled water temperature and flow rate affect the heat transfer in the evaporator, and keeping the chilled water flow temperature as high as operating conditions permit is beneficial in reducing compressor energy consumption.
(2) Lowering the condensing pressure reduces compressor energy consumption and is possible at reduced load and lower ambient air temperatures. This can only be fully exploited by using microprocessor-controlled electronic expansion valves. These give a marked increase in COP under most operating conditions. Chillers only operate at full load for very short periods of the year. The part load COP is therefore more important. Extra energy is used to start chillers and it is therefore an advantage to reduce the starts/hour to a minimum. This is achieved by increasing the water capacity of the system. In the case of chiller with positive displacement compressors, the more steps of control the better; this is achieved by multiple compressors or sometimes by incremental speed changes.

$$\text{starts/hour} = \frac{900q}{s\,m\,t} \tag{1.5}$$

where:

q is the average cooling capacity of the last step of refrigeration (kW)
s is the specific heat of the fluid (4.18 kW kJ^{-1} for water)
m is the mass of fluid in the primary refrigeration circuit (kg)
t is the allowable change of primary fluid temperature: This is usually determined by the refrigeration control system and is typically around 1°K.

1.11 AIR SYSTEM SUPPLY TEMPERATURES

Fan energy is a major proportion of the electrical energy input to a building.

Where air is the medium used to remove room sensible heat

gains, a reduction in supply air temperature of 1°K can reduce the fan energy by about 10% (12°C supply air temperature for a room temperature of 22°C is fairly typical for mixing type air distribution). The choice of air supply temperature needs careful consideration as it affects the system energy consumption in many ways, e.g. reducing the supply air temperature for cooling application, has the following effects.

(i) reduces fan size and fan energy;
(ii) requires increased cooling coil surface for given chilled water or evaporating temperatures;
(iii) may require lower chilled water temperatures with possible reduction in chiller COP and consequent increase in refrigeration compressor power and energy consumption;
(iv) psychrometric process may show an increase in the air latent heat load due to a lower room humidity;
(v) duct temperature rise may increase because the air is at a lower temperature and also because the ducts are either smaller or operate at a lower velocity. This effect can be offset by improved insulation, but needs evaluating.

1.12 WATER SYSTEM TEMPERATURES

The energy consumption of pumps associated with heating and cooling systems is a relatively small proportion of the electrical energy input to building services. The choice of system temperatures does, however, merit consideration as it also affects distribution thermal losses and primary plant efficiencies.

1.12.1 Heating systems

In systems with extensive mains such as district or multi-building installations, the heat loss from the mains may be a high proportion of the heat delivered. Remember that while the mains loss only reduces slightly with increasing ambient temperature, the mean annual rate of heat delivered is probably less than half of the design maximum. For example, a system with a design mains loss of 20 per cent of the design heat load may have a seasonal mains loss of about 50 per cent of the heat delivered. Mains losses can be reduced by:

(i) Designing water distribution systems with as high a water temperature differential as is economic, having regard to a practical selection of equipment, particularly control and regulating valves.
(ii) Using the lowest temperatures that are compatible with

economic terminal selection.

(iii) Using higher standards of insulation, particularly at pipe supports.

1.12.2 Chilled water systems

The mains energy loss (heat gain) with chilled water distribution systems is usually very small and unlike heating, is not a significant factor in water temperature selection. Chilled water temperatures are usually determined by cooling coil selection to meet psychrometric requirements and are dealt with in Chapter 8.

1.13 BUILDING AIR LEAKAGE

Building and other regulations have progressively increased the standards of building insulation over the years, but at the time of writing there has been no standard for the maximum leakage acceptable for a given differential pressure between outside and inside a complete building. The incidence of excessive air leakage causing problems on recently erected buildings is quite high, in some cases affecting occupant comfort, and resulting in a failure of the heating system to provide design internal temperatures under windy conditions. The infiltration air loss is a major part of heat requirement, and it is up to the services designer to highlight its significance to the rest of the building design team.

EXAMPLE 1.2
See Figure 1.8.

Take a one metre length of an intermediate floor of a building. Assume the following:

Floor to ceiling height is 2.7 m.
Depth from outside wall is 6 m.
A window 1 m wide and 1.5 m high, U value 3.4 W $m^{-2}K^{-1}$.
External wall 1 m wide and 1.2 m high, U value 0.45 W $m^{-2}K^{-1}$.
Infiltration rate 1 air change per hour (CIBSE Table 4) [3].

The steady-state fabric heat loss would be

(Wall: 1.2 $m^2 \times 0.45$ W m^{-2} K) + (window: 1.5 $m^2 \times 3.2$ W m^{-2}K) = 5.34 W K^{-1}

The infiltration heat loss would be

(1.0 ac/h $\times$ 2.7 m $\times$ 6.0 m $\times$ 1.0 m $\times$ 0.33 W m^{-3}K) = 5.35 W K^{-1}

i.e. almost 50 per cent of heat loss in this example is due to infiltration.

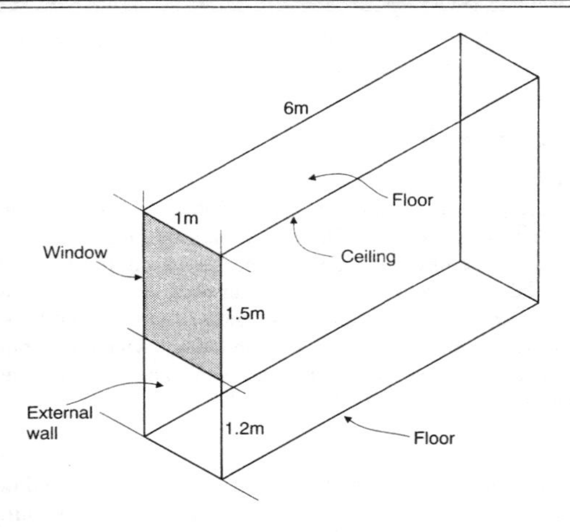

Figure 1.8 Typical building module

1.14 COMFORT FACTORS

Comfort is a complex subject about which much has been written and it is dealt with in the CIBSE guide in Section AI. The main factors appear to be air temperature, mean radiant temperature (i.e. the mean surface temperature of the room enclosure) and air movement. Relative humidity is a less important factor and many occupants are unable to detect quite large variations in humidity. What is important is the need of the occupant to change the air temperature and/or air movement to suit his or her preference and this need may change from day to day. The range of adjustment of a thermostat by an occupant should, however, be restricted to avoid overheating or overcooling the structure beyond the possibility of quick recovery. 19°C to 24°C is a suitable range. Noise is an important factor the mechanics of which are dealt with in Chapter 3. It is important to remember that it is possible to have rooms too quiet for comfort so that the occasional intermittent noises that occur in most buildings become disturbing and speech privacy is not obtained. NR 35 is a typical sound pressure level for good quality cellular offices, with NR 42 for open plan offices. It is usual to rely on the air conditioning or ventilation system to provide this level of background noise. Alternatively, the noise may be electronically generated and transmitted over the PA system. Rooms requiring sound pressure levels below NR 30 require specialist design for the structure, finishes, partitioning and services.

Other factors affecting comfort are colour and lighting levels which are outside the scope of this book.

1.14.1 Fresh air supply

Need

A constant introduction of fresh air (often described as outside air) is necessary for all buildings during occupied hours. It is necessary to dilute body odours and gases emitted by some foams, adhesives and other modern finishes. A useful measure of the adequacy of fresh air introduction in an occupied building is the measure of CO_2 concentration, which generally should not exceed about 0.2 per cent and certainly not approach the threshold limit value of 0.5 per cent.

How

Fresh air introduced naturally through cracks around windows and doors used to be entirely adequate for heated-only buildings of traditional construction and proportions. However, present construction methods often use relatively impervious external cladding panels with sealed windows. This tight construction combined with deeper building plans means that normal infiltration of fresh air is inadequate for the needs of the occupants, and fresh air must be introduced by mechanical ventilation or purpose-designed openings and shafts.

Amount

The minimum amount of fresh air required is about 8 to 12 $l\ s^{-1}$ per person or 1.4 $l\ s^{-1}\ m^{-2}$ floor area, whichever results in the greater air quantity. There is, however, considerable pressure to increase this standard, particularly where significant smoking is likely. It should be borne in mind that the air quality at breathing level is also very much a function of the efficiency of the air distribution in the occupied space.

1.14.2 Distribution efficiency

With conventional air mixing terminals their ability to disperse and mix contaminants uniformly to avoid local high concentrations is important. With displacement or low velocity terminals their ability to envelope occupants in supply air flow and to avoid disturbing natural convection and stratification is important if maximum efficiency is to be achieved. The efficiency of the air distribution is probably as important a factor as the fresh air quantity, and merits more attention than it gets. Examples of this are the problems that occurred with many of the earlier VAV systems using fixed geometry supply diffusers and high turn-down ratios in air flow.

1.14.3 Fresh air source

The position of the fresh air intake in relation to sources of external contamination is important. Generally, intakes at low level, near busy roads should be avoided and intakes at the top of buildings are preferred, but care is necessary to minimize the risk of picking up contaminants and odours from other roof level exhausts, possibly from adjacent buildings.

1.14.4 Exhaust outlets

Exhaust discharges which are hot, smelly, humid, toxic or contaminated should be jetted vertically upwards at high velocity (6 m s^{-1} to 12 m s^{-1} or higher) dependent upon the volume and proximity of obstructions above the discharge level. The use of any form of cowl to prevent the ingress of rainwater must be avoided and rainwater should be drained at the base of the discharge duct. Normal air conditioning or ventilation exhausts should also be made at significant velocity to minimize the risk of them being carried to intakes by the wind and by the surface effect of walls and roofs.

1.15 THERMAL BEHAVIOUR OF BUILDINGS

1.15.1 Buildings

Building regulations already enforce high standards of building insulation and the trend is for these to improve. However, more thought needs to be given to the other thermal characteristics of buildings such as the mass, the time constant and the admittance factors of the internal surfaces, as these can have a considerable affect on energy consumption, summertime temperatures, the need for mechanical cooling and external design criteria. For example, buildings with a high mass and consequently a high thermal capacity on the room side of the insulation will change temperature at a slow rate for a given energy gain or loss. A similarly sized building with lightweight construction on the room side of the insulation will be subject to much larger temperature changes under the same conditions.

1.15.2 Time constant

The time taken for a mass to cool to its ambient temperature if exposed continuously to its initial cooling rate is called its time constant. For example, if a building heated to 21°C when the ambient temperature is 0°C has an initial cooling rate when the heating is switched off of 0.5°K h^{-1} then it may be said to have a time constant of

$$\frac{21°C}{0.5} = 42 \text{ hours}$$

In practice, buildings do not cool or heat uniformly and the cyclic heat flow in and out of a structure is a complex matter. Nevertheless, the simplified time constant described above is a useful measure for comparing the thermal performance of buildings.

1.15.3 Heating energy effects

Generally, a building with a smaller time constant will cool to lower temperatures and consequently lose less heat when the heating system is off. Such a building will, for the same reason, recover temperature at a higher rate when preheating and, because of its lower heat loss during the off period, less heat energy is required when subject to intermittent heating. For continuously heated buildings the time constant make little difference to the heat energy required (see Figure 1.9).

Thermal mass outside the insulation will only have a small effect on the time constant but will act as a flywheel on reducing the effects of swings of outside air temperature.

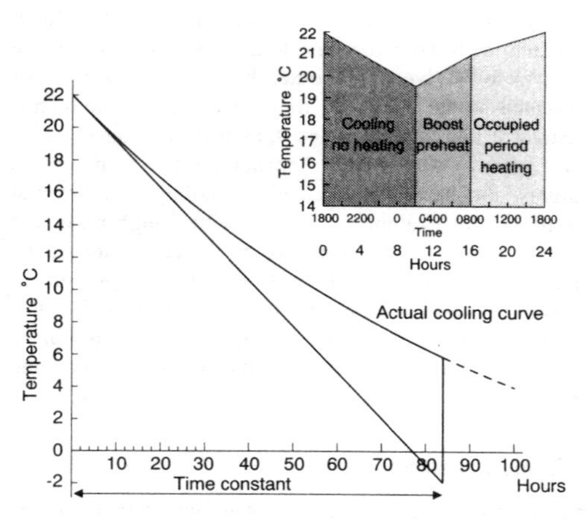

Approximate mean structure temperature for a building with an 84-hour time constant when subject to continuous heat loss, and inset when subject to normal intermittent heating regime

Figure 1.9 Building thermal characteristics

1.15.4 Cooling energy effects

A building with a high thermal capacity and long time constant will store more cooling capacity if ventilated with outside air at night, than will a similarly sized building with a low thermal capacity and a short time constant. The cooling capacity or negative cooling energy stored will reduce the mechanical cooling energy required the following day, or alternatively reduce the room temperature if not mechanically cooled. This method of exploiting the thermal storage of buildings has not been fully appreciated until recently.

Higher insulation standards

(1) reduce heating energy requirements;
(2) increase the importance of reducing uncontrolled air infiltration;
(3) do not significantly reduce cooling energy requirements. Under some circumstances it may increase them;
(4) increase time constants.

Higher building air tightness

(1) reduces heat energy requirements significantly;
(2) does not significantly reduce cooling energy requirements;
(3) increases the need for effective control of ventilation rates, whether mechanically, or by natural wind and stack effect;
(4) enables more effective control of comfort and cleanliness to be obtained;
(5) reduces the design margins needed on the size of terminal and plant equipment.

Higher building mass inside insulation

(1) increases heating energy requirements of intermittently heated buildings;
(2) reduces cooling energy requirements, especially where controlled night cooling ventilation is used.

Higher building mass outside insulation

(1) gives a flywheel effect which enables less rigorous external design temperatures to be used;
(2) has minimal effect on heating or cooling energy requirements.

Glass, area, shading and orientation

has a significant effect on comfort, cooling and heating energy requirements.

Room height

Increasing the room height generally increases the heat loss and heat energy requirement but it also increases the effectiveness of stratification as a means of reducing cooling energy when the appropriate type of air distribution is applied. This is usually low level, low velocity supply, coupled with a high level extract. This system is discussed in greater detail in Chapter 3.

1.16 MAINTENANCE AND HYGIENE

Maintenance is a major part of the cost of operating HVAC systems. Lack of maintenance is also a frequent cause of problems, poor environmental conditions, high energy costs and sometimes illness. It is therefore important that HVAC systems are designed to minimize the amount of maintenance required, and especially maintenance in the occupied space. Adequate access must be provided to all equipment requiring regular maintenance.

(i) Terminal equipment requiring regular maintenance which is sited in congested ceiling voids is likely to be neglected, and its maintenance become costly.
(ii) Terminal equipment which recirculates room air via the ceiling void needs frequent cleaning and the room dirt deposited in the ceiling void makes maintenance a messy operation.
(iii) Terminal cooling equipment which dehumidifies involves cooling coils which run wet and need drains. The drain trays and drain pipes can become a source of odours and possible infection, unless regularly cleaned.
(iv) All air ducts, particularly extract ducts, accumulate dirt. Even supply air carries a certain amount of dirt which has bypassed the filter media. Ductwork should have access opening for regular inspection and cleaning.
(v) Humidifiers, particularly of the steam injection type, are a frequent source of internally wet ducts. The control, design and arrangements should ensure that:

- steam and air are mixed before contact with cool surfaces. The control of steam injection always ensures that the mix does not approach saturation.
- care is taken with the insulation of the duct and air handling unit, to ensure that there are no cold bridges. These might produce local internal surface temperatures close to the dew point of the air.

The combination of internal duct deposits and moisture will inevitably lead to mould and bacteria growths. Serious consideration should be given to minimizing the amount and period of humidification required. Indeed, for many comfort applications the need for any at all is questionable.

1.17 THERMAL OVERLAP

Thermal overlap is the term used to describe controlling the output of a terminal unit by cancelling cooling energy with heat energy or vice versa. This is obviously a very wasteful control method but unfortunately it is still very commonplace in many air conditioning systems and even some heating systems. It may be reduced or eliminated by using more sophisticated controls and/or recommissioning control set points to match the actual requirements of the building, rather than those given in the original design brief.

The following are examples of common air conditioning systems which are inherently wasteful under most operating conditions, with some suggested modifications which would reduce the energy consumption.

EXAMPLE 1 CONSTANT VOLUME REHEAT

All-air, constant volume reheat systems of air conditioning. These are probably the most wasteful. The system supplies the sum of the maximum cooling requirements of every zone which is cancelled as required under partial load conditions by reheat locally at the zone.

Suggested improvements

(1) Convert to VAV.

(2) Control supply air temperature within limits to just meet the maximum requirement of the least favourable zone.

Any reheat control valve reaching a fully closed position initiates a progressive decrease in supply air temperature until the valve starts to reopen. *All* reheat control valves opening more than say 10 per cent of their full movement initiate a progressive increase in supply air temperature until any valve backs off. This type of control reduces reheat to a minimum.

EXAMPLE 2 TWO-PIPE NON-CHANGEOVER AIR WATER SYSTEMS

Two-pipe fan coil or induction systems. Heating is provided by primary air at constant volume and a variable temperature that is adjusted automatically with changes in external temperature, according to a predetermined schedule. This invariably means cancelling some of the heating capacity with cooling by the terminal cooling coil under its local control. This type of system is likely to exist only in older office buildings.

Suggested improvements

(1) Control primary air temperature to suit the requirements of the least favourable terminal control by removing the outside compensator winter control resetting the supply air temperature and substituting control from the status of the terminal control valves.

Any single valve reaching a fully closed position initiates a progressive increase in supply air temperature until it starts to reopen.

All valves opening more than say 10 per cent of their full movement initiate a progressive reduction of supply air temperature until any valve backs off. This type of control ensures that the primary air temperature is continually adjusted to the lowest acceptable temperature.

(2) If possible zone primary air on an aspect basis with separate zone reheaters and distribution ductwork. Each zone controlled as described above.

EXAMPLE 3 VAV PERIMETER HEATING

VAV with individually controlled diffuser terminals and perimeter heating under single zone control, scheduled against outside temperature. This common VAV system suffers from thermal overlap in that the perimeter heating must be controlled at a temperature high enough to offset the heat losses of the least favourable module, plus reheating the minimum airflow from the VAV diffusers. This means that in most modules where the net heat loss is lower, the overheating at the perimeter is cancelled out by increased air flow and cooling from the VAV diffuser.

Suggested solutions

(1) Perimeter heating should, in addition to overall outside temperature compensation, be fitted with some form of local modular control which will reduce the output of a perimeter heating module before cooling overlap occurs. This usually takes the form of a direct action thermostatic radiator valve with a control band between say 20°C and 22°C.

(2) The thermal overlap is significantly reduced by ensuring that the supply air temperature is set as high as possible without causing any overheating. This may be achieved automatically by a control system which resets the supply air temperature within suitable limits from the status of terminal diffuser throttling devices.

EXAMPLE 4 VAV WITH REHEAT BOXES

These systems are arranged so that, with reducing cooling load at the terminal, the primary airflow reduces down to a minimum predetermined by ventilation requirements. Upon further reduction of cooling load, the air supply is reheated at constant volume.

Suggested solutions

As for improvement 2, given in Example 3.

EXAMPLE 5 HEATING SYSTEMS

Heating systems. All LTHW heating systems should have the flow temperature reset lower as the outside temperature increases, to a schedule suited to the terminal characteristic. This reduces mains losses and improves boiler efficiency but it is not sufficient as the only control. Terminal control is essential to avoid overheating and thermal overlap caused by occupants opening windows.

Suggested solutions

Radiators and natural convectors should always be fitted with terminal control. This usually takes the form of thermostatic radiator valves.

References

1. Building Research Establishment Environmental Assessment Method Report 1993.
2. Building Services Research and Information Association TN 2/77 Coincidence of dry and wet bulb temperatures.
3. Chartered Institution of Building Services Engineers Guide Book A Table A 4.12 Empirical Values for air infiltration.

2 Heating System Design

2.1 BUILDING HEAT LOADS

2.1.1 Heat loss components

Fabric loss

Heat losses through the fabric of a building are usually calculated on the basis of steady-state heat flow through all the externally exposed surfaces of the building plus the heat flow through internal walls where the room the other side is at a lower temperature.

Air Infiltration Loss

Outside air infiltrates through apertures in the building fabric under the combined effects of wind and stack effect which cause a pressure difference between inside and outside the wall or roof. These pressure differences may be positive or negative, but any pressure difference causes air flow through the building.

Heat is required to raise the temperature of the infiltrating outside air to room temperature.

Ventilation air heat load

Outside air has to be introduced into the building to provide oxygen for occupants, and to dilute, disperse and remove contaminants and odours. This air may be introduced naturally by opening windows or mechanically by fans; either way it needs to be considered separately from infiltration and requires heating to room temperature.

Humidification heat load

In some cases it is necessary to increase the moisture content of the air in order to raise the relative humidity of the room air in winter. Heat is required to raise the temperature of the water and to change its state from liquid to vapour. In the case of steam injection this heat is provided at the steam generator but in the case of water spray and extended wetted surface humidifiers the heat is usually provided at the air preheater upstream of the humidifier.

Hotwater supply

Generally it is preferable to provide independent local systems for dealing with hotwater supply (HWS) rather than serving the hotwater service from the main space heating system. Separating the space heating system from the hotwater service systems has the following advantages:

- The space heating system can be designed to operate at lower and variable water flow temperatures giving reduced mains losses and higher boiler efficiencies.
- The space heating system can be shut down in summer reducing mains and standing losses as well as improving the annual boiler plant load factor and seasonal efficiency.
- Decentralized HWS systems have lower mains and standing losses than central HWS systems fed from space heating boiler plant.

The heat load required for HWS can be calculated by the procedure and curves given in CIBSE Guide B [1], but for combined heating and HWS systems consideration should be given to the fact that the HWS load peaks only for short periods which do not coincide with space heating load peaks. The HWS load to be dealt with by the boilers can be reduced by the use of low capacity, fast recovery hotwater storage heat exchangers combined with a control system which sheds part of the space heating load to give priority to HWS recovery.

Material Stock

An unusual heat loss which must be considered for warehouse and industrial buildings having a significant throughput of materials; these may enter the building at external temperatures and will absorb heat in the building. The additional heat load can be calculated:

$$\text{heat load kW} = \text{mean intake of goods (kg/s)} \times \text{specific heat (kJ/kg K)} \qquad (2.1)$$

The mean intake of materials may be the maximum over a period of a few hours for materials with a high heat conductivity and exposed surface area or over several days for materials in insulated packing boxes.

2.1.2 Fabric heat loss calculations

These are computed from the sum of (the area × U values × the difference in the air temperature inside and outside):

$$\text{watts} = \text{area (m}^2\text{)} \times \text{U value (W/m}^2\text{ K)} \times (t_i - t_o) \qquad (2.2)$$

Areas should be calculated from internal dimensions taken finished floor to finished floor and wall to wall (see Figure 2.1).

U values for different types of construction, exposure and direction of heat flow can usually be obtained from the building designer or CIBSE A3[2]. With highly insulated buildings it is very important to investigate and calculate the effect of

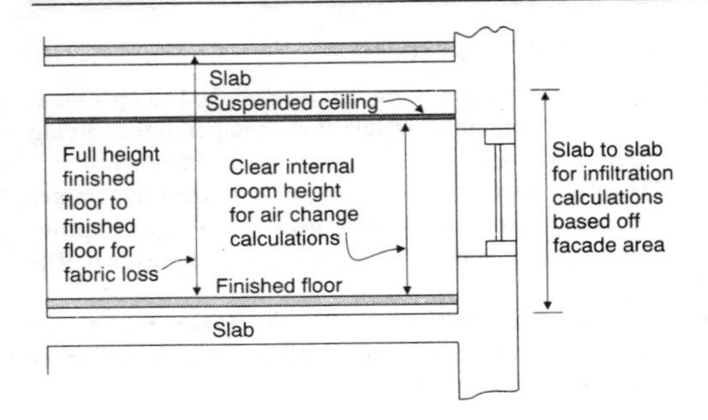

Figure 2.1 Room dimensions for heat loss and air change calculations

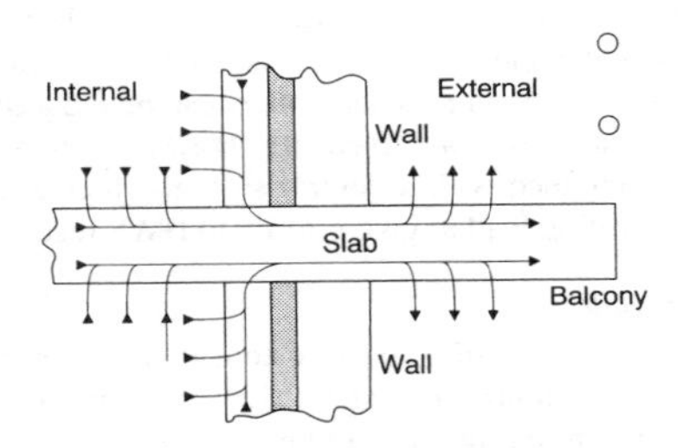

Figure 2.2 Example of lateral heat flow at heat bridge fin

heat bridges or fins as they give rise to a higher heat flow than would be obtained by considering the individual elements separately. This is due to the lateral heat flow which occurs either side of the heat bridge, (Figure 2.2). Example calculations are given in CIBSE A3.

U values for uniform non-bridged walls and roofs are simply calculated:

$$U = 1/R_{si} + R_1 + R_2 \ldots + R_a + R_{so} \qquad (2.3)$$

Where R_{si} = internal surface resistance (m^2 K/W)
R_1, R_2, etc. = the thermal resistances of structural layers m^2 K/W
R_a = air space resistance (m^2 K/W)
R_{so} = external surface resistance (m^2 K w^{-1})

2.1.3 Internal temperature

For most normal construction buildings with single-storey height rooms the internal air temperature for fabric and infiltration head losses may be assumed to be the comfort resultant temperature for occupants. However, the air temperature corresponding to the resultant temperature may need to be adjusted in the following cases:

(a) Rooms with an unusually high proportion of cold surfaces, such as windows, and/or poorly insulated walls, floors or ceilings, require a higher internal air temperature to compensate for the lower mean radiant temperature.
(b) Rooms heated by radiant surface require a lower internal air temperature to compensate for the higher mean radiant temperature (MRT).
(c) Rooms of buildings which are heated occasionally will require a higher room air temperature to compensate for the lower MRT. In many cases admittance factors will be more appropriate than U values (which are based on steady state heat flow). Buildings which are unheated for long periods of time will have lower internal surface temperatures and will absorb more heat during the building heat input period. (See CIBSE A5 - 6, Cyclic Conditions.)
(d) The occasional need for internal air temperatures higher than usual is discussed above. These apply to both fabric and air filtration heat losses.

In those cases where it is judged that the air temperature (t_{ai}) may need to be significantly above or below the comfort dry resultant temperature (t_{res}) a more complex procedure for determining fabric and air heat losses is required (See CIBSE A1.3 and A9 - 3 to A9 - 6) [3] Briefly:

Calculate for each room values for

$$\frac{NV}{3\Sigma(A)} \quad \text{and } \Sigma(AU)/\Sigma(A) \qquad (2.4)$$

where N = Number of air changes/hour (h^{-1})
V = Room volume (m^3)
A = Area (m^2)

Refer to tables A9.1 to A9.2 in the CIBSE Guide to obtain values for F_1 and F_2 [4].
calculate the fabric heat loss $Q_u = F_1 (t_{ai}-t_{ao}) \Sigma AU$ (2.5)
and the air heat loss $Q_a = F_2 (t_{ai}-t_{ao}) \frac{1}{3} NV$ (2.6)

Note that in these formulae t_{ai} is assumed to be equal to t_{res}. F_1 and F_2 correct for actual differences between t_{ai} and t_{res} and t_{ei}

where t_{ai} = internal room air temperature (used for calculating air loss)
t_{ao} = outside air temperature

t_{ei} = environmental room temperature (used for calculating fabric loss)
t_{res} = dry resultant temperature (used for assessing comfort) (see CIBSE Table A1.3).

Note
This complex procedure is only justified for unusual buildings.
Examination of Tables A9.1 to A9.6 indicates when F_1 and F_2 become significant corrections.

(e) Vertical temperature gradients are significant in high rooms resulting in higher temperatures and heat loss in the higher levels to achieve the required comfort resultant temperature at occupant level. The increased heat loss due to temperature gradients, is considerably affected by the type of heating system, convective heating with high terminal leaving air temperatures giving the worst temperature gradients and low level radiant heating having the least effect. CIBSE Table A9 gives an indication of allowances to be added to normally calculated heat losses for different room heights and heating systems.

Temperature gradients in high spaces may be considerably reduced by the use of anti-stratification fans which draw air from close to the ceiling and redistribute it at low level.

Although for reasons given above, steady state heat flow is not really appropriate for today's well-insulated long-time constant buildings (see Figure 3.7) experience has shown that it gives safe results for most buildings. In practice most heating problems are usually due to excessive infiltration and/or unsuitable control systems.

2.1.4 Outside design temperatures

These need to be carefully considered taking into account the use of the buildings, its thermal capacity, the heating system operating programme, the overload capacity of the system when relieved of its air heating load, and the location of the building.

CIBSE gives data to make a judgement of the outside temperatures to be used. Bear in mind the following considerations:

- The thermal capacity of the building structure damps out the effect of daily cycles of external temperature so that a 24-hour or longer lowest mean temperature is appropriate and CIBSE Table A2.2 [5] is normally used for calculating fabric heat losses.

- For ventilation air heating loads there is no such storage effect and air heating coils need to heat outside air from the lowest outside temperature that occurs while the plant is operating. This temperature can be obtained from Kew from CIBSE Table 2.8 [6] which suggests a figure of –6.6°C. Correction for other locations can be assessed from Table A2.2.
- For infiltration air heating loads it could be argued that the internal thermal capacity of the building also damps the effect of outside temperature cycles, but this must be to a lesser extent than for fabric losses. CIBSE Table A2.2 is usually used and in conjunction with the usual assessment of air change rates gives satisfactory results for tight buildings. When more accurate procedures for calculating air infiltration rates become available it will be necessary to use outside air temperatures lower than the CIBSE Table A2.2 24-hour mean outside air temperatures for calculating the infiltration air loads.

2.1.5 Infiltration air flow

The air flow into a building through leaks in the external envelope is very difficult to assess, yet with present insulation standards it may account for a greater heat loss than the conduction heat loss through the external envelope.

It is normal to use the air change rates given in CIBSE Table A4.12 as a basis for calculating the infiltration air flow. Experience has shown that this data is safe to use for well-constructed buildings and is probably excessive for some types of external sealed cladding now commonly used for commercial buildings.

It would be more logical to base infiltration rates on the external facade area, type and exposure than on the volume of the room but unfortunately data in this form is not readily available at the time of writing.

A procedure for calculating air flow through building apertures due to wind pressure and stack effect is given in CIBSE A4-4 to 13 [7]. Window and cladding manufacturers often publish leakage performance data obtained under laboratory test conditions.

The problem with using this data is that it often gives an infiltration air flow which is too low in practice because at design stage it is not possible to assess the additional leakage due to other construction joints. These can vary considerably depending on the quality of building workmanship and the detailing. These factors are not known at the design stage. Until more experimental data is available winter infiltration air flow rates must be based on CIBSE A4.12[3].

Infiltration due to wind pressure occurs only on the windward faces of a building and not on the leeward faces. The full infiltration heat load has to be allowed for the heater emitters but for the whole building the infiltration heat load may be about half the sum of the module infiltration loads. The whole building heat load should be used for assessing boiler power not the sum of the module heat loads or terminal outputs.

CIBSE formulae A4.13 and Figure 4.5 give procedures for calculating the whole building infiltration rate.

2.2 HEAT LOAD PROFILES AND INTERMITTENT HEATING

Graphical presentations of the building heat losses and gains over 24 hours are useful for selecting the optimum boiler sizes and heat input programme. Graphs to show both the peak winter design condition and the mean winter conditions are necessary as it is more important to deal with the mean winter condition as efficiently as possible. The peak design condition rarely occurs consequently a lower full load efficiency is acceptable (see the weather tape load diagram, Figure 2.3).

The heat load profiles shown in the following example are simplified to make them easier to construct and are subject to the following qualifications:

(i) They assume the external air temperature remains constant over the 24-hour period whereas it usually cycles through several degrees (see CIBSE Figure A2.2).[8]
(ii) The infiltration rate is assumed as constant whereas it may be lower when the building is unoccupied.
(iii) The internal gains are assumed to be constant over the occupied period whereas in practice a proportion of the gains are absorbed by the building mass during the day and released after the occupied period.
(iv) Solar gains are ignored.

EXAMPLE 2.1

Consider a modern commercial building mechanically ventilated or air conditioned with the following characteristics expressed per m^2 of heated floor area for peak winter design conditions:-

Infiltration air heat loss	10 W/m^2 (1 air change/hour in half the building)
Fabric heat loss	20 W/m^2 (current insulation standards)
Mechanical ventilation	38 W/m^2 (1.4 l/m^{2}s)

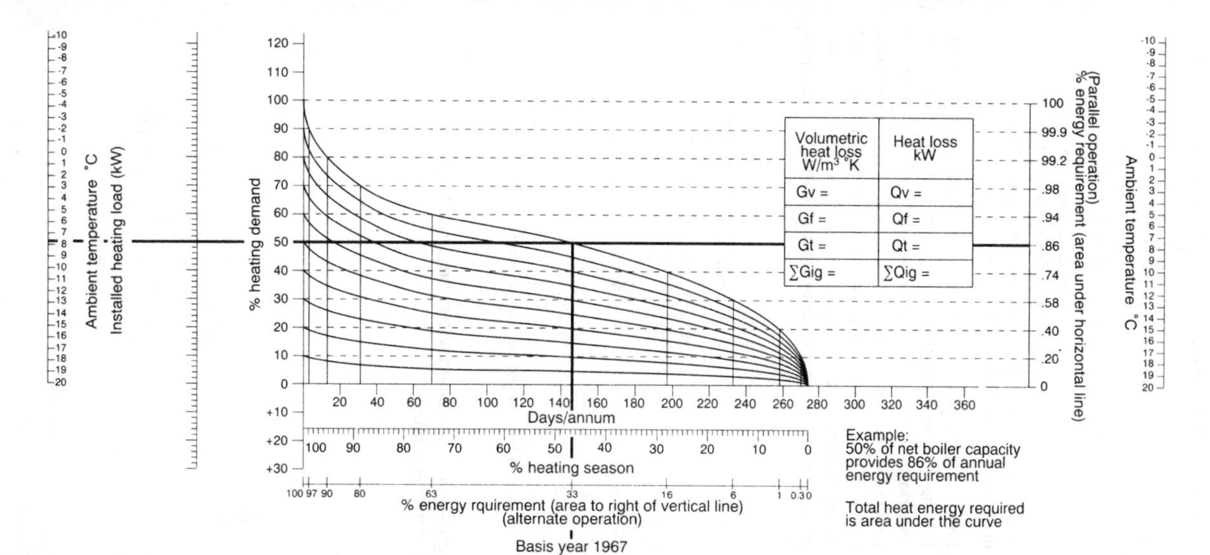

Figure 2.3 Weather tape — heating analysis chart

Average internal gains (min)	20 W/m² (lights, people, small power)
Building cooling rate	0.25°C/h (well-insulated heavy structure)
Occupation period	0800 h to 1800 h

Figure 2.4 shows the profile of heat loads imposed at full winter design conditions with fabric and infiltration heat loss dealt with by continuous heating. Figure 2.5 is a similar diagram at mean typical winter conditions.

Figure 2.6 is similar to Figure 2.4 but with an intermittent heating regime switching the heating off at 1800 hours requiring 48 W/m² heat input starting at 23.25 hours to restore the heat lost during the off period before occupation. It would be possible to extend this boost into the occupied period thereby reducing the maximum heat load but at the penalty of higher internal temperatures during the occupied period.

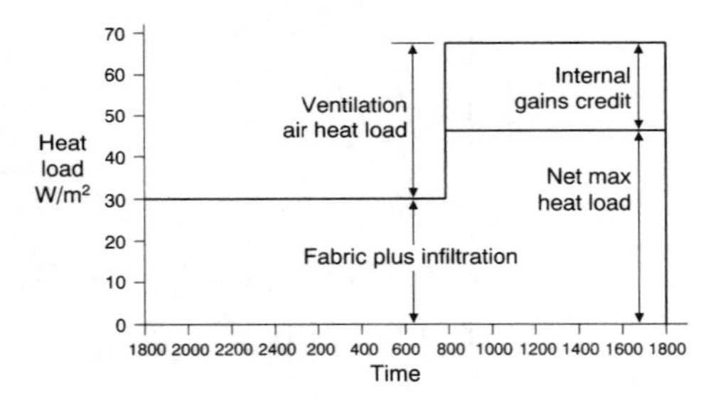

Figure 2.4 Simplified winter design load profile continuous heating

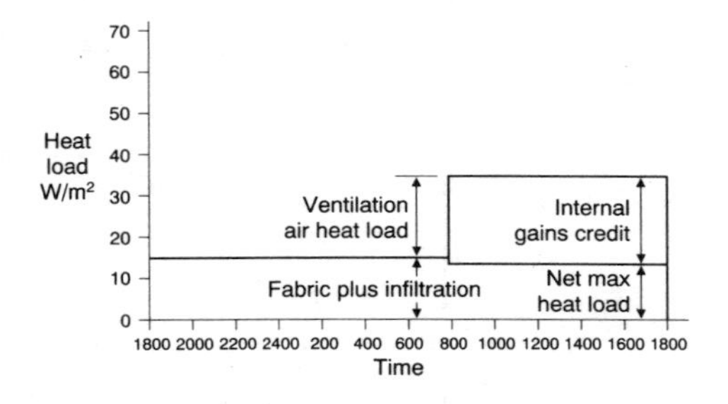

Figure 2.5 Simplified typical winter load profile continuous heating

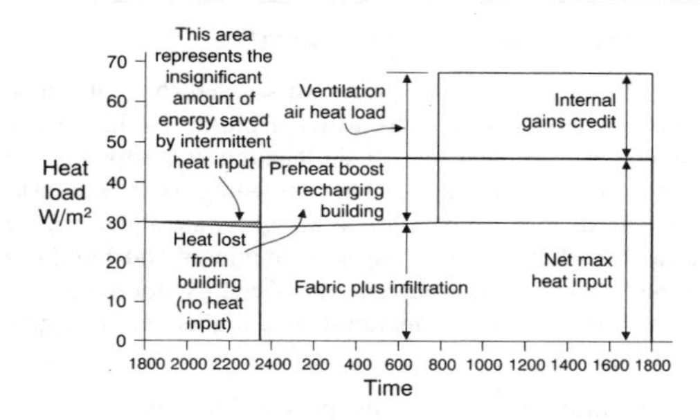

Figure 2.6 Simplified winter design load profile: Intermittent heating with 48 W/m² available input

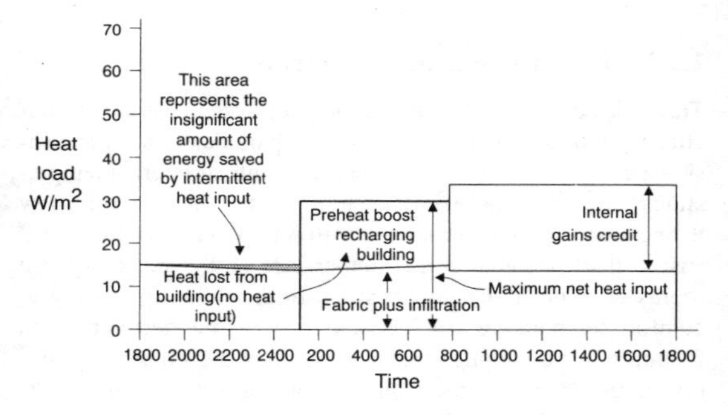

Figure 2.7 Simplified typical winter load profile: Intermittent heating with 30 W/m² available input

Figure 2.6 demonstrates that there is little or no benefit in operating intermittent heating at near design winter conditions. The very small saving in heat loss achieved by allowing the building temperature to fall is offset by the higher infiltration air loss during the boost period when the internal temperature is increased to restore the heat lost from the structure.

Figure 2.7 shows the heat load profile obtained by operating an intermittent heating regime at a mean winter condition. This requires no increase in the required heating capacity. Although this operation is viable, the energy savings compared with continuous operation are very small.

2.2.1 Intermittent heating comments

From the simplified heat load profiles discussed above it can be seen that for the present well-insulated buildings the energy savings obtained by shutting off the heating and allowing the building to cool are very small. Older buildings with low insulation standards lose temperature at a higher rate and will give larger savings. Low thermal capacity buildings and buildings where the internal mass is partially insulated by floor and ceiling voids will also lose temperature at a higher rate and give larger savings.

Intermittent heating is worth while provided it does not involve increasing the heating capacity required to deal with the peak design winter condition. This means that at near design conditions the system offsetting the fabric loss should operate continuously.

2.2.2 Intermittent heating controls

These decide at what time the heating system needs to start after night or weekend shut down. The parameters for this decision are complex, and simple measurements of internal temperatures are not necessarily appropriate. The room low temperature limit control is equally important as this must ensure that the internal air temperature in the most exposed rooms does not fall below a value from which a recovery to comfort temperature can be achieved before occupancy. In Example 2.1, assuming a preheat period of 5.25 h and a recovery at 0.25°C/h the low limit should be set at about 20°C $-(5.25 \times 0.25) = 18.69$°C. In practice due to the complexity and lack of information about the thermal behaviour of a building, the settings of optimum start and low limit controls should be found experimentally from thermographs in the most exposed rooms under cold weather conditions.

2.2.3 Design implications

The ventilation air plant, if separate from the system offsetting the fabric and infiltration losses, should be controlled to give the lowest supply air temperature that can be distributed in the room without causing discomfort (about 11°C). The fabric heating system (with local terminals controlled from room temperature) should be sized to have capacity to deal with the peak design fabric heat loss plus reheating the ventilation air up to room temperature. In this way some limited cooling is available from the ventilation air in mild weather when the internal gains exceed the fabric heat loss. The weather tape Figure 2.3 and load profile Figure 2.7 suggest this will be the case for a major part of the winter period.

The heat load profiles demonstrate that the internal gains offset a major part of the heat requirement during the occupied period and to ignore them (as is customary) must result in oversized boiler plant with consequent loss of efficiency. Care must be taken to correctly assess the minimum internal gain that is likely to occur (this will usually be from the lights) and also to ensure that they are switched on early enough for their gain to become effective before starting the ventilation plant. The minimum internal gain can be credited to the required boiler power but not to the heat emitters (some rooms may not have internal gains). The assessment of boiler power should take into account the effect of one boiler unit failing although the risk of this happening on the rare occasion of design full load is remote. A margin of about 20 per cent should be added to the net heat requirement to allow for some fall off in boiler performance, the possibility of heat losses being higher than those calculated, due to higher U values, excessive air infiltration and/or severe external conditions and also the possibility of internal gains being less than estimated.

2.2.4 Air heat recovery

The heat load profile shows that by adding a 50 per cent efficient air-to-air heat recovery device such as a run-around coil system the net maximum heat load could be reduced by 19 W/m^2 (39.6 per cent).

2.3 HEAT EMITTER CHARACTERISTICS

2.3.1 Direct gas-fired air heaters

These comprise air intake filter, combustion chamber with natural gas burner, capacity and safety interlock controls, fan chamber discharge plenum for either long throw air grilles or duct connections. These units do not use a heat exchanger, the combustion products (mainly water vapour and CO_2) mix with the incoming fresh air. These units are only suitable for make-up fresh air applications in large buildings such as factories, warehouses and shopping centres. They cannot be used in a recirculation mode because of the build-up of CO_2 and humidity, but some units are designed to deliver relatively small amounts of high temperature, high pressure air via specially designed high induction nozzles which entrain and mix large volumes of secondary room air.

These units have very high efficiencies and a fast control response. They must be combined and interlocked with a matching extract system.

2.3.2 Indirect air heaters

These are similar but use a heat exchanger with a flue which avoids contamination from products of combustion. They may be oil or gas fired. Similar units may have finned heat exchangers for steam or hot water instead of oil or gas firing. Indirect air heaters have a wider application as they can be used with air recirculation.

2.3.3 Warm air heaters

Warm air heating units for large rooms have to a large extent taken over from centralized steam and water systems with multiple terminals, largely because of their lower initial cost, reduced design time, fast response and simplicity. The main drawback with warm air heating compared with radiant heating in high rooms is the higher temperature gradient and heat loss.

2.3.4 Air distribution

With warm air heating it is very important to select the air outlets correctly taking into account supply air temperature, throw, height above floor level and sound power.

Unless the supply air mixes thoroughly with the room air before it loses momentum, its buoyancy will take over and it will rise, increasing the temperature gradient and failing to achieve comfort temperatures at occupant level. This phenomenon is a common cause of problems with warm air systems, and may be caused by a combination of the following:

(a) Leaving air temperature too high.
(b) Air outlets too large (outlet velocity too low).
(c) Air outlets mounted too high.
(d) Extract (return air) position too high.

If the air outlet velocity is too high discomfort many result from excessive air movement (see Figure 2.8.)

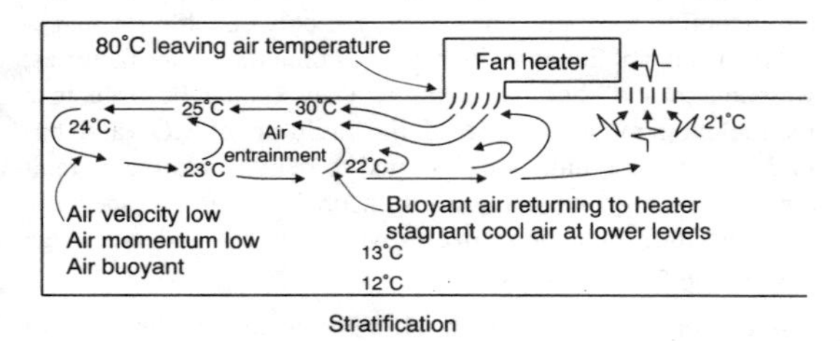

Figure 2.8 Typical air movement pattern obtained with high leaving air temperature and inadequate mixing

2.3.5 Radiant heating low temperature panels

These take the form of metal sheets heated by conduction from hot water, steam or direct gas flue pipes which are bonded to them. The units are usually mounted a high level and used for industrial type applications.

A major part of the total heat emission is radiant but the proportions of radiant and convected heat vary depending on the plane in which the panels are mounted and the efficiency of the back insulation (see Figure 2.9).

The convected heat is usually not desirable as it contributes to high temperature gradients and high heat loss.

Care must be taken in the configuration and mounting height of the radiant strip. High radiation intensity on the head of occupants can lead to discomfort depending on duration of occupancy. Most manufacturers give guidance on spacing and mounting height. CIBSE A1 - 7 [9] gives comfort guidance for radiant heat. It is also necessary to avoid mounting close to external walls as the radiant heat transfer will increase the wall heat loss by raising the internal surface temperature of the wall or window.

Areas of high air infiltration such as those close to external doors are not effectively dealt with by radiant heat and require local warm air heaters. In high rooms low temperature radiant heating is usually more efficient than warm air heating because of the higher MRT. Lower air temperature and reduced temperature gradient significantly reduce the heat losses. However, the capital cost of radiant heating is usually higher than warm air heating and the fans in warm air heaters can, in some cases, be used in the summer to improve comfort by increasing air movement.

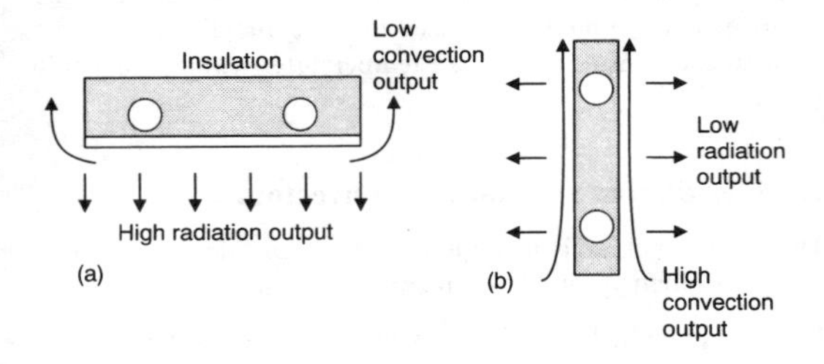

Figure 2.9 (a) Insulated radiant panel. (b) High total output non-insulated panel

2.3.6 Embedded panel radiant heating

This is another form of low temperature radiant heating which is becoming more popular. Low temperature hot water pipes in sinuous coils are cast into the floor screed over insulation: 15 mm steel or copper pipes used to be used at about 200 mm to 300 mm centres but today crosslinked polypropylene or similar plastic tube is used in conjunction with special reinforced screeds over insulation. A diffusion barrier coating on the tube prevents the passage of oxygen and other gases through the pipe wall, thus avoiding potential corrosion problems elsewhere in the system.

The main application for embedded panel heating is for continuously heated rooms which are not subject to high intermittent heat gains and where the heating needs to be entirely unobtrusive.

Warm floors give increased comfort provided they are not too warm. The main disadvantage of embedded panel heating is its high thermal capacity and hence slow response to deal with sudden load changes.

2.3.7 High temperature radiant heating units

These are either gas-fired refractory or infra-red electric elèments backed by a reflector. Their output is mainly radiant but some convection does occur. The gas-fired units require minimum ventilation rates to provide combustion air and dilute the products of combustion, which are usually exhausted via ridge ventilators. Spacing, mounting height and angle are critical to avoid overheating, fire risk and discomfort.

They are useful for providing local comfort in unheated areas. Other applications are for buildings such as churches and community halls which are used intermittently. In these cases it is often not possible or economic to warm the building structure to any extent and the heaters need to shine directly at the occupants to compensate for low air temperatures and building cold surfaces.

2.3.8 Radiators and natural convectors

The steel panel radiator although often considered unacceptable on aesthetic grounds has many advantages.

(a) It provides heat at the perimeter where the heat loss occurs.
(b) Its emission is partly radiant offsetting cold surfaces of the window.
(c) It is individually controllable either automatically, if thermostatic rad valves are used, or manually.

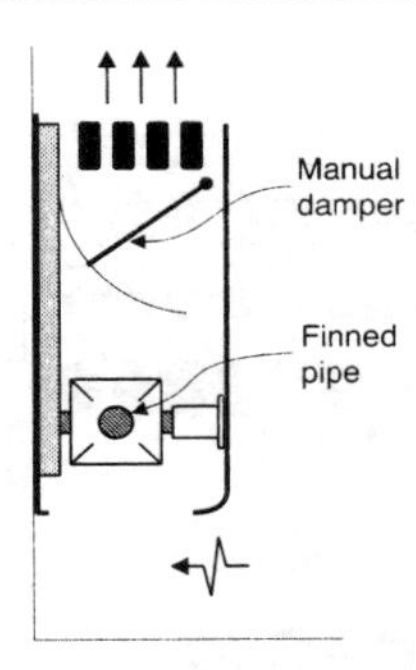

Figure 2.10 Damper controlled convector

(d) It has no moving parts to generate noise.
(e) It requires no maintenance apart from occasional re-painting.

When noise problems occur they are usually due to free air in the water or cavitation at the control valve due to excessive pressure drop. Radiators can be unobtrusive if carefully planned and selected. Natural convectors usually comprise a steel casing housing an aluminium finned copper tube. They can be arranged as separate units or in a continuous line wall to wall. Their output is mainly convective. Control is either water side via control valves or air side by means of a manual damper which when closed stops air circulation through the finned heat exchanger. Unfortunately the air side control does not stop the heat output completely. The damper is rarely a tight shut off and the steel casing temperature increases with the damper closed resulting in waste and overheating (see Figure 2.10).

2.4 HEAT PRODUCER CHARACTERISTICS

These are mostly so-called boilers. Water heaters would be a more correct term unless they generate steam. Other heat sources are heat recovery from exhaust air, condenser heat rejection from refrigeration plant, heat pumps using outside air or water as the low grade heat source, direct electric resistance heating, combined heat and power where heat is used from the exhaust gas and jacket of the engine driving the generator for electrical power.

2.4.1 Water heaters

Cast iron sectional boilers ranging an output from about 9 kW to 1500 kW (see Figure 2.11).

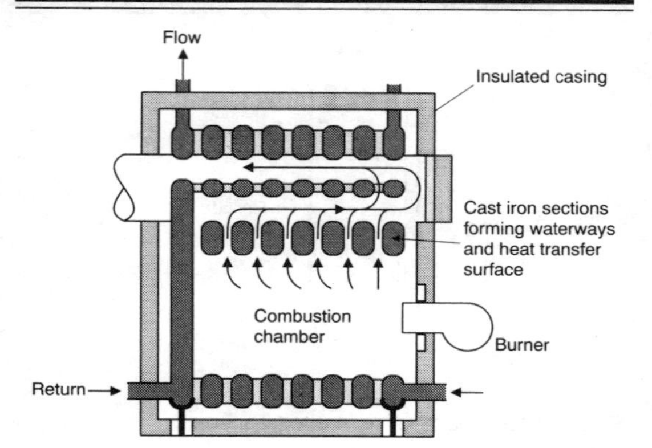

Figure 2.11 Typical cast iron sectional boiler

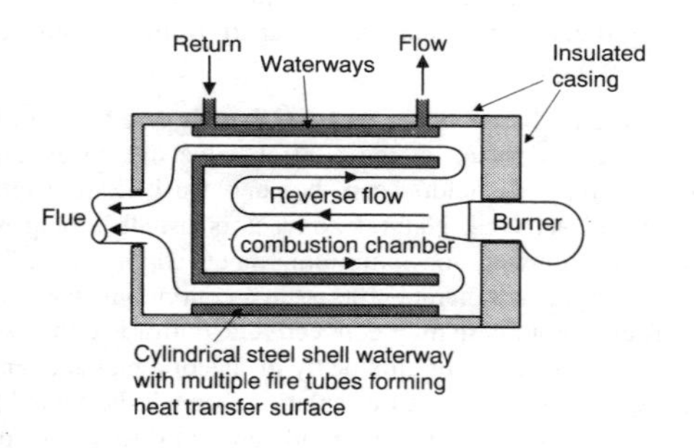

Figure 2.12 Typical steel shell boiler

Steel shell boilers in which the main convective heat transfer surface takes the form of multiple steel tubes carrying the products of combustion to the flue, these range in output from about 50 kW up to over 12 000 kW (see Figure 2.12).

Water tube boilers in which the main convective heat transfer surface is in the form of forced circulation water tubes, these range from domestic size units which usually use copper tubes with fins, to power station sized units with nests of steel water tubes. The main characteristic of water tube boilers is their small water content which gives them a fast response to load changes and reduces their standing heat losses. The small water content also makes them sensitive to water flow rate, if the flow drops below a critical rate for a short time local overheating, steam production and possible failure of the tubes may occur (see Figure 2.13).

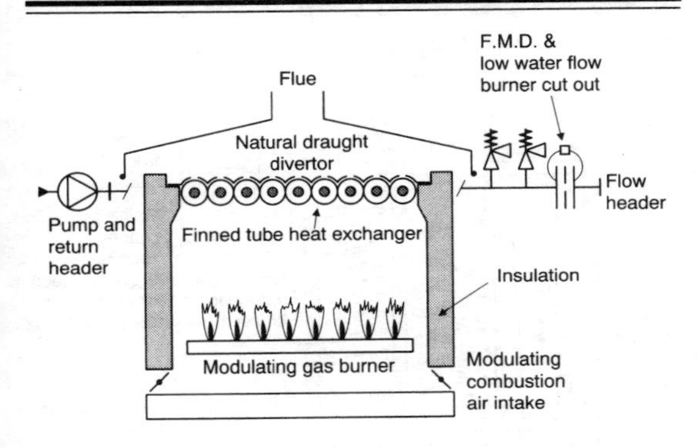

Figure 2.13 Typical water tube boiler

Most of the boiler types described above may be fired by gas, oil or solid fuel. The output when fired by gas is often slightly lower than when fired by oil or solid fuel which burn with a more luminous flame giving a higher radiant heat transfer in the combustion chamber.

Condensing boilers

Unfortunately due to their higher capital cost condensing boilers are rarely used in the United Kingdom yet they can provide very significant savings in gas consumption especially when matched to a suitable heating system. They are confined to clean burning fuels like natural gas.

The construction of the main boiler convective heat transfer surface is similar to conventional boilers described above but the products of combustion are then passed through a separate additional heat exchanger constructed from corrosion resistant material such as stainless steel or glass-lined cast iron. This final heat exchanger cools the product of combustion below 100°C thereby recovering the latent heat and some sensible heat from the flue gas and increasing the boiler efficiency by approximately 10 per cent (see Figure 2.14).

To obtain the maximum benefit the system design should include the following features:

(a) The operating water temperatures should be kept as low as possible; in particular the return temperature at full load should not exceed about 50°C.
(b) The control system should ensure that the return water temperature reduces with falling load. Note that the normal bypass three-port terminal control increases the return water with falling load.

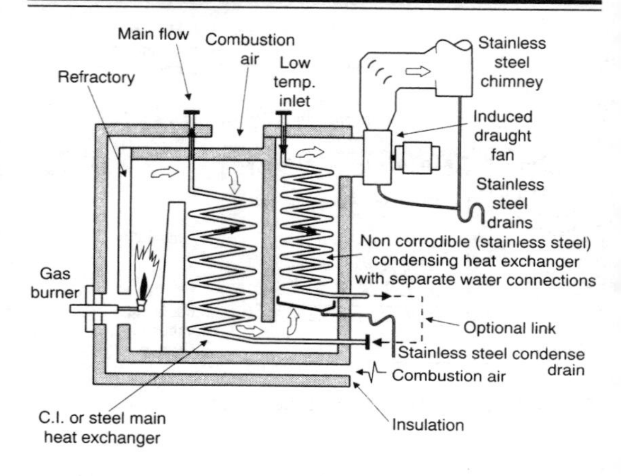

Figure 2.14 Typical condensing boiler

(c) The boiler selection and circuitry should ensure that the lead condensing boiler units operate at the lowest water temperature and carry the annual base heating load, with lower cost non-condensing boilers dealing with the peak lopping (see Figure 2.15). In practice if 50 per cent of the installed boiler capacity is in condensing units they will handle more than 86 per cent of the annual heat energy requirement leaving only 14 per cent to be handled by the less-efficient non-condensing units (see weather tape Figure 2.3).

2.4.2 Heat pumps

These usually operate on a vapour compression cycle described in section 8.1. The evaporator cools a low grade heat source such as outside air or water and the condenser becomes the heat source. Some are equipped with refrigerant reversing valves which reverse the roles of evaporator and condenser when the unit is required to cool instead of heat. The use of heat pumping is most economical when heating and cooling are required at the same time which is often the case with air conditioned buildings during mid-season.

2.4.3 Electric thermal storage

Electric resistance heating is more economical if the electrical energy is used at night or other times when it is available at a special low tariff. The heat is stored either by heating water in

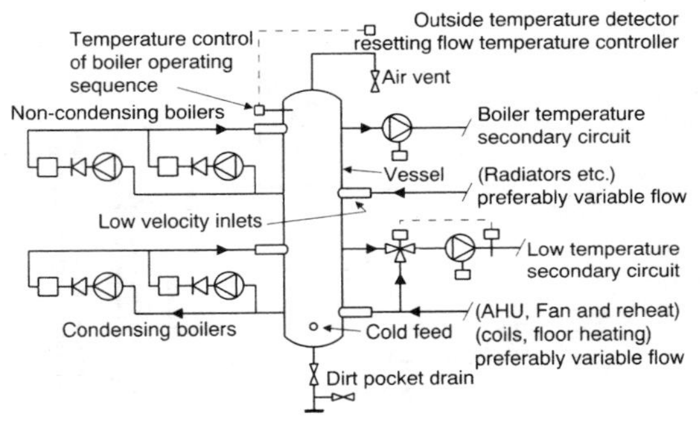

NOTES:
1 Vessel ensures lowest temperature water inlet to condensing boilers
2 Vessel provides water velocities below 0.2m/s for air and solids separation
3 Vessel provide additional water capacity to reduce temperature swings and boiler cycling
4 Vessel inlet connections via sparge pipes to maximize stratification
5 Water temperature controller sequences boilers to provide lowest acceptable flow temperature
6 Condensing boilers always lead in sequence
7 Variable flow secondary circuits provide lowest return temperature

Figure 2.15 Condensing and non-condensing boiler circuit arrangement using stratification vessel

pressurized cylinders or by melting eutectic salts or by heating blocks of masonry. The heat is recovered from the store when needed either by water circulation or by direct heat exchange to air. Electric resistance heating is relatively expensive and produces approximately three times the amount of CO_2 per useful kilojoule as an efficient gas boiler, but it is relatively cheap to maintain.

2.4.4 Combined heat and power

Where electrical power is site generated by engine driven alternators heat is available from the engine cooling system and from exhaust gas heat exchangers, this heat may be used for space heating and hotwater supply. High grade heat from the exhaust gas may be used to power absorption water chillers.

The rate of heat production is a function of the electrical power required and there may be times when it is insufficient to meet the needs, conversely there will be times when more heat is produced than is required. Supplementary boiler plant and heat rejection equipments are therefore required. These complications lead a complex arrangement of plant and controls that is generally better suited to large scale applications (see Figure 2.16).

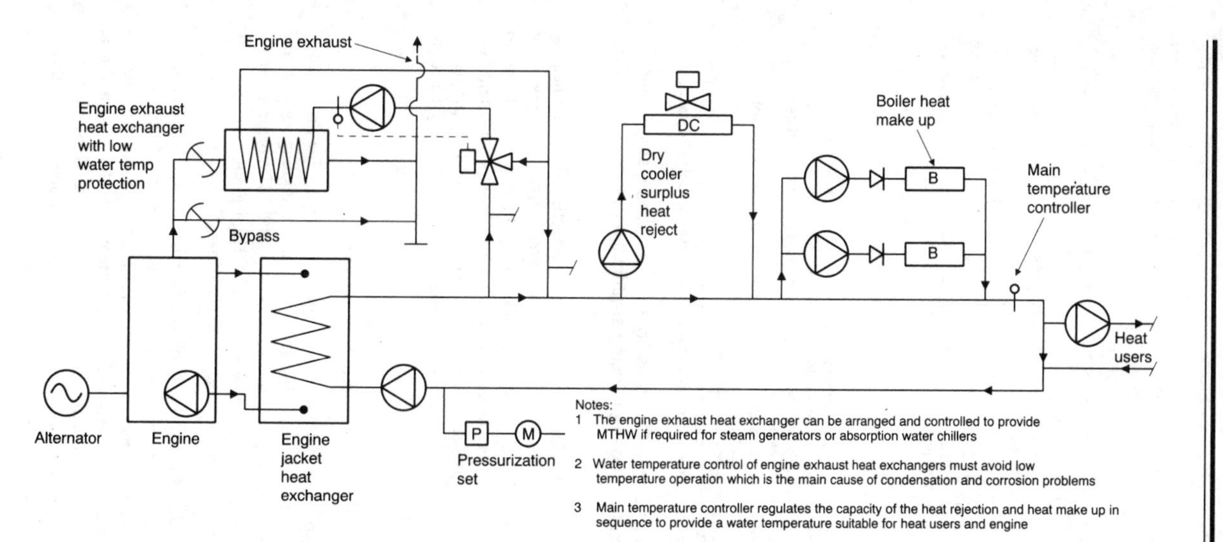

Figure 2.16 Typical CHP heating circuit diagram

2.5 HEATING SYSTEM CONTROL FUNCTIONS

Heating systems for buildings are required to:

(a) Maintain comfort conditions over specific periods of the day.
(b) Operate with the lowest possible energy consumption.
(c) Operate in such a manner that the heating system and building finishes are not damaged.

2.5.1 Time control and optimized start (see Section 2.2)

Heating systems if shut down need to start early enough to ensure that the rooms are at comfort temperature at the start of occupancy. This is often achieved by an optimum start device which varies the start time dependent on the building temperature. The mechanical ventilation system needs only to operate during the occupied period and can be controlled to fixed start and stop times. A CO_2 concentration detector in the extract duct may be used to control the proportion of fresh air introduced by the ventilation system in winter, thereby giving significant savings in air heating energy. In winter the ventilation system should normally be stopped when occupants leave.

High building insulation standards and consequent small heating capacities mean that the heating system will recover a fall in the temperature of the building fabric only at a very slow rate (see Figure 1.9). It is therefore important to include room low-limit thermostats in the more exposed rooms to override the time control and restart the heating system to ensure than any room temperature does not fall below a predetermined level, usually over 16°C. Under near full design external temperatures the heating should operate continuously.

2.5.2 Room temperature control

It is now important to have individual temperature control over each room and the simple way of achieving this is by means of thermostatic radiator valves or electrically operated valves controlling each terminal. These controls should be resettable by the occupant within limits (say 18°C to 20°C). Many older systems are only zoned or whole building controlled by resetting the flow temperature lower as the outside temperature increases, this being achieved by an automatic compensator control. These controls perform a useful function in reducing mains losses and increasing boiler efficiency at part load but as a sole means of room temperature control they are not appropriate for modern buildings where the internal heat gains often meet the heat loss at quite low external temperatures; hence both compensator and individual room temperature controls are essential to avoid wasteful and uncomfortable overheating.

2.5.3 Damage avoidance

Damage to buildings due to frost or condensation is usually eliminated by the room low limit control described in Section 2.5.1. Provided rooms are kept above 10°C and are not subject to significant moisture gains they will be kept free of condensation damage.

Heating systems and air conditioning systems are vulnerable to frost damage, particularly heating and cooling coils in air handling units. Control functions to give these protection are:

(a) Automatic tight sealing of fresh air inlet and exhaust air dampers when the air handling unit fans stop in winter.
(b) Heating and cooling coil control valves arranged to automatically fully open to coils when air handling unit fans stop in winter.
(c) Heating and chilled water circulating pumps arranged to operate when the outside air temperature drops below 0°C under plant shutdown conditions.
(d) Air handling unit fans switched off in the event of low supply air temperature. This may be caused by control malfunction or heat source failure.
(e) Restart boiler plant if heating return water falls below 20°C or room low limit temperature reached. Boiler continues to operate until normal flow temperature achieved.

Boilers can be damaged by condensation corrosion if allowed to operate with low water temperatures for long periods, oil or solid fuel fired steel boilers being particularly prone to corrosion damage. With large water content systems it is necessary to arrange for the boiler water circulation to bypass the system so that on start-up the boiler water temperature reaches normal operating levels as quickly as possible and the boiler circulates water from and to the system only when it achieves normal operating temperatures. Boilers may also risk damage if their water flow drops below a minimum value. Protection is given by a low-flow detection burner lock-out. Boilers need to have the water circulation maintained for a short period after burner shutdown to remove residual heat, this is usually achieved by a time delay controlled pump run on.

2.6 FUEL CHARACTERISTICS

The following notes give a brief summary of the general features of the more common boiler fuels including handling, storage, ash disposal, flue gas treatment, pollutants and plant maintenance. Except for refuse, detailed fuel specifications are to be found in CIBSE Guide Section C5 [10].

2.6.1 Refuse

Now that landfill disposal of refuse is becoming increasingly difficult there will be greater pressures for disposal of combustible refuse by incineration with heat recovery via waste heat boilers. This is only viable on a large scale usually in conjunction with steam power generation and district heating. Refuse combustion requires complex storage, mechanical handling, combustion control, ash disposal and flue gas treatment: apart from this mention as a possible heat source it is outside the scope of this book.

2.6.2 Coal

Coal as a fuel is similar in many respects to refuse, also requiring storage space, mechanical handling plant, ash disposal and flue gas treatment. However, the fuel and ash volumes are much lower due to its higher calorific value and coal is normally burnt in the boiler combustion chamber, unlike refuse which usually requires a large separate combustion chamber. Coal contains sulphur which gives rise to corrosion and flue gas emission problems. Combustion efficiencies are usually lower than with oil or gas. Coal-fired plant is complex and difficult to control at low heat outputs requiring skilled operating and maintenance staff.

2.6.3 Oil

Oil is generally available in five grades ranging from the most expensive light distillate kerosene suitable for vaporizing domestic boilers to the four blended grades which as they become heavier have higher sulphur contents, require higher temperatures for storing, pumping and atomizing and become cheaper. Steel oil storage tanks require a bund oil-proof wall to contain the contents of the oil tank in the event of leakage.

Oil-fired boilers can operate fully automatically under any load conditions, but because oil does not burn as cleanly as gas, deposits form on burners and boiler heat transfer surfaces which require regular cleaning to maintain efficiency. The dirtier flue gas also requires higher chimneys to disperse the pollutants.

The significant sulphur content gives rise to corrosion hazards unless the products of combustion are kept above critical temperatures, and as a consequence boiler return water temperatures have to be kept above critical levels whenever the boilers are operating for significant periods.

2.6.4 Natural gas

Competes with oil and has a number of advantages:

(a) It burns cleanly with high combustion efficiency and high boiler seasonal efficiency.
(b) No storage is required.
(c) Boilers and burners require less maintenance and last longer.
(d) It is the least polluting of all common fuels.
(e) The ancillary handling equipment is minimal.

2.6.5 Electricity

Needs to be considered as a possible alternative to the fuels considered. Electricity can be used to generate heat by direct electric resistance and by driving heat pumps. Compared with gas it has the following features:

(a) Although non-polluting at the point of use its contribution to CO_2 and other atmospheric pollution at the generating power station is much higher than that of an equivalent gas fired boiler plant.
(b) If cheaper off peak electricity is to be used some form of thermal storage is usually required.
(c) Electric resistance heating requires virtually no maintenance at the point of use.
(d) The capital cost of resistance heating is low but heat pump equipment is usually much more expensive than gas boiler plant.
(e) The cost per useful kilowatt hour is likely to be higher than obtained from fossil fuels.

2.7 BOILER HOUSE ANCILLARIES

This section deals briefly with various items related to boiler house operation which need to be considered at planning stage.

2.7.1 Ventilation

Natural or mechanical ventilation is required to provide combustion air and to avoid excessive temperatures in the boiler house. Combustion air requirements are given in CIBSE Guide C5 [10] and British Gas regulations give ventilation requirements for gas-fired plant. It is important that mechanical ventilation is interlocked to prevent boilers being fired unless it is operating.

Mechanical ventilation must be balanced to ensure a slight positive air pressure in the boiler house and some low level ventilation is necessary to prevent possible build up of CO_2.

2.7.2 Heating

Some form of heating is often required in large exposed boiler houses to provide some degree of comfort for operators and maintenance staff and frost protection in the event of shutdown.

2.7.3 Drainage

Floor gullies and sometimes blowdown cooling tanks are necessary to enable draining of high or medium temperature boilers or other major equipment for repair or inspection. A standpipe is useful for washing down floors, etc. Oil interceptors are required for oil-fired plant.

2.7.4 Water treatment

Make-up water under some circumstances is either softened or demineralized to avoid scale formation. This treatment entails storage of chemicals, vessels for mixing, drainage and emergency showers if acids or alkalis are involved. Chemical dosing of closed circuits is usually required for corrosion inhibiting, pH correction, and sometimes biocides are required for control of organisms.

2.7.5 Storage space

A lockage room is required for storing tools, spares and servicing equipment.

2.7.6 Fire and safety precautions

Current regulations concerning health and safety should be consulted and complied with, including:

(a) Alternative escape routes from any part of the boiler house.
(b) Break glass initiated plant shut down and alarm at all exits.
(c) Thermal high temperature cut-outs over boiler burners and other potential fire hazards initiating fuel shut-off and alarm.
(d) Separation and security of dangerous chemicals.
(e) Separation of boiler plant from refrigeration plant.
(f) Termination of safety valve and other discharge pipes in a manner least likely to cause injury.
(g) Ventilation of voids containing gas pipes.

2.7.7 Administration

A large boiler house should include an office and usual amenities with space for records, drawings, operating and maintenance instructions as well as control, instrumentation, monitoring and alarm panels.

References

1. CIBSE Guide Book B. Figures B4.7 to B4.17, HIVS Plant Sizing curves.
2. CIBSE Guide Book A. Section 3 Examples 1 to 5, Calculation of U values.
3. CIBSE Guide Book A. Table A1.3. Dry Resultant Comfort temperatures and A9-3 to A9-6, Heat losses.
4. CIBSE Guide Book A. Tables A9.1 to A9.7, Correction Factors.F_1 F_2
5. CIBSE Guide Book A. Table A2.2, Summary of winter external design temperatures.
6. CIBSE Guide Book A. Table A2.8, Weather data banded for Kew (Jan, Feb, March).
7. CIBSE Guide Book A. Pages A4-4 to A4-13, Calculations of infiltration and natural ventilation.
8. CIBSE Guide Book A.. Figure A2.2, Typical winter weather over 24 hours.
9. CIBSE Guide Book A. Page A1.7, Radiant Heating.
10. CIBSE Guide Book A. Pages C5-8 to C5-10, Fuel Combustion Data.

3 Air conditioning and ventilation

3.1 HEAT GAINS

3.1.1 Sensible heat gain components

To maintain a room temperature less than that outside a cooling capacity that matches the sensible heat gain must be provided in the room. Sensible gain arises from: transmission through the building envelope and natural infiltration — abbreviated as *T*; solar gain through windows (*S*); the heat emitted by people (*P*); lights (*L*); and business machines (*M*). A simple load diagram can be drawn (Figure 3.1) showing how the maximum gains vary as the outside temperature changes.

The transmission gain depends on the difference between the outside and room air temperatures and the load line is easily established: point 1 (transmission gain for the summer design outside temperature) is joined to point 6 (heat loss for the winter design outside temperature).

Gains from people, lights and machines are constants and so points 2, 3 and 4 are identified and joined to points 7, 8 and 9, giving load lines for: $T + P$, $T + P + L$ and $T + P + L + M$.

Solar gain through windows is not related to outside air temperature, but maximum solar gains occur in summer, when the outside temperature is high, and minimum gains in winter, when the outside temperature is low. Hence, for the case of a west-facing window, the design solar heat gain is associated with the outside design temperature in July to establish the point 5 and the winter design outside temperature linked with the solar gain in January to identify the point 10.

For windows facing south the maximum solar gain is at noon in the spring or autumn and it is reasonable to link the midday outside temperature in March or September with the solar heat gain. This gives a cranked load line (shown broken) through the points 5', 11 and 10', in Figure 3.1.

East-facing rooms have a peak solar gain at about 0800 h or 0900 h, suntime, in June or July, when the outside air temperature is lower than the summer design value. A similar procedure can be adopted.

There is likely to be a net sensible gain, requiring cooling for the conditioned room, over much of the year. It must be remembered that there will be diversification of the heat gain components. Population will change and there will be variation in the use of lights and machines. Furthermore, cloud cover

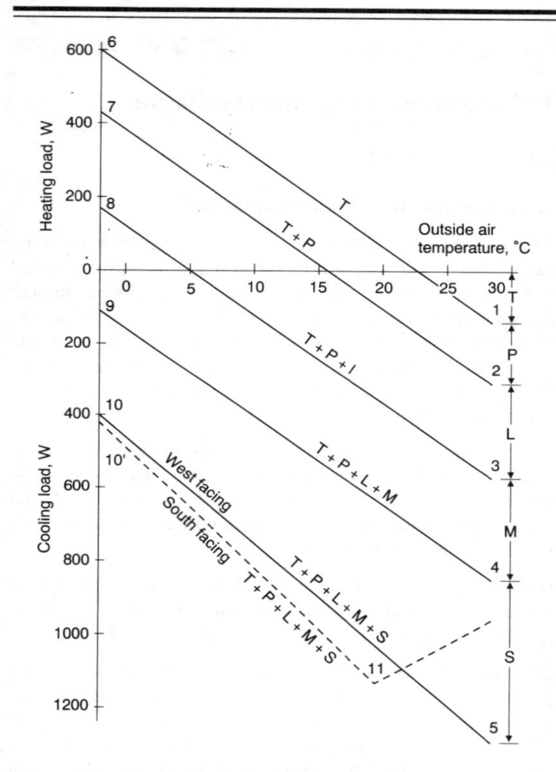

Figure 3.1 Load diagrams for west facing (————) and south facing (– – – – – –) modules

will obscure the sun for some of the time. Hence load diagrams show the maximum heat gains likely, but average gains will be less and, for some of the time in winter, there will be no heat gains at all and the load will be entirely heating, particularly at start-up in the morning. Sensible gains must not be included when calculating design heating loads and boiler powers.

3.1.2 Outside design state in summer

The warmest temperatures in the UK are in July or August, at about 1500 h suntime and the coolest are then approximately nine degrees less than the maximum, occurring at about 0300 h suntime. The average moisture content of the air tends to remain fairly constant in a particular month and hence the humidity rises as the temperature falls after 1500 h, with the possible formation of dew if the temperature drop is enough. After sunrise, a reverse process occurs: the dew evaporates and the humidity subsequently falls as the temperature increases in the approach to 1500 h. (See Figure 3.2).

Meteorological data are available for many places in the UK and throughout the world and they may be analysed to yield a suitable design condition for 1500 h suntime in July. It is customary to record maximum daily air temperatures and relative humidities, usually at the same time in the afternoon. Over a given month the maximum temperatures on each day are noted and this is repeated for several years. The average of these temperatures is established and termed the mean daily maximum dry-bulb, for the period of years considered. The greatest temperature in each month is also noted and the average calculated over the same period of years. This is called the mean monthly maximum dry-bulb temperature. Thus mean daily maximum temperatures can be regarded as referring to typical weather in the month while mean monthly maxima refer to spells of warm weather in the month.

Since the moisture content of the air does not change very

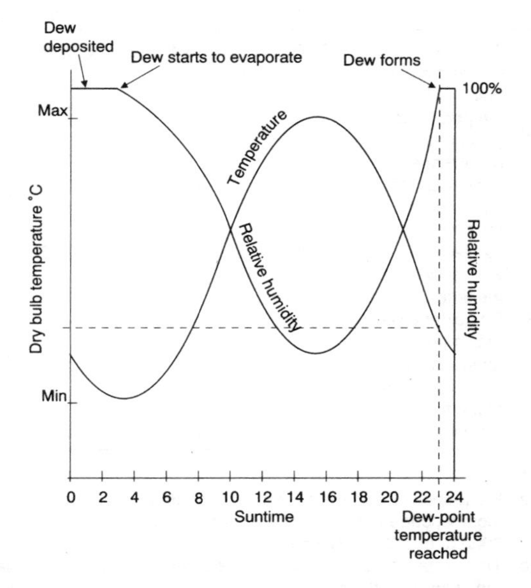

Figure 3.2 Daily variation of dry-bulb temperature and relative humidity

much in a particular month it is determined from the mean daily maximum dry-bulb (in the month), expressed at the same time of the day as the measurement of relative humidity. Knowing the moisture content and the mean monthly maximum dry-bulb temperature for the month allows the design state to be expressed, usually as a dry-bulb and wet-bulb temperature. See Section 3.4.1 for a clarification of the meaning of wet-bulb and dry-bulb temperatures.

If the meteorological station from which the data are obtained is in a rural district the design value of dry-bulb temperature, obtained as above, may be used directly. If the location of the air conditioned building is in a city then an addition of about 1° or 1.5° should be made. This accounts for the solar radiation absorbed by the building surfaces during the morning and later convected into the air, increasing its temperature. On this basis the outside design state chosen for London is 28°C dry-bulb, 19.5°C wet-bulb (sling) (see Table 3.1).

Table 3.1 A summary of outside summer design states for London

Basis	Dry-bulb (°C)	Wet-bulb (sling) (°C)	Moisture content (g/kg)
1. Analysis of mean daily and monthly temperatures	26.9	18.4	9.556
2. Frequency of occurrence			
Exceeded for 1 per cent of the time	27.0	19.7	11.28
Exceeded for 2.5 per cent of the time	25.4	18.9	11.13
3. Coincident with maximum solar radiation intensity for 2.5 per cent of the time	24.5	–	–
4. Common usage	28.0	19.5	10.65

Notes:

(a) Frequency of occurrence data are for the four summer months, June to September, during the occupied period from 0800 h to 1700 h suntime.

(b) The temperature coincident with 2.5 per cent maximum solar radiation is for the month of July.

(c) Common usage includes about 1° for the effect of buildings in the city centre.

(d) When rating plant performance it is prudent to add one or two degrees to the design value. Thus an air-cooled condenser would be selected for 29°C or 30°C dry-bulb if the outside design value was 28°C dry-bulb.

The frequency with which different temperatures occur is also obtainable and may be used as an alternative for expressing an outside summer design state.

Solar gains through windows play a dominant part in heat gain calculations and another approach is to choose an outside design state that is linked with periods of high solar intensity. The CIBSE Guide [1] tabulates dry-bulb temperatures for the summer months coincident with solar radiation intensities exceeded for 2.5 per cent of the time.

3.1.3 Inside design state

There are two forms of air conditioning: industrial and comfort. The condition required for industrial conditioning is specified by the needs of the process and verified by measurement. On the other hand, the room state suitable for comfort conditioning is directly dependent on the feeling of comfort experienced by the occupants and is more difficult to express. A study of human comfort is therefore relevant [2].

The human body is a fairly inefficient machine (maximum efficiency about 20 per cent) and it must establish a heat balance with the environment, for reasons of both health and comfort:

$$M - W = E + R + C + S \tag{3.1}$$

where
M = metabolic rate (W)
W = rate of doing work (W)
E = rate of heat loss from the body, by evaporation (W)
R = rate of heat loss from the body, by radiation (W)
C = rate of heat loss from the body, by convection (W)
S = heat storage rate in the body (W) (zero when in health).

The body usually has to lose surplus heat to the environment and there are four environmental factors that influence this:

(1) Dry-bulb temperature, affecting convection.
(2) Air velocity, affecting convection and evaporation.
(3) Mean radiant temperature of the surrounding surfaces, affecting radiation.
(4) Relative humidity, affecting evaporation.

Dry-bulb temperature is the most important of these factors and is under the direct control of the air conditioning system. Suitable temperatures in the UK lie between 22°C and 23°C but in hotter climates values up to 26°C may be adopted, depending on the duration of the occupancy and economic factors. The duration of occupancy is relevant in all countries. For example, in the UK the foyer of a theatre might have short-term occupancy and a suitable inside design dry-bulb could be 25°C with an outside design temperature of 28°C. On the other hand, the auditorium, would be at 22°C for its long-term occupancy.

Air velocity is next in importance and its value should generally not exceed about 0.15 m/s, provided that the other factors are comfortable. The part of the body on which the air movement is directed is relevant. Although the air movement is not automatically controlled by the system to give comfortable conditions, a proper selection of the supply air distribution terminals and system for the treated space must be made to ensure comfort. This is particularly so for variable air volume systems where the minimum supply airflow rate, as well as the design rate, must be carefully considered when selecting the supply terminals and designing the air distribution system.

Mean radiant temperature is not under the control of the air conditioning system (except for systems using chilled ceilings) but high intensity solar radiation through windows must be excluded. This is done by the provision of suitable solar control methods for windows that can be exposed to direct solar radiation. It usually takes the form of internal Venetian blinds for commercial buildings but the use of external shading is best although not always practical in the UK. Solar reflective glazing may be satisfactory but glazing that absorbs a large amount of solar radiation is not (see Section 3.1.6).

Relative humidity is no longer considered to be of great significance in the provision of human comfort. Taking account of comfort, skin dryness, respiratory health and mould growth it has been shown [3] that dew-point is the significant factor and can lie between 1.7°C and 16.7°C. At 20°C the corresponding humidities are 30 per cent and 80 per cent and, at 22°C, about 26 per cent and 72 per cent.

Human beings lose heat through their surfaces areas and, since these are different, comfort conditions will vary slightly from person to person. The clothing worn also exerts an influence and it follows that comfort cannot be expected for everyone in a mixed population of men and women.

Research [2] shows that if account is taken of all the variables involved, namely, the metabolic rate related to the activity, body surface area, the clothing worn, air dry-bulb temperature, air velocity, mean radiant temperature and relative humidity, satisfying more than 95 per cent of a mixed population is impossible.

Several synthetic scales of comfort have been developed over the years, with mixed success. In Europe and the UK dry resultant temperature (t_{res}) is often taken as an index of comfort. This is defined by

$$t_{res} = [(T_{rm} - 273) + t_a - \sqrt{10v}\,]/[1 + \sqrt{10v}\,] \qquad (3.2)$$

where
T_{rm} = mean radiant temperature of the surrounding surfaces (K)
t_a = dry-bulb temperature (°C)
v = air velocity (m/s).

In equation (3.2), if v equals 0.1 m/s the dry resultant temperature is the average of the mean radiant temperature and the dry-bulb and this is a common way of expressing it:

$$t_{res} = [T_{rm} - 273) + t_a]/2 \qquad (3.3)$$

Floor temperature, the vertical temperature difference between the feet and the head, the asymmetry of radiant temperature and the carbon dioxide content of the air (see Section 3.2.1) are also significant. In summary, the following conditions are desirable for human comfort in a room:

(1) The dry-bulb should exceed the mean radiant temperature in summer but be less than it in winter.
(2) The mean air velocity should not exceed 0.15 m/s, unless the dry-bulb is greater than 26°C.
(3) Relative humidity should lie between about 20 per cent and 60 per cent.
(4) The foot-to-head temperature difference should be as small as possible, normally less than 1.5° and never exceeding 3°.
(5) Floor temperatures should be within the range 17°C to 26°C for people who are standing.
(6) Radiant temperature asymmetry should be not more than 5° vertically or 10° horizontally.
(7) The concentration of carbon dioxide should not exceed 0.1 per cent.

In the UK, for long-term occupancy, inside design conditions are often taken as 22°C dry-bulb with 50 per cent saturation (virtually the same as relative humidity — see Section 3.4). There is a reasonable view that the design dry-bulb could be 22.5°C to 23°C, with adequate comfort and a benefit in capital and running costs.

3.1.4 Sensible transmission heat gain through glass

There is no thermal inertia in window glass, hence:

$$Q_g = A_g\, U_g\, (t_o - t_r) \qquad (3.4)$$

where

Q_g = sensible heat gain through glass (W)
A_g = surface area of glass (m^2)
U_g = thermal transmittance coefficient of glass (W/m^2 °C)
t_o = outside air dry-bulb temperature (°C)
t_r = room air dry-bulb temperature (°C).

U-values are found from the CIBSE Guide or manufacturers' data.

3.1.5 Sensible heat gain by natural infiltration [4]

Air infiltrates through openings in the building fabric by two mechanisms: wind effect and stack effect.

Wind effect

This occurs because of the pressure difference exerted by a wind across opposite faces of a building, but the airflow through the building is influenced by the present of partitions. Wind speeds increase with altitude and are higher in winter than in summer, with corresponding effects on the seasonal infiltration rate.

Stack effect

In summer, the air outside an air conditioned building is warmer and less dense than the air within. Consequently air tends to enter openings in the upper part of the building fabric and to leave through openings in the lower parts. The effect is influenced by the use of doors in the entrance lobby and the movement of lifts inside the building. The presence of many open windows, one above the other, in multi-storey buildings, makes calculation difficult.

The CIBSE Guide [4] provides tabulated data and equations to calculate combined wind and stack effects but in practice this is difficult because there is uncertainty about the tightness of the building structure. In the UK, with modern, well-made buildings infiltration may be small, but in countries where the standards of building construction are poorer, the infiltration rate can be large. In a hot humid climate this may cause an incalculable and significant proportion of the total heat gain.

The sensible heat gain by the infiltration of warm outside air is defined as the product of the mass flow rate of air, its specific heat capacity and the outside-to-inside air temperature difference. A practical equation can then be developed by assuming an air density of 1.2 kg/m³ and a specific heat capacity of 1.0 kJ/kg K:

$$Q_{si} = n\,V\,(t_o - t_r)/3 \qquad (3.5)$$

where

Q_{si} = sensible heat gain by infiltration (W)
n = infiltration rate (h^{-1})
V = volume of the treated space (m^3)
t_o = outside air dry-bulb temperature (°C)
t_r = room air dry-bulb temperature (°C).

In the UK a value of 0.5 is often taken for *n* in summer, but some designers allow nothing. The proportion of the total sensible gain, represented by infiltration, is then quite small (about 3 per cent) but, in winter it is large and can dominate the heat losses, particularly in entrance halls where air change rates may exceed 4.

3.1.6 Solar gain through glass

Direct, diffuse and ground radiation

Direct radiation. The intensity of direct solar radiation reaching a place on the surface of the earth depends on its path length through the atmosphere and is related to the position of the sun in the sky. In turn, this depends on the month of the year, the latitude of the place on the surface of the earth and the time of the day. Figure 3.3 shows how the path length is different for the months of December and June, in the northern hemisphere. Atmospheric clarity and cloud cover introduce local fluctuations.

The position of the sun in the sky is expressed by two coordinates: solar altitude, *a*, and solar azimuth, *z*. Values of these angles are tabulated [5].

The intensity of direct solar radiation is expressed in W/m² on a surface at right angles to the radiation and the symbol

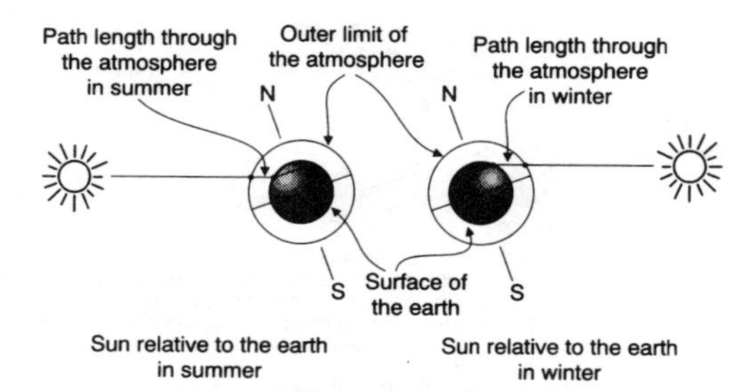

Figure 3.3 The path length of the rays of the sun through the atmosphere in summer and in winter, for the northern hemisphere

commonly used is *I*. Numerical values of such intensity are given in reference [6].

The angle of incidence of direct radiation on an actual wall, roof, or window varies with the position of the sun in the sky and it is customary to resolve the direct radiation in a direction at right angles to the actual receiving surface. Multiplying the resolved intensity (W/m²) by the area of the receiving surface (m²) then yields a value for the radiation normally incident upon the surface (W). Figure 3.4 shows that the following equations emerge:

$$I_h = I \sin a \tag{3.6}$$
$$I_v = I \cos a \cos n \tag{3.7}$$

where

I_h = intensity of direct solar radiation normally incident on a horizontal surface (W/m²)

a = solar altitude (degrees)

I_v = intensity of direct solar radiation normally incident on a vertical surface (W/m²)

n = wall-solar azimuth angle (between the horizontal component of the rays of the sun and the normal to the vertical receiving surface) (degrees).

For the case of sloping surfaces it can be shown [7] that the value of the resolved radiation is given by:

$$I = I \sin a \cos \delta + I \cos a \cos n \sin \delta \tag{3.8}$$

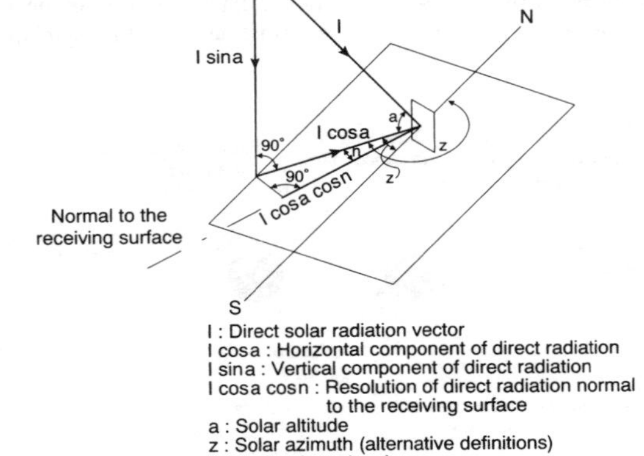

Figure 3.4 Direct solar radiation resolved in a direction normal to the receiving surface

where

I = intensity of direct radiation normally incident on a surface tilted at an angle δ to the horizontal (W/m^2).

Diffuse radiation. In its passage through the upper atmosphere the total solar radiation is scattered by the molecules of nitrogen, oxygen and water vapour. Some of the radiation is also absorbed, mostly by molecules of carbon dioxide, ozone and water vapour, which re-radiate thermal energy in all directions. Part of this scattered and re-radiated energy reaches the surface of the earth and is termed diffuse, scattered or sky radiation. In very approximate terms about 10 per cent of the solar radiation reaching the surface of the earth on a clear day is diffuse, the remainder being direct. The amount of cloud cover has an obvious influence, reducing the direct and increasing the diffuse radiation received as the cloud cover increases. Diffuse radiation is more intense when coming from the part of the sky in the vicinity of the sun and is stronger for higher solar altitudes but is not strong enough to cast a shadow. References 6 and 7 provide numerical values. Increases in height above sea level give a reduction in the strength of the scattered radiation but the direct radiation is correspondingly greater.

Ground radiation. Direct and scattered solar radiation is reflected from the ground and from building surfaces. Its intensity can be significantly large and should be taken into account when calculating the solar energy incident upon nearby windows. The reflectances of the surfaces play an important part [6, 7].

Shading and solar control glass

Direct solar radiation must be prevented from passing through a window into an air conditioned room. The intensity of direct radiation from a surface temperature of 6000°C at the sun causes discomfort, even if the other environmental factors in the room are correct. In order of preference, the measures that may be taken to prevent such discomfort are as follows:

(1) External, motorized, adjustable blinds
(2) Light coloured, reflective, Venetian blinds between the clear glass panes of double glazing.
(3) Light coloured, reflective, Venetian blinds on the inside of single clear glass.
(4) Light coloured, reflective vertical slat blinds, on the inside of clear single glass, with a shading coefficient not less than that of item (3), above.
(5) Sealed unit double glazing with an outer pane of heat-reflecting glass and an inner pane of clear glass, having a shading coefficient not exceeding 0.27.
(6) Double frames of wood or metal with a thermal break,

an outer pane of heat-reflecting glass and an inner pane of clear glass, with a shading coefficient not exceeding 0.27.

(7) Double frames of wood or metal with a thermal break, an outer pane with a heat-reflecting film on its inner surface and an inner pane of clear glass, with a shading coefficient not exceeding 0.27.

(8) Single frame of wood (preferably) or metal with a thermal break and a pane of heat reflecting glass, with a shading coefficient not exceeding 0.27.

(9) Single frame of wood (preferably) or metal with a thermal break and a clear glass pane with a reflective film on its inner surface and a shading coefficient not exceeding 0.27.

Shading coefficient is defined as the ratio of the total thermal solar radiation transmitted through a particular glass and shading combination to the total thermal solar radiation transmitted through single clear 4 mm glass. For Venetian blinds on the inside of single clear, 6 mm float glass the total shading coefficient is about 0.54, but this is acceptable because the heat radiated from the blinds into the room is from a surface at a temperature of about 40°C (as a result of the blinds having absorbed some of the solar radiation) and is not intense enough to cause discomfort in the way direct solar radiation would.

Motorized external shades are excellent but not a practical proposition unless there is adequate access for maintenance. This is seldom the case in the UK.

Fixed external shades are of little use in the UK and in northern, high latitudes, because the sun is low in the sky for most of the day and year, allowing direct radiation to penetrate into the depth of the room for much of the time. The maximum solar altitude of the sun at noon in June in London is only 62° and at noon in December it is only 15°. Fixed external blinds are best suited to the tropics where the maximum solar altitude is near 90° for much of the day.

Heat absorbing solar glasses are of little use: their shading coefficients exceed 0.27 and they do not exclude enough of the high intensity direct radiation from the sun.

It is not prudent to fit reflective film or shades on the inner face of solar control glass without the written agreement of the manufacturers. The presence of the inner film or shade reflects direct radiation back through the glass and increases its temperature, setting up thermal stresses that can be dangerous [8].

Double glazing

Whereas the shading coefficient of single clear glass is 1.0, that of double clear glass is about 0.85 and none of the high intensity direct solar radiation is excluded. Double glazing is unnecessary for air conditioning unless relative humidity is provided at a value that would give condensation on single glass in winter. A thermal break must be provided in metal frames when used with double glazing. A calculation should always be carried out to establish the inner surface temperature of the double glass, in relation to the room dew-point and the coldest expected outside condition.

There is no economic case for double glazing, either for heating or for air conditioning, in a commercial building occupied for nine hours a day, five days a week. However, there may be a case for hospitals and the like, occupied continuously. The only reason for double glazing (other than dealing with condensation in special circumstances), is acoustic. The mean attenuation provided by a single glazed, openable window without weather stripping, is about 20 dB over the range of frequencies from 160 to 3150 Hz. If an openable, double glazed window is provided having a 200 mm air gap, the mean attenuation rises to 40 dB. It is not possible to attach an economic value to the acoustic benefit but, in noisy urban districts, the noise from traffic entering a room through single glass is very intrusive, often making telephone conversations difficult.

The effect of building mass

Solar radiation entering a room (termed the instantaneous gain) does not immediately cause the air temperature to rise and provide a load for the air conditioning system. Some heat is convected from the warm blinds over the sunlit windows and this causes an immediate rise in the air temperature. However, the radiation that does enter, either directly from the sun or from the blinds, is incident upon the room surfaces, principally the floor, although some is also reflected from the floor onto the walls, ceiling and furniture, all of which play a part. The radiation is absorbed by the upper thickness of the solid surfaces on which it is incident and warms the material. Time is taken for this to occur and some of the incident energy is stored in the mass of the material but is eventually convected into the room to provide a load on the air conditioning system. Figure 3.5 illustrates what happens and shows that there is a time lag and a decrement factor. Hence the building mass plays an important part in the calculation of solar heat gain through windows.

For the purpose of calculating solar heat gains buildings are usually classified as:

(i) Heavyweight. These have bare concrete floors or floors covered with hardwood blocks, linoleum or vinyl, with no suspended ceilings and few partitions.

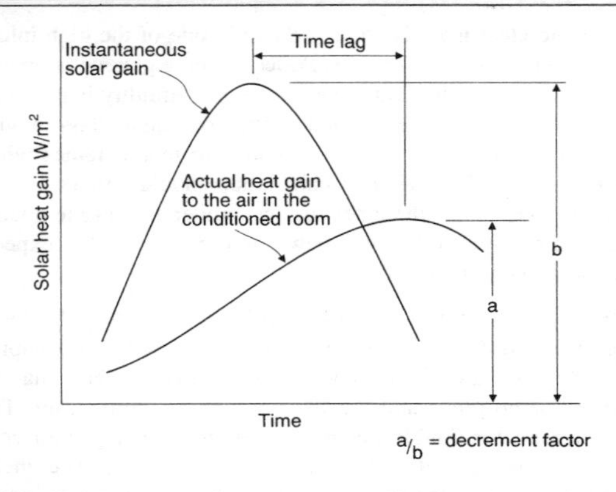

Figure 3.5 Instantaneous heat gain through glass and the cooling load on the air conditioned room some time later, causing the room air temperature to rise

(ii) Lightweight. These are provided with supported or carpeted floors with suspended ceilings and multiple partitions.

For intermittently heated buildings a thermal response factor (f_r) is sometimes used [8] to define the weight of the building:

$$f_r = \frac{[\Sigma (AY) + nV/3]}{[\Sigma (AU) + nV/3]} \qquad (3.9)$$

where

A = surface area of a building structural element (m^2)
Y = thermal admittance of the surface of a building structural element (W/m^2 °C)
U = thermal transmittance of a building structural element (W/m^2 °C)
n = infiltration rate (h^{-1})
V = volume of the space (m^3).

The response factor is unsuitable for expressing the weight of a building when calculating solar heat gains through glazing. It gives the wrong answer according to the definition of building weight given above. Referring to equation (3.9) it can be seen that a building with a large number of partitions (which will increase $\Sigma(AY)$ but will not affect $\Sigma(AU)$) gives a large numerator, implying a large response factor and hence a heavy building. In fact, a building with many partitions is regarded as a lightweight building when calculating solar heat gain through glazing.

The practical calculation of solar heat gain

CIBSE tables are available [9] that take account of all the relevant variables and give the actual solar heat gain to the room as a cooling load on the air conditioning system. The tables cover orientations through 360° at 45° intervals from north. at hourly intervals from 0800 h to 1800 h suntime, for each month of the year. Cooling loads are tabulated for bare glass windows and for windows fitted with light coloured, slatted blinds, which are assumed to be adjusted by the occupants to exclude the entry of the direct rays of the sun. Separate tables are given for heavyweight and lightweight buildings.

The tables are based on a theoretical analysis [10] and assume that the air conditioning system maintains a constant dry resultant temperature within the room and that the plant runs for 10 h each day. No corrections are given for other periods of plant operation but factors are provided to take account of different forms of shading and systems that maintain a constant air (dry-bulb) temperature — as all air conditioning systems do. Latitudes dealt with are: 0°, 10°N, 20°N, 30°N, 35°N, 40°N, 45°N, 50°N, 51.7°N, 55°N and 60°N. The latitude of London is about 51.5°N and the table for 51.7°N is always used for London. The latitude of Aberdeen is approximately 57°8'N so the tables adequately cover most of the UK.

The actual cooling load on the air conditioning system due to solar heat gain through vertical glazing is given by:

$$Q_{sg} = F_c F_s q_{sg} A_g \qquad (3.10)$$

where

Q_{sg} = the actual cooling load due to solar heat gain through vertical glazing (W)
F_c = air point control factor
F_s = shading factor
q_{sg} = tabulated cooling load due to solar heat gain through vertical glazing (W/m^2)
A_g = the area of glass if wooden frames are used, or the area of the opening in the wall in which the frame is fitted, if metal frames are used (m^2).

EXAMPLE 3.1

Determine the cooling load due to solar heat gain at 1500 h suntime in July for a west-facing window using CIBSE Table A9.15 [9], for a lightweight building with intermittently used, internal, Venetian blinds. The relevant window area is 2 m^2.

Answer

From CIBSE Table A9.15, the cooling load at 1500 h suntime in July, though a west-facing window is 270 W/m^2. At 1500 h suntime in July a west face will be in sunlight so the blinds are drawn. From Table A9.15, F_c is 0.91 for a lightweight building with closed internal blinds. F_s is 0.77 for a lightweight building with 6 mm clear glass and closed, internal Venetian blinds. Hence, by equation (3.10):

$$Q_{sg} = 0.91 \times 0.77 \times 270 \times 2 = 378 \text{ W}$$

Other tabulated data exist for the calculation of the cooling load due to solar heat gain through glass and yield answers similar to the CIBSE tables.

3.1.7 People

Sensible, latent and total emissions

The total rate of emission depends on the activity, but the split between the sensible and latent proportions depends on the dry-bulb temperature: at higher temperatures the body does not lose heat as readily by convection and radiation and hence the losses by evaporation must increase. Table 3.2 shows this.

Population densities

Studies have been carried out on the density of population in different types of commercial building and some details are reported in reference 11. Table 3.3 summarizes results.

Diversity factors

When calculating the maximum sensible heat gain for a particular room or module, a diversity factor of 1 is used because the cooling capacity provided must be able to deal with the maximum heat gain. On the other hand, when calculating the refrigeration load for an entire building not all the components of the heat gain (Section 3.1.1) will be at a maximum value, simultaneously. This is only allowable if the system to be used permits it. For example, a constant volume re-heat system cancels reductions in sensible heat gain by warming the air supplied and so the load on the refrigeration plant never reduces, but stays at its peak duty virtually all the time. On the other hand, a four-pipe fan coil system modulates its cooling capacity to match variations in the sensible gain and permits diversity factors to be used, when calculating the design duty of the central refrigeration plant.

Diversity factors may be applied to the heat gains from people (both sensible and latent), lighting and business machines. A typical value for people is 0.75.

Table 3.2 Heat emission from people

		Heat emitted in watts at various dry-bulb temperatures							
	Metabolic rate	20°		22°		24°		26°	
Activity	(watts)	S	L	S	L	S	L	S	L
Seated at rest	115	90	25	80	35	75	40	65	50
Office work	140	100	40	90	50	80	60	70	70
Standing	150	105	45	95	55	82	68	72	78
Eating	160	110	50	100	60	85	75	75	85
Light work in a factory	235	130	105	115	120	100	135	80	155
Dancing	265	140	125	125	140	105	160	90	175

Table 3.3 A summary of typical population densities

Application	Density of population (m^2/person)
Offices:	9
Department stores:	
typical over the total area	1.7 – 4.3
basement sales ares	3
ground floor sales areas	4
upper floor sales areas	6
Hotels:	
bars	1.8
restaurants	1.9
banqueting suites	1.9
reception area	7.5
foyers	3.6
Shopping centres:	
malls	10 (approx.)
shop units	3.3
Supermarkets:	3 (over the gross area but about half the population is concentrated over the front third of the floor area in the vicinity of the tills)
Museums, art galleries and libraries:	
typical	10
occasional peaks	2–3
Theatres, cinemas and concert halls:	
lobbies	2–3
auditorium	0.75–1.0 (over the total area, including aisles)

3.1.8 Electric lighting

Illuminance and power dissipation

All the electrical energy drawn from the mains is liberated as a sensible gain to the conditioned space. The power quoted on the bulb of a tungsten lamp is the total rate of heat dissipation but, for fluorescent tubes, the gain to the room is about 25 per cent more than the power quoted on the tube because of the heat liberated by the control gear.

A typical relationship between illuminance and heat dissipation is given in Table 3.4.

Table 3.4 Illuminance and heat dissipation

Illluminance (lux)	Total rate of heat liberation, (W/m² floor area)					
	Tungsten lamps		65 W white fluorescent tubes			58 W poly-phosp0hor 1.5 m tubes
	Open reflector	Diffusing fitting	Open plastic trough	Diffusing fitting	Louvred ceiling panel	
100	19–28	28–36	4–5	6–8	6	4–8
200	28–36	36–50	6–7	8–11	9–11	2–10
300	38–55	50–69	9–11	12–16	12–17	10–16
500	66–88	–	15–25	24–27	20–27	14–16
1000	–	–	32–38	48–54	43–57	30–58

Supply and extract ventilated luminaires

If supply air distribution fittings are combined with luminaires the lighting tube may be overcooled and the illuminance reduced. On the other hand it is sometimes claimed that the life of the lighting tube is lengthened. The manufacturers must always be consulted before using combined supply fittings.

Extract ventilated luminaires have been successfully used in the past and Figure 3.6 illustrates the arrangement. About 40 to 50 per cent of all the heat emitted by the recessed light fitting can be taken away by the extract ventilation system [12] and returned to the central air handling plant for discharge to waste or re-cycling, as expedient. This is even after an allowance has been made for the heat transmission from the warmed ceiling void to the room above the slab and to the room beneath the suspended ceiling. It is unnecessary to make duct connections directly to each luminaire but it is essential that extract spigots are provided for each module/room and connected into the

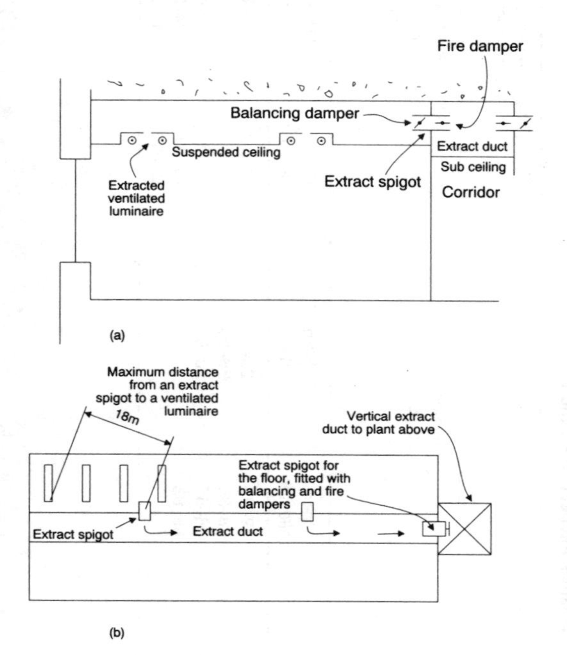

Figure 3.6 The use of extract ventilated luminaires. (a) Section of a room with extract-ventilated luminaires. (b) Plan of an open-plan office fitted with extract-ventilated luminaires

central ducted extract system that leads back to the air handling plant. Otherwise there will be preferential extract ventilation from luminaires nearest to the spigot. Each extract spigot should be fitted with both balancing and fire dampers. With large treated areas the distance from an extract spigot to the most remote luminaire should not exceed about 18 m.

Extract fittings must not be used with polyphosphor tubes – the illuminance reduces.

Diversity factors

The diversity allowance to be made for lighting depends on several factors: the depth of the room, the proportion of the exterior wall occupied by windows, the colour scheme of decoration within the room and the method of control adopted for the lighting. If all these factors are taken into account the value obtained for the diversity factor is sometimes very low – too low, in fact, for a prudent design engineer to adopt. It is suggested that, in the absence of reliable information to the contrary, a diversity factor of 0.75 to 0.8 is adopted when calculating the refrigeration load for an entire building.

3.1.9 Small power and business machines

Heat emissions from equipment

These vary according to the ratings of the business machines used but have been greatly over-estimated in the past. This has led to excessive design cooling capacity being installed with air conditioning systems that have never run at full duty. Automatic control at reduced actual load tended to be poor. With variable air volume systems this may give rise to complaints of draught at partial load.

Nameplate powers on machines are not operating powers. A survey [13] of powers for typical business machines used in offices shows that the average rate of heat dissipation over a half-hour period lies between 20 per cent and 80 per cent of the nameplate power. A typical allowance for heat dissipation appears to be about 20 W/m^2, referred to the floor area, in the absence of other, firm information to the contrary.

Diversity factors

A reasonable diversity factor, based on information from reference 13, is 0.65 to 0.70.

3.1.10 Transmission through walls and roofs

Sol-air temperatures

Transmission heat gain through walls and roofs for multi-storey buildings usually amounts to only about one or two per-cent of the total sensible gain. This is to be borne in mind because it does not make sense to spend excessive time in cal-culations for a heat gain of such small significance. The excep-tion is when heat gains are calculated for the roof of a large plan, low-rise building where thermal inertia of the roof is very small. Nevertheless, heat gains through walls and roofs should not be ignored.

The method advocated by the CIBSE involves the use of sol-air temperatures. Sol-air temperature (t_{eo}) is defined as that notional outside air temperature that gives the same rate of sensible heat entry into the outer surface of a wall or roof as the actual com-bination of air temperature and radiation exchanges does. This is shown by the following:

$$(t_{eo} - t_{so})\, h_{so} = (t_a - t_{so})\, h_{so} + \alpha I + \alpha I_s + R$$

The value of R, a term to cover long-wavelength radiant exchanges with the surroundings is unknown and, for all practi-cal purposes, can be ignored.

In the case of vertical walls the sol-air temperature is then defined by

$$t_{eo} = t_a + \frac{\alpha\, I \cos a \cos n + \alpha\, I_s}{h_{so}} \qquad (3.11)$$

and, in the case of horizontal roofs, it is defined by

$$t_{eo} = t_a + \frac{\alpha\, I \sin a + \alpha\, I_s}{h_{so}} \qquad (3.12)$$

where

t_a = outside air dry-bulb temperature (°C)
α = absorption coefficient for a surface
I = intensity of direct solar radiation on a surface at right angles to the rays of the sun (W/m^2)
a = solar altitude (degrees)
n = wall-solar azimuth angle (degrees)
I_s = intensity of scattered radiation normal to a surface (W/m^2)
h_{so} = outside surface heat transfer coefficient (W/m^2 °C).

Sol-air temperatures can be calculated and are tabulated [1] for southern England and the months of March to October, inclu-sive. Dry-bulb temperatures are quoted at hourly intervals over 24 hours and are for 2.5 per cent highest solar radiation (see Section 3.1.2). 24-hours mean values are given and eight verti-cal wall orientations and a horizontal roof are considered.

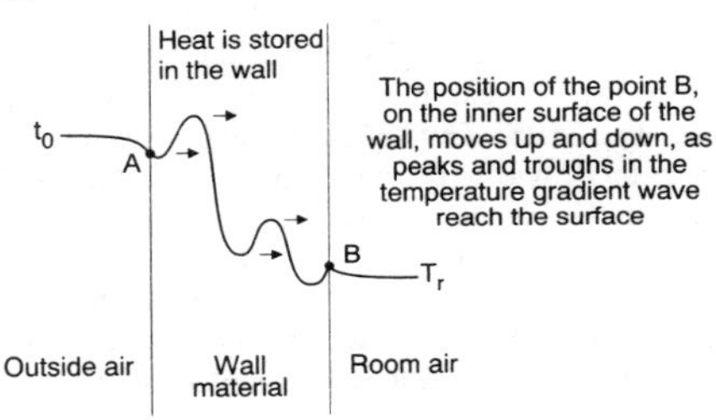

Figure 3.7 The temperature gradient for unsteady-state heat flow through a wall

Time lags and decrement factors

Outside surfaces of walls or roofs are subject to daily variations in air temperature and solar radiation. These cyclic changes mean that the shape of the temperature gradient through the wall is very complicated, even though the air conditioning system may be holding the room dry-bulb at a constant value. Figure 3.7 illustrates this. Heat is stored in the wall and as the cyclic waves in the gradient proceed through wall, the inside surface temperature fluctuates. The first consequence is that the heat gain to the room is less than the rate of heat entry to the outside surface and this is expressed by the application of a decrement factor. The second consequence is that a time lag is involved. Suitable decrement factors and time lags are quoted for various practical building constructions in reference [14].

Corrections to outside air and sol-air temperatures

The outside air design dry-bulb temperature (Section 3.1.2) may not be the same as that tabulated in reference 1 and corrections must be applied to the tabulated values before they can be used to calculate heat gain through a wall or roof.

Equations (3.11) and (3.12) show that the sol-air temperature is developed by an addition to the outside air dry-bulb temperature. Hence the difference between the chosen design value of the outside air dry-bulb at 1500 h suntime and the tabulated value for the same time, in the particular design month, can be applied as an addition (or subtraction) to all the tabulated sol-air temperatures and mean sol-air temperatures. For example, if the tabulated [1] outside air temperature at 1500 h suntime in July is 24.5°C but the design temperature is 28°C then 3.5° must be added to all tabulated sol-air and mean sol-air temperatures, for the purpose of calculating heat gain through a wall or roof. The application of this is clarified in Section 3.1.11.

Heat gains through walls and roofs

The principle adopted [15] is that the heat gain at a particular time is the 24-hour mean gain plus the variation about the mean. The CIBSE Guide[15] introduces various correcting factors to these two terms which have the effect of reducing the value of the calculated heat gain through a wall or roof by about 10 percent. In view of the relative unimportance of the heat gain and the uncertainty about the accuracy of the correcting factors, it is scarcely worth while using them. Hence it is reasonable to use the following simpler equation:

$$Q_{\theta + \phi} = AU[t_{em} - t_r) + f(t_{eo} - t_{em})] \tag{3.13}$$

where

$Q_{\theta + \phi}$ = heat gain to a room through a wall or roof at a time $(\theta + \phi)$ (W)
ϕ = time lag of a wall or roof (h)
A = area of a wall or roof (m^2)
U = thermal transmittance coefficient of a wall or roof ($W/m^2\ °C$)
t_{em} = 24-hour mean sol-air temperature (°C)
t_r = constant dry resultant temperature at the centre of the room (°C)
f = decrement factor
t_{eo} = sol-air temperature at 0 (°C).

In practice, t_r is taken as the room dry-bulb temperature and, of course, it is not constant over 24 hours, as the equation assumes, but this is ignored.

Equivalent temperature differences

The following equation gives the sensible heat gain through a wall or roof in terms of a notional equivalent temperature difference, outside air to room air, Δt_{eq}, that takes account of the effects of temperature difference, solar radiation, etc.:

$$Q_{\theta + \phi} = AU\, \Delta t_{eq} \tag{3.14}$$

Values of Δt_{eq} can be established at different times of the day and different orientations for various building structures, by combining equations (3.13) and (3.14). If the results are tabulated, equation (3.14) provides a simple way of calculating the heat gain through walls or roofs. The method is used in the USA[16] but has not been extensively adopted in the UK.

Floors exposed on the underside

Floors over open car parks and similar spaces have their underside exposed to the air temperature but are not irradiated by the sun and do not lose heat by radiation to the black sky at night. The sensible heat gain is given by

$$Q_f = A_f U_f (t_o - t_r) \quad (3.15)$$

The equation should be used to determine the surface temperature of the floor in a room, when occupied, during summer and winter. This is not as easy as it appears: if the system is operated intermittently it will be shut down over weekends and, in cold weather, the mass of the floor slab will cool. On start-up on Monday morning the floor surface temperature will be less than the value predicted by equation (3.15) until the system has operated for sufficient time to bring the mass of all the floor slab up to the steady-state temperature. This difficulty can be pre-empted by providing sufficient thermal insulation on the floor slab. The best place for the insulation is on the upper surface of the slab but this can pose problems arising from the point loads imposed by furniture legs and people's heels on the upper surface of the insulation. If a sufficiently strong and durable insulation material cannot be found a possible solution is to place thicker insulation under the slab. Sometimes heating coils are placed beneath the slab, in conjunction with adequate insulation, to deal with the downward heat loss.

Excessively warm floors are seldom a problem in summer, except if a boiler room, or the like, is located beneath. Adequate insulation under the slab, finished on its lower surface with reflective foil, is then a satisfactory answer, possibly in conjunction with mechanical ventilation in the room beneath. Floor surface temperatures should lie between 17°C and 26°C.

3.1.11 Practical heat gain calculations

Design sensible heat gain for a room or module

This is the maximum sensible heat gain that will occur, within the limiting conditions imposed by the design brief. The maximum gains must be calculated for each room or module so that the air conditioning system can provide enough sensible cooling capacity to offset such gains. The method is shown by means of an example.

EXAMPLE 3.2

Calculate the maximum sensible heat gains at 1500 h suntime in July for the module shown in Figure 3.8, making use of the following design data:

Latitude	51.7°N
Outside design state:	28°C dry-bulb, 19.5° wet-bulb (sling)

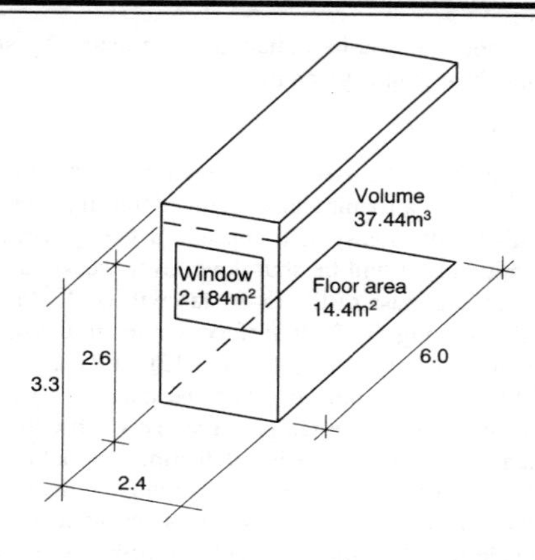

Figure 3.8 Modular dimensions for Example 3.2

Room design state:	22°C dry-bulb, 50 per cent saturation (air point control over room temperature)
Wall:	time lag (φ) 5 h, decrement factor 0.65 U-value 0.45 W/m² °C, light-coloured surface
Glass window:	metal frame, 6 mm single clear glass U-value 5.6 W/m² °C internal white Venetian blinds
Building structure:	lightweight for the purpose of calculating solar gains through glass
Population:	2 people engaged in office work 90 W/person sensible heat emission (Table 3.2)
Illuminance:	500 lux, 17 W/m² (Table 3.4), not extract ventilated
Business machines:	20 W/m² (Section 3.1.9)
Infiltration rate:	0.5 air changes/hour (Section 3.1.5)

Answer

Referring to Table A8.3(e) in the CIBSE Guide[1] the relevant sol-air temperatures and corrections can be determined:

Design outside air temperature at 1500 h:	28°C
Tabulated outside air temperature at 1500 h:	24.5°C
Correction:	+3.5°
Time of heat gain to room:	1500 h
Time lag:	5 h
Time of relevant sol-air temperature:	1000 h

Tabulated sol-air temperature at 1000 h:	23.0°C
Correction:	3.5°
Actual sol-air temperature to use (t_{eo}):	26.5°C
Tabulated 24 h mean sol-air temperature:	22.5°C
Correction:	+3.5°
Actual 24 h mean sol-air temperature to use (t_{em}):	26°C

If, to be on the safe side, it is assumed that the wall area through which sensible heat gains occur is the gross area (based on the floor-to-floor height) minus the opening in the wall for the window, then equation (3.13) can be used to calculate the transmission gain through the wall into the room at 1500 h suntime.

Referring to Table A9.15 in the CIBSE Guide [9] the cooling load due to solar gains through windows for a lightweight building with intermittently used blinds and 10 h plan operation is 270 W/m². This is q_{sg} in equation (3.10). The footnote to the table gives the following correction factors:

The sun is on the west face of the building at 1500 h so the shade correction factor for a lightweight building with light slatted blinds closed is 0.77. This is F_s in equation (3.10). The factor for air point control for a lightweight building with blinds closed is 0.91. This is F_c in equation (3.10). Using equations (3.4), (3.13), (3.5) and (3.10), the following sensible heat gains are calculated:

	watts	proportion
Glass: 2.184 × 5.6 × (28 – 22) =	73	6%
Wall: (3.3 × 2.4 – 2.184) × 0.45 [(26 – 22) + 0.65 × (26.5 – 26)] =	11	1%
Infiltration: 0.5 × 37.44 × (28 – 22)/3 =	37	3%
Solar (glass): 0.91 × 0.77 × 270 × 2.184 =	413	33%
People: 2 × 90 =	180	14%
Lights: 17 × 14.4 =	245	20%
Business machines: 20 × 14.4 =	288	23%
Total design maximum sensible heat gain =	1247	100%

Solar gain through glass is the dominant element and this can be used as an indicator if there is doubt as to the time of the day and month of the year when gains are a maximum. For a room with windows facing in different directions multiple calculations would establish the maximum gain and the solar gain through glass would be of help in choosing the times and months for which to do such calculations.

For the top floors of buildings the transmission gain through the roof is of some importance but its significance is decreasing with the reducing U-values adopted in accordance with the Building Regulations.

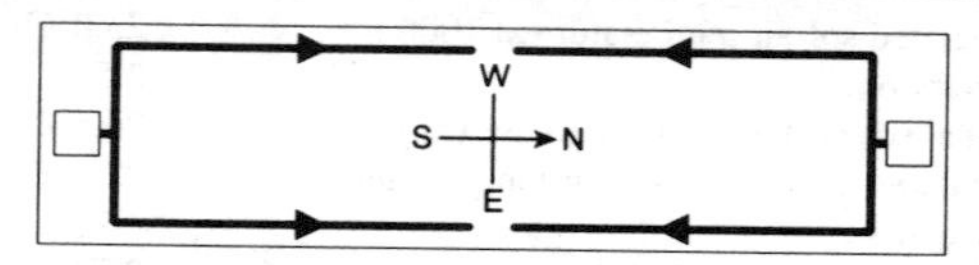

Correct arrangement: takes account of the natural diversity of solar gain as the sun moves round the building

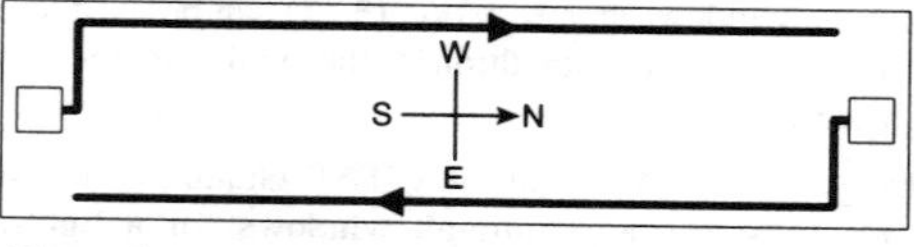

Wrong: does not take into account solar diversity

Figure 3.9 Air handling plants used for VAV systems should be arranged to feed opposite faces of a building

Maximum simultaneous sensible heat gain for the part of a building treated by a VAV air handling plant

Figure 3.9 shows part of a building conditioned by a variable air volume (VAV) system. It should be arranged that an air handling unit deals with opposite faces of the building so that, as the sun moves round the building, the corresponding natural diversity in the solar heat gain through glass is exploited. The sensible heat gains are calculated as a whole for the part of the building treated by the air handling plant and diversity factors for people (Section 3.1.7), lights (Section 3.1.8) and business machines (3.1.9) are applied. This is to size the air handling unit correctly: the plant will never have to handle the sum of the maximum individual VAV air supply rates because of the diversification of the solar gain and the gains from people, lights and machines.

Similarly, the duct system for that part of the building dealt with by the air handling plant will never have to carry the sum of the maximum individual airflow rates of the VAV terminals. Some skill is necessary in applying diversity factors for sizing the ducts: the section of ducting for the most remote part of the building dealt with by the air handling plant will obviously have to handle the sum of the maximum duties of the individual VAV terminals, at certain times. So, for this section, diversity factors of unity must be applied for people, lights and machines. Maximum solar heat gain must also be used for the windows.

Maximum simultaneous heat gains for the calculation of the refrigeration load for the whole building

The refrigeration load for the entire building must be calculated when a central refrigeration plant is used. A large part of the total refrigeration load is due to the sensible heat gains and, when determining these, diversity factors have to be applied if the air conditioning system used can accept such diversification. (For example a VAV system can but a constant volume re-heat system cannot.) The refrigeration load should be calculated for the time when it is a maximum and this usually at about 1500 h suntime in July, for the UK. The matter is dealt with in Section 3.4.6.

Typical values

The heat gains calculated in Example 3.2 were for a typical office module having a construction conforming with the Building Regulations and realistic loads for people, lights and machines.

For comparative purposes it is common practice to express duties and loads per unit of treated floor area and for the example the specific sensible heat gain is 1247 W/14.4 m^2, namely, 87 W/m^2. Specific sensible heat gains in offices will largely depend on the size of the windows and the internal loads from people, lights and machines. For the core area of an office, gains by transmission, infiltration and solar radiation through glass are missing and, using the results from Example 3.2, the specific sensible heat gain is 713 W/14.4 m^2, which is only 50 W/m^2. In some applications, such as theatres, the load will be almost entirely from people but in others, such as TV studios, the lighting will be dominant.

3.1.12 Latent heat gains

The two sources of latent heat gain are from people (Q_{lp}) and natural infiltration (Q_{li}). The total latent heat gain (Q_l) is then given by:

$$Q_l = Q_{lp} + Q_{li} \tag{3.16}$$

It follows that if the humidity in the conditioned space is not to rise to unacceptably high levels the room must be provided with a dehumidifying capacity that matches the latent heat gain. The air supplied must be dry enough to absorb the moisture gains corresponding to the latent heat gains.

People emit moisture by the exhalation of humid air from the lungs and by the evaporation of moisture from the skin. Table 3.2 gives the latent heat gains from people in environments at different air temperatures and for different activities.

In summer design weather the moisture content of the air outside an air conditioned building is higher than it is inside. Hence wind and stack effects (Section 3.1.5) cause a latent gain from infiltration (Q_{li}), expressed by

$$Q_{li} = 0.8\, n\, V\, (g_o - g_r) \tag{3.17}$$

where

n = number of air changes per hour of infiltration (h^{-1})
V = volume of the room or building (m^3)
g_o = moisture content of the outside air (g/kg)
g_r = moisture content of the room air (g/kg).

EXAMPLE 3.3

Calculate the latent heat gains for the module and design conditions used in Example 3.2, given that the design moisture contents (from psychrometric tables or a psychrometric chart — see Section 3.4) are 10.65 g/kg for the outside air and 8.366 g/kg for the room air.

Answer

From Table 3.2 the latent emission from people is 50 W each and hence, for two people, Q_{lp} is 100 W. The infiltration rate is 0.5 air changes per hour. Hence, by equation (3.17)

$$Q_{li} = 0.8 \times 0.5 \times 37.44\,(10.65 - 8.366) = 34 \text{ W}$$

By equation (3.16) the total latent heat gain is

$$Q_l = 100 + 34 = 134 \text{ W}$$

In the UK, with reasonably good building construction, the uncertainty about the air change rate is not of great significance when calculating sensible and latent heat gains but, in a hot humid climate, particularly when building construction is unreliable, the infiltration rate can be large and the latent heat gain by infiltration then assumes considerable importance.

3.1.13 Sensible heat gain to ducts

For most practical purposes the temperature rise between two duct sections, 1 and 2, one metre apart (Figure 3.10), resulting from heat gain to the ducted airstream [7] is given by:

$$(t_2 - t_1) = \frac{(t_r - t_1)}{KDV} \tag{3.18}$$

where

t_1 = upstream air temperature (°C)
t_2 = downstream air temperature (°C)
t_r = temperature in the room through which the duct runs (°C)

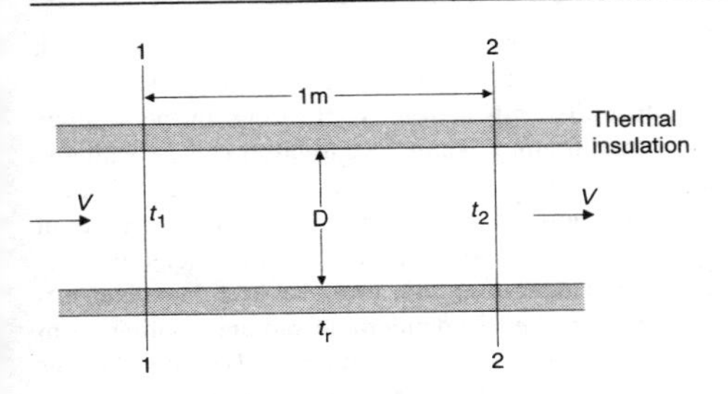

Figure 3.10 Duct heat gain

K = 200 for 25 mm lagging thickness, 363 for 50 mm lagging thickness and 523 for 75 mm lagging thickness (s/m²)
D = internal equivalent duct diameter (m)
V = mean air velocity in the duct (m/s).

Equation (3.18) must be used for successive, short duct lengths (say 4 m) because t_1 and t_2 increase as heat gains occur. For example, if D is 300 mm, V is 10 m/s, t_1 is 12°C and t_r is 22°C, equation (3.18) gives a temperature rise of 0.017° per metre, and t_2 is 12.07°C after 4 ms, with 25 mm lagging.

3.2 FRESH AIR ALLOWANCES

3.2.1 The need for fresh air

Fresh air is needed for four reasons

(i) For breathing. This is quite small: 0.1 to 1.2 l/s for a person, depending on the activity.
(ii) For CO_2 control. Fresh air contains approximately 0.3 to 0.34 per cent of CO_2 and people each produce CO_2 at a rate of 0.00472 litres/s. Hence to prevent the concentration of CO_2 in an occupied space from rising to an unacceptably high level fresh air must be introduced. An acceptable concentration in an occupied space is about 0.1 per cent. The threshold limit value (TLV) for an 8 h exposure is 5 per cent but concentrations exceeding 2 per cent are not acceptable. Beyond this value increasing human discomfort is experienced [17]. CO_2 is a narcotic poison, fatal to humans.
(iii) To control odours, The most important reason for supplying fresh air is to dilute odours to a socially acceptable level. If this is achieved the other requirements will be more than satisfied. Smoking has a very considerable

effect on the odour content of air in an occupied room and CIBSE recommendations are given in Table 3.5.

(iv) To reduce discomfort from overheating, in the absence of air conditioning. Natural ventilation can be valuable but is unpredictable and mechanical ventilation is necessary if a defined airflow rate is desired. Up to about eight air changes per hour can be beneficial but the inside air temperature can never be less than that outside, ignoring any imponderable radiant cooling from the mass of the building structure. Beyond this air change rate diminishing returns occur (Figure 3.11). Noticeable cooling by air movement is better provided by supply air ventilation rather than by extract, the cooling benefit of which escapes notice. See section 3.3.

3.2.2 Practical allowances

Table 3.5 Recommended outdoor air supply rates for sedentary occupants [18]

Condition	Recommended outdoor air supply rate for each person (l/s)
No smoking	8
Some smoking	16
Heavy smoking	24
Very heavy smoking	32

Reproduced by kind permission of the CIBSE.

Fresh air supply rates are conveniently expressed per square metre of floor area, when the population density is predictable, as in offices, and not particularly dense (see Section 3.1.7).

In office buildings, a typical population is 9 m^2/person. On this basis, for a whole building, the fresh air allowance from Table 3.5 would be 0.9 l/s m^2 with no smoking, 1.8 l/s m^2 with some smoking and 2.7 li/s m^2 for heavy smoking. An allowance for an entire office building might be 1.4 l/s m^2 but, where a head count is possible (as in a theatre), the fresh air supply should be per person, not per square metre of floor area.

The American recommendation [19] is 7.5 l/s for each person, regardless of whether there is smoking or not, but this standard has been criticised. The German standard [20] requires 13.9 l/s for each person in open plan offices without smoking, increased to 19.4 l/s when there is smoking.

3.3 VENTILATION

3.3.1 Natural ventilation

There are two sources of natural ventilation: wind effect and stack effect (see Section 3.1.5). The determination of the ventilation rates achieved [4,21] is as follows.

Wind effect — openings on opposite faces of the building

$$Q_{we} = C_d A_w U_z C_p \tag{3.19}$$

where

Q_{we} = volumetric airflow rate due to wind effect (m^3/s)
C_d = coefficient of discharge (usually taken as 0.61)
U_z = mean wind speed at height z above the ground (m/s)
C_p = difference in the pressure coefficient between the windward and leeward sides of the building, typically taken as 1.0
A_w = equivalent area for ventilation (m^2).

The mean wind speed is given by

$$U_z = U_m K_s Z^a \tag{3.20}$$

where

U_z = mean wind speed at a height z above the ground (m/s)
U_m = meterological mean wind speed at a height of 10 m above ground level in open country (m/s)
K_s = parameter relating wind speed to the nature of the terrain
a = exponent relating wind speed to the height above ground
Z = height above ground (m)

Equation 3.19 is for the case when the area of the inlets equals that of the outlets. When the two areas, A_1 and A_2, are unequal the value of A_w is given by

$$A_w = (A_1 A_2)/(A_1^2 + A_2^2)^{0.5} \tag{3.21}$$

Values of K_s and a are shown in Table 3.5.

Table 3.6 Parameters related to wind speed

	Open flat country	Country with scattered windbreaks	Urban	City
K_s	0.68	0.52	0.35	0.21
a	0.17	0.20	0.25	0.33

Reproduced by kind permission of the CIBSE [21]

Stack effect — openings on the same side of the building

An approximate equation [4] that can be used for determining the airflow rate due to thermal forces (stack effect) is

$$\Delta_p = 3462\, h[1/(t_o + 273) - 1/(t_r + 273)] \quad (3.22a)$$

whence, if $(t_r - t_o) \infty 10$ K,

$$Q_{se} = 0.827\, [\,(A_1 A_2) / (A_1^2 + A_2^2)^{0.5}]\, (\Delta_p)^{0.5} \quad (3.22)$$

where

Q_{se} = volumetric airflow rate due to stack effect (m^3/s)
h = the stack height: the vertical distance between the centre-lines of the inlet and outlet openings (m)
t_o = mean outside air temperature over the stack height (°C)
t_r = mean inside air temperature over the stack height (°C)
Δ_p = pressure difference due to the stack height (Pa)

Reference 4 gives other arrangements of inlets and outlets with graphs that allow airflow rates to be estimated, but ventilation is unpredictable and accurate calculation impossible

3.3.2 Mechanical ventilation

Practical values and limitations

In commercial applications mechanical ventilation can only ever be a partial substitute for air conditioning. To deal with sensible heat gains air must be supplied to a room at a temperature less than the temperature desired in the room. Air conditioning achieves a comfortable room temperature of 22°C to 23°C (in the UK) by cooling a mixture of fresh and recirculated air to about 11°C and this requires mechanical refrigeration for much of the year. However, refrigeration plant is not needed when the outside dry-bulb is less than about 10°C or 11°C (for about half the year in London) because such a temperature can be obtained by mixing cooler outside air with warmer recirculated air. For the other half of the year, when it is warmer outside, comfortable temperatures within a room may not be possible by mechanical ventilation alone.

The case is a little worse than this: to distribute the air to the places where it is needed a system of supply ductwork and a fan is necessary. The pressure drop caused by ductwork friction and any related plant is dealt with at the fan by compressing the airstream. This adiabatic compression causes a temperature rise [7] of approximately 1° for each kPa of fan total pressure when the fan and motor are not in the airstream and 1.2°/kPa when they are in the airstream. The fan in a low velocity ventilation system could be developing about 0.25 kPa to 0.35 kPa of fan total pressure and the supply air will then be about 0.25° to 0.42° warmer than the outside air.

Overheating is inevitable in warm weather and openable windows are essential, to let the occupants obtain actual and psychological relief. A strong reason for not opening the windows in an urban environment is that doing so admits noise from traffic, and dirt. Traffic noise varies considerably, and depends on the density of traffic, its mean speed, the angle of the gradient climbed and the distance from the window. It can be highly objectionable to the extent that telephone conversations are impossible when windows are open.

Table 3.7 Typical mean reductions of noise provided by windows over the frequency range 160–3150 Hz

Window type	Mean noise reduction in (dB)
Open window (35% of the inner window-wall area)	14
Fixed single glass or openable single glass with a weatherstrip	25
Sealed unit double glazing	30
Openable double glazing with 200 mm air gap and a weatherstrip	40
Openable double glazing with 400 mm air gap, a weatherstrip and sound absorbing lining on the reveals	45

Theoretical considerations [1, 20, 22–25] suggest that mechanical air change rates up to 15 per hour can give acceptable room temperatures: a good standard would be 24 ± 2°C during a sequence of several warm days. The temperature obtained depends on the heat gains, the building mass and the extent to which it can assist cooling by storing unwanted heat. The temperatures quoted above are for typical modern office blocks and typical sensible heat gains from people, lights and business machines (see Section 3.1.11).

Providing more than about ten air changes per hour yields diminishing returns (Figure 3.11) and if calculation suggests that higher rates are needed, air conditioning is advisable.

From the foregoing it is evident that mechanical ventilation is a partial solution to the provision of comfort, without adopting air conditioning, provided that:

- (i) the building is not in an urban location where dirt and noise will be objectionable; and
- (ii) some measure of overheating in summer is accepted, even though the windows are opened.

Conventional methods of air distribution in rooms for commer-

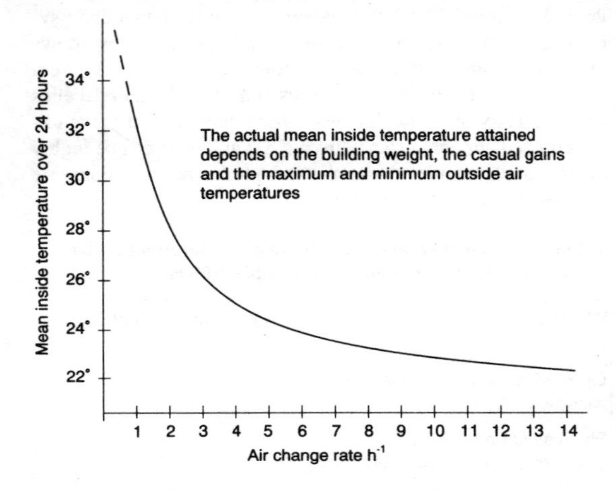

Figure 3.11 Diminishing returns from mechanical ventilation

cial applications involve the supply of air from high level, either through ceiling diffusers or from side-wall grilles. Air is usually extracted mechanically through grilles, the exact location of which is comparatively unimportant because the pattern of the supply air delivery dictates the air movement in the occupied space. Conventional air supply terminals can handle up to about 20 air changes per hour. Beyond that it is impossible to select fittings that will provide comfort in the occupied space, in terms of air movement and noise. A possible alternative is to supply the air at low level and allow it to diffuse upwards, for extraction at high level. A vertical temperature gradient prevails and the temperature in the occupied space will be lower than that of the air removed at high level. There may then be some advantage for rooms treated by mechanical ventilation where the prevailing temperatures could be a little lower than those suggested earlier, although still above the outside air temperature in warm weather. See section 7.8.

The later addition of refrigeration plant to mechanical ventilation systems

This is possible but it may not always give the performance that would have been obtained if air conditioning had been installed at the outset. The following points must be considered:

(i) Even if the system can have a cooler coil added it must handle enough air to deal with the sensible and latent heat gains in the rooms treated.

(ii) If the system is unable to do this, then auxiliary cooling must be provided in the treated rooms — for example as fan coil units.

(iii) The air supplied may have to deal with all or part of the latent heat gains in the rooms.

(iv) If the auxiliary room units deal with any of the latent heat gain then a condensate drainage system must be provided for them.

(v) The noise level in the treated rooms may be increased by the additions proposed.

(vi) The air distribution and hence the air movement and comfort experienced in the rooms may alter.

(vii) Independent thermostatic control must be provided for each treated room.

(viii) The space available in the air handling plant for the addition of the cooler coil must be of adequate size and in a suitable position — preferably on the suction side of the supply fan.

(ix) The face velocity over the cooler coil must not be so high that condensate carryover takes place.

(x) The fan and driving motor must be able to deliver the correct airflow rate after the addition of the extra frictional resistances of the cooler coil, any terminal heater batteries and any necessary air distribution terminals, silencers and ductwork.

(xi) A suitable location must be available for the location of the additional refrigeration plant.

(xii) If an air-cooled, direct-expansion, air cooler coil is added to the air handling plant then the air-cooled condenser, the compressor and the air cooler coil should not be very far apart.

(xiii) If an air-cooled direct expansion system is used then it must be properly controlled to run safely at partial load.

(xiv) If a water chiller is used then it must be properly controlled to run safely at partial load, chilled water storage being provided if necessary.

(xv) The location of the air-cooled condenser or cooling tower must be such as to permit adequate airflow, without short-circuiting.

(xvi) The location of the air-cooled condenser or cooling tower must not cause a noise nuisance to neighbouring property or the environment.

(xvii) The air discharged from the air-cooled condenser or cooling tower must not be directed towards the air intake of the ventilation system or to any neighbouring, openable windows or other air intakes.

(xviii) The location of the refrigeration plant and cooling tower must be carefully considered, so that noise or vibration is not transmitted into any occupied spaces.

(xix) It must be possible to run piping between the water chiller and the cooler coil and between the refrigeration plant and any cooling tower.

(xx) The electrical supply installation for the building must be able to handle the extra loads imposed by the addition of the refrigeration plant and any other necessary equipment.

Industrial ventilation

Mechanical extract ventilation is often provided to remove waste material (such as sawdust and shavings in woodworking machine shops) and to exhaust objectionable or dangerous vapours or solid pollutants from industrial processes. The air quantities handled are directly related to the process and the rate of pollution produced. There may also be a system of supply ventilation to assist the extract process and make good the air removed from the room/building in a balanced and controlled manner. The design of the systems must be primarily concerned with the industrial needs, rather than the comfort of people.

Where it is only necessary to reduce the effects of overheating, mechanical ventilation can be used and in this case supply ventilation is more effective than extract, because it gives a measure of spot cooling, by its directional properties, whereas extract ventilation cannot do this. Both supply and extract are necessary to give a balanced air distribution.

It is also possible to provide the air change rate needed by natural means (without the benefit of spot cooling) and this is helped by the large floor-to-roof heights in some industrial buildings, that assist stack effect.

However the air is removed from a workplace it must be made good by fresh air from outside and this requires heating in winter.

Polluted air cannot be discharged anywhere and treatment may be necessary before the air can be discharged to outside. The requirements of the Health and Safety Executive should be considered and the local Factory Inspector consulted as necessary.

3.4 AIR CONDITIONING

3.4.1 The psychrometric chart

An air conditioning system handles a mixture of a large amount of dry air (mostly nitrogen and oxygen) with a small amount of

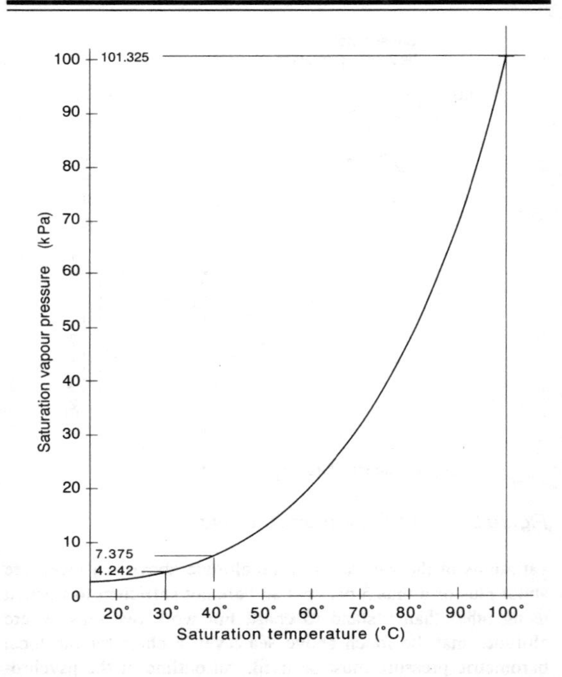

Figure 3.12 The relationship between saturation vapour pressure and saturation temperature

water vapour. The water vapour (steam) is at the same temperature as the dry air but exerts only a small partial pressure because there is not much water vapour present. At this low pressure (Figure 3.12.) the water vapour exists in the superheated or saturated state at the same temperature as the dry air with which it is mixed. There is a large amount of dry air in the mixture so it exerts a large partial pressure. The sum of the partial pressures of the dry air and water vapour is the total pressure of the mixture (the barometric, or atmospheric pressure) and psychrometric charts and tables are published for a quoted barometric pressure. The international standard adopted is 101 325 Pa, which is the mean value at sea level and 45°N latitude. The psychrometric chart shows the relationship between the amount of water vapour in the atmosphere and the temperature and other properties of the mixture, as various processes are carried out by an air conditioning system. The values of the relevant properties and the way in which they change during a process is different if the barometric pressure alters. However, the changes of barometric pressure within the UK, due to

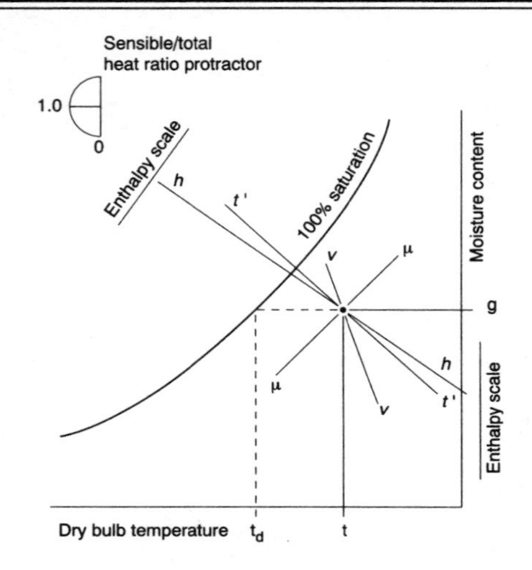

Figure 3.13 The psychrometric chart

variations in the weather and the altitude above sea level, are about plus or minus 5 per cent and are not sufficient to warrant using other than a standard chart. For work overseas, where altitudes may be much above sea level, a chart for the local barometric pressure must be used. An outline of the psychrometric chart is illustrated in Figure 3.13 and the relevant properties are defined as follows:

Dry-bulb temperature (symbol t, °C).
The equilibrium temperature indicated by a dry thermometer, shielded from radiation, over which the velocity of airflow is not less than 4.5 m/s.

Wet-bulb temperature (sling or aspirated) (t', °C).
The equilibrium temperature indicated by a thermometer with a wetted bulb, shielded from radiation, over which the velocity of airflow is not less than 4.5 m/s. This is the wet-bulb temperature generally used and adopted for the CIBSE psychrometric chart. Lines of sling wet-bulb temperature are closer together in the bottom left-hand corner of the chart and further apart in the top right-hand corner.

Wet-bulb temperature (screen) (t'_{sc}, °C).
The equilibrium temperature indicated by a thermometer with a wetted bulb, partially shielded from radiation in a louvred housing, over which the air velocity is unlikely to be above 4.5 m/s. This is the wet-bulb measured by meteorologists. Its value is about 0.5° higher than the sling wet-bulb. It is not shown on a psychrometric chart.

Moisture content (g, kg/kg dry air or g/kg dry air)
The mass of moisture, as dry saturated or superheated steam, mixed with 1 kg of dry air, in an air–water vapour mixture.

Vapour pressure (p_s, Pa)
The partial pressure of the steam mixed with 1 kg of dry air in an air–water vapour mixture. This is not shown on a psychrometric chart.

Relative humidity (ϕ, %)
The ratio of the partial pressure of the steam in an air–water vapour mixture at a given dry-bulb temperature to the partial pressure of saturated steam in an air–water vapour mixture at the same dry-bulb temperature. This is not shown on the CIBSE psychrometric chart. Percentage saturation, which is almost the same, is shown instead.

Percentage saturation (μ,%)
The ratio of the moisture content of an air–water vapour mixture at a given dry-bulb temperature to the moisture content of saturated air at the same dry-bulb temperature. Saturated air is at a relative humidity or a percentage saturation of 100 per cent. Relative humidity and percentage saturation are equal at 100 per cent and at 0 per cent.

Dew-point temperature (t_d, °C)
The temperature at which an air–water vapour mixture becomes saturated. This is when the steam mixed with the dry air is saturated and any further reduction of the mixture temperature will cause condensation (as dew or frost) to form.

Specific volume (v, m³/kg dry air)
The volume of 1 kg of dry air at a given temperature and partial pressure in an air–water vapour mixture.

Enthalpy (h, kJ/kg dry air)
The energy content of 1 kg of dry air together with its associated moisture content, above a datum of 0°C for dry air and liquid water. A useful equation for enthalpy that gives good agreement with values published in the CIBSE tables and shown on the psychrometric chart, over the range from 0°C to 60°C, is

$$h = (1.007t - 0.026) + g(2501 + 1.84t) \qquad (3.23)$$

where
h = enthalpy of an air–water vapour mixture (kJ/kg dry air)
t = dry-bulb temperature (°C)
g = moisture content (kg/kg dry air).

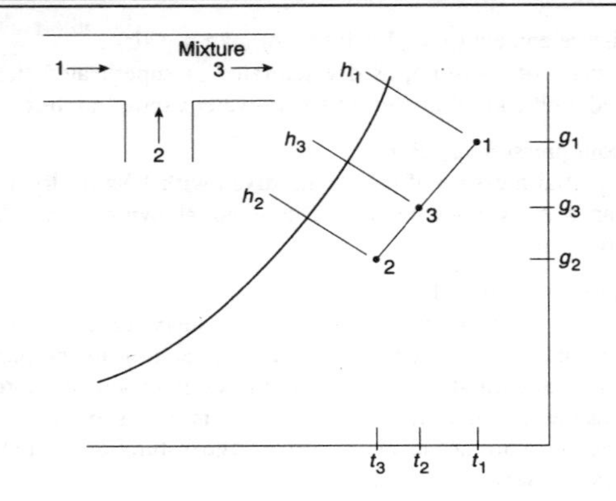

Figure 3.14 A mixing process

3.4.2 Psychrometric processes

Mixing two airstreams

The mixture state, 3, is found by joining the component states, 1 and 2, by a straight line and locating the state 3 on it according to the masses of the mixing components (Figure 3.14).

Sensible heating

Air is warmed, at constant moisture content, from dry-bulb t_1 to dry-bulb t_2. (see Figure 3.15). The process is shown on the psychrometric chart by a straight line joining the entering state, 1, to the leaving state, 2. The mass flow rate entering the heater battery is the same as that leaving it but the volumetric flow

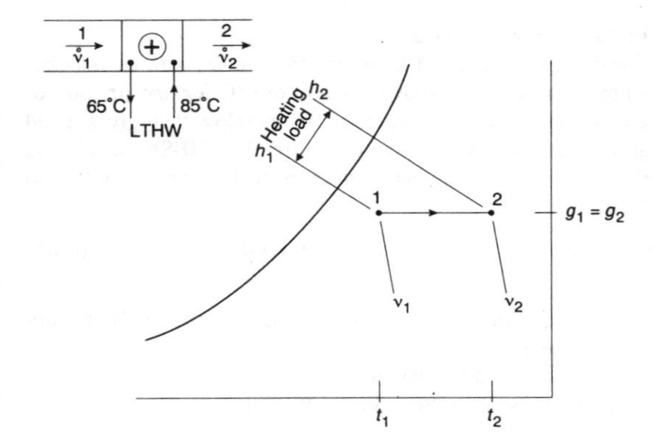

Figure 3.15 Sensible heating

rates are different, the leaving flow rate exceeding the entering flow rate because its temperature is higher and hence its specific volume is greater. If the volumetric flow rate is known, the mass flow rate is expressed by using the correct specific volume.

In general, a cooling or heating load (kJ/s or kW) is given by the product of the mass flow rate of air (kg dry air/s) and the enthalpy change (kJ/kg dry air). In this case, the sensible heating load, Q_h, in kW, is given by

$$\begin{aligned} Q_h &= m\,(h_2 - h_1) \\ &= (\dot{v}_1/v_1)\,(h_2 - h_1) \\ &= (\dot{v}_2/v_2)\,(h_2 - h_1) \end{aligned} \tag{3.24}$$

where
m = mass flow rate of dry air (kg dry air/s)
h = enthalpy of the air (kJ/kg dry air)
$\dot{v}$ = volumetric flow rate of air (m^3/s)
v = specific volume of the air (m^3/kg dry air).

Cooling and dehumidification

See Figure 3.16. Air is cooled from state 1 to state 2. The process is shown on a psychrometric chart by a straight line joining the entering and leaving states, 1 and 2. If extended, this line must be able to cut the saturation curve at a point 3. The point 3 is termed the apparatus dew point and its temperature is also the mean surface temperature of the cooler coil.

The cooling load (kW) is given by the product of the mass flow rate of air (kg dry air/s) and the enthalpy drop $(h_1 - h_2)$ (kJ/kg

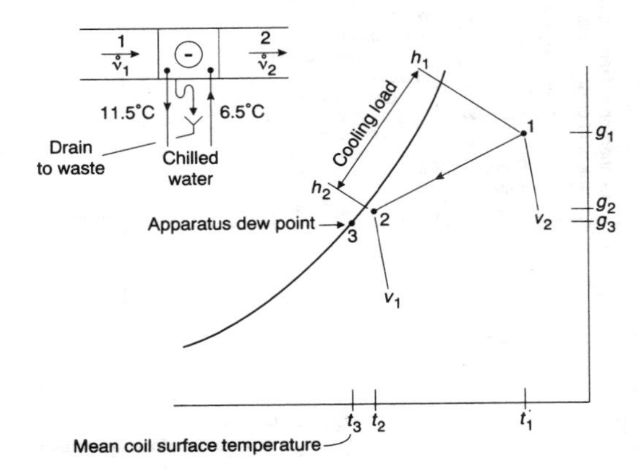

Figure 3.16 Cooling and dehumidification

dry air). The mass flow rate may be expressed in terms of the entering volumetric airflow rate and the entering specific volume or in the corresponding leaving properties.

The effectiveness of a cooler coil is defined by its contact factor, β :

$$\beta = (h_1 - h_2)/(h_1 - h_3) \quad (3.25)$$
$$\beta = (g_1 - g_2)/(g_1 - g_3) \quad (3.26)$$
$$\beta = (t_1 - t_2)/(t_1 - t_3) \quad (3.27)$$

The by-pass factor equals 1 – β.

Equations (3.25) and (3.26) are exact but equation (3.27) is approximate, although quite accurate enough for all practical purposes. This is because the only linear properties on the psychrometric chart are moisture content and enthalpy. All the other properties are non-linear, some obviously so, such as percentage saturation. Dry-bulb temperature is not quite linear but for most cases it can be regarded as so. The only temperature line that is at right angles to the zero moisture content line is that for 30°. All the other dry-bulb lines diverge slightly from it.

Sensible cooling

Figure 3.17 shows a process of sensible cooling from state 1 to state 2. If the process line from 1 to 2 is extended to 3, this point cannot lie on the saturation curve. The point 4, is the dew-point, t_{d1}, of the entering air state 1. If 3 coincided with 4

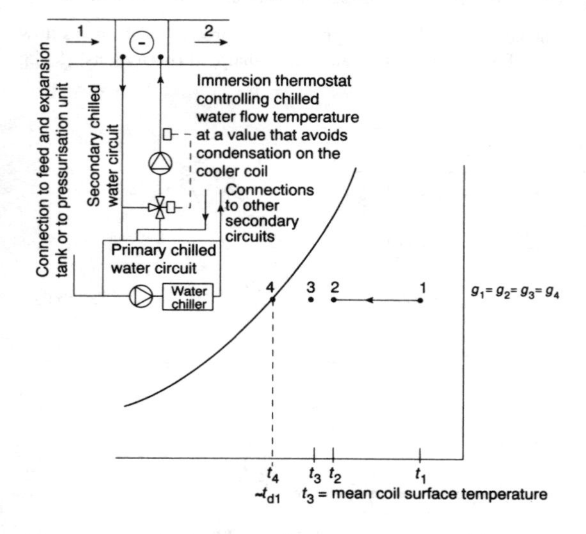

Figure 3.17 Sensible cooling

there would be a contradiction because the temperature of 3 is the mean coil surface temperature which, being a mean, implies that some of the coil surface is at a temperature less than the mean and this would be less than the entering air dew-point. Some dehumidification would then occur and the process would not be one of sensible cooling.

The consequence is that the chilled water flow temperature onto the cooler coil must be controlled at a value that will ensure that none of the surface of the coil is at a temperature less than the entering air dew-point. Hence it is necessary to have a secondary chilled water circuit as well as a primary circuit, as Figure 3.17 shows.

Humidification

In the past, humidification has been achieved by passing air through a spray chamber handling water recirculated from a sump or by injecting spray water directly into the airstream. This is no longer acceptable because of the hygienic risk to people in the occupied space. The method now adopted is to inject dry steam into the airstream (Figure 3.18). The process line is almost a dry-bulb line unless the temperature of the steam is very high. Even then the inclination to the dry bulb line is only about four degrees of angle. If the steam is superheated the inclination to the dry-bulb line will, of course, be greater. The steam must be injected at a place in the air conditioning system where the air can accept the moisture added. For example, if steam is injected after the cooler coil the air, being almost saturated, cannot accept any more moisture so the steam will collect as condensate on the floor of the ducting and

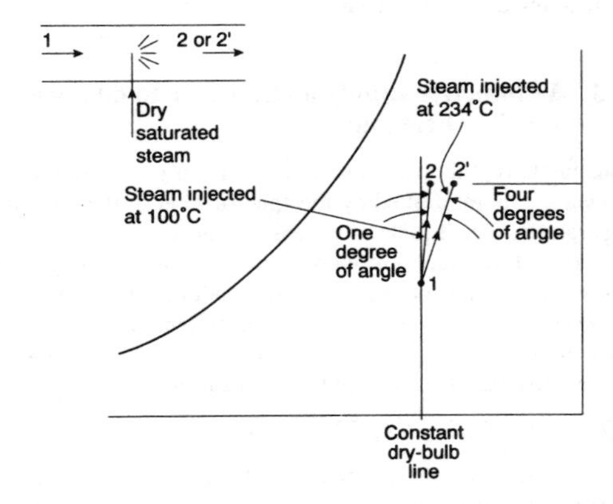

Figure 3.18 Humidification by dry steam injection

cause problems. Dry saturated steam is best injected as far as possible from the plant, preferably very close to the final supply point to the conditioned room. Sometimes, in industrial applications, the steam is injected in the room itself.

As with other psychrometric processes, the humidification load is the product of the mass flow rate of the airstream (volumetric flow rate divided by the appropriate specific volume) and the enthalpy change.

Pre-heating and re-heating

Pre-heating. Systems that handle 100 per cent outside air and include chilled water cooler coils must have pre-heaters to warm the air before it flows over the cooler coil, for frost protection in cold winter weather. If there is a risk of freezing fog then the air filter should also be protected by a frost protection pre-heater. Even if there is no cooler coil, it is often necessary to warm the air to a temperature that will avoid condensation on duct walls and will not cause discomfort if delivered to an occupied space. When a mixture of outside and recirculated air in handled it is usually unnecessary to pre-heat the fresh air component, unless this is large and the mixture state would have a low temperature. Poor mixing and stratification may also sometimes require the outside air component to be pre-heated. The psychrometric process occurs along a line of constant moisture content.

Re-heating. Re-heater batteries are used to warm the air leaving a cooler coil to a higher temperature and the re-heater is usually controlled from the air temperature of the room to which the air is being supplied. The psychrometric process occurs along a line of constant moisture content.

3.4.3 Volumetric supply airflow rate to deal with a sensible heat gain

Fundamentally, the mass flow rate of the air supplied to a conditioned space, multiplied by the specific heat of the air and its temperature rise, equals the sensible heat acquired by the airstream. However, it is inconvenient to deal in mass flow rates of airstreams because all air handling and distribution equipment is expressed in terms of volumetric flow rates. Hence it is useful to develop an equation relating volumetric flow rate to sensible heat gain (Q_s). This is as follows:

$$Q_s = [\dot{v}_t \, \rho_0 \, (273 + t_0)/(273 + t)] \, c \, (t_r - t_s)$$

The expression in square brackets is the mass flow rate, involving the volumetric flow rate, $\dot{v}_t$, at temperature t, a standard density ρ_0 at a standard temperature t_o, and a density correction term $(273 + t_o)/(273 + t)$, according to Charles' law. The specific heat of air is c, and the room and supply air temperatures are t_r and t_s, respectively.

The following standard values are chosen:

$$\rho_o = 1.191 \text{ kg/m}^3,\ t_o = 20°\text{C and } c = 1.026 \text{ kJ/kg °C}$$

Inserting these values and re-arranging:

$$\dot{v}_t = \frac{Q_s}{(t_r - t_s)} \times \frac{(273 + t)}{358} \qquad (3.28)$$

In the above equation note that:

(i) If Q_s is in kW, $\dot{v}_t$ is in m³/s but if Q_s is in W then $\dot{v}_t$ is in l/s.

(ii) The symbol t is the temperature at which the volumetric airflow rate, $\dot{v}_t$, is to be expressed.

3.4.4 Volumetric supply airflow rate to deal with a latent heat gain

The moisture picked up by an airstream as it flows through a room is expressed by the product of the mass flow rate of air (kg dry air/s) and the difference between its initial and final moisture contents (kg moisture picked up/kg dry air). Each kg of moisture acquired by the airstream represents a latent heat gain corresponding to its latent heat of evaporation (h_{fg}).

As before (Section 3.4.3), it is convenient to convert the mass flow rate into a volumetric flow rate with a Charles' law temperature correction:

$$Q_l = [\dot{v}_t \rho_o (273 + t_o)/(273 + t)](g_r - g_s)\, h_{fg}$$

The term in square brackets is the mass flow rate of air, g_r and g_s are the moisture contents of the room air and the supply air, and h_{fg} is the latent heat of evaporation of water.

Taking 1.191 kg/m³ as the density of air at a temperature of 20°C and adopting 2454 kJ/kg moisture as the latent heat of evaporation of water, the following equation is developed for a relationship between the volumetric airflow rate, $\dot{v}_t$ at a temperature t, and latent heat gain, Q_l:

$$\dot{v}_t = \frac{Q_l}{(g_r - g_s)} \times \frac{(273 + t)}{856} \qquad (3.29)$$

In the above equation note that:

(i) The moisture contents of the room air and the supply air, g_r and g_s, are in g/kg dry air (not kg/kg dry air).

(ii) t is the temperature at which the volumetric airflow rate is expressed.

(iii) If the latent heat gain is in W the volumetric air flow rate is in l/s but, if it is in kW, the airflow rate is in m³/s.

3.4.5 The choice of a suitable design supply air state

If the sensible gain is known and a value chosen for the supply air temperature, t_s, the necessary volumetric supply airflow rate can be calculated from equation (3.28). Equation (3.29) is then used to determine the supply air moisture content required to deal with the latent heat gain, using the same volumetric airflow rate. The air supplied to the conditioned room does two things simultaneously: it absorbs the sensible gain as its temperature rises from t_s to t_r and absorbs the latent gain as its moisture content rises from g_s to g_r. The air supplied must be cool enough to deal with the sensible heat gain and dry enough to deal with the latent heat gain.

The ratio of the sensible to the sensible plus latent heat gains in a conditioned room will follow a process line having this slope as it diffuses through the room, absorbing the sensible and latent gains simultaneously. For the summer design heat gains a line drawn on the psychrometric chart through the room state point and having such a slope is termed the design room ratio line and a protractor appears in the top left-hand corner of the chart giving the ratios of sensible to total heat gains. Use is made of this when choosing a suitable supply air temperature.

The practical considerations when choosing a design supply air temperature are: the lowest safe air temperature possible from an air cooler coil, the contact factor of the coil, and the rise in temperature from fan power (Section 3.3.2) and duct heat gain (Section 3.1.13). An allowance must also be made for the minimum fresh air to be supplied.

A suggested procedure is as follows (see Figure 3.19):

(1) Identify the design room state, *R*, and design outside state, *O*, on a psychrometric chart. This is usually for 1500 h suntime in July in the UK.

(2) Knowing the design sensible and latent gains, calculate the slope of the design room ratio line. Draw a line through the origin of the sensible-total heat ratio protractor with this slope.

(3) Draw the design room ratio line through the room state, *R*, parallel to the line through the protractor.

(4) Knowing the design occupancy of the room or module and the minimum design fresh air allowance, calculate the minimum fresh airflow rate to be supplied.

(5) Knowing the type of system to be installed, make an allowance for the probable air temperature rise through the supply air fan and air duct, and the extract fan and duct.

(6) Add the temperature rise through the extract fan and duct to the room temperature, t_r, and identify the

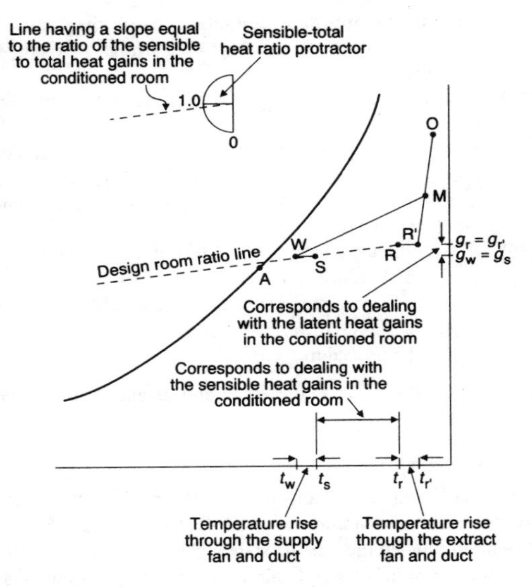

Figure 3.19 The choice of a suitable supply air state, *S*

recirculation air state, *R'*, on the chart, with the same moisture content, g_r, as the room state, *R*, but having a higher dry-bulb temperature, $t_{r'}$.

(7) Make a first choice of supply air dry-bulb temperature, t_s, about 8° or 10° less than the room dry-bulb, t_r, and identify the corresponding supply air state, *S*, on the design room ratio line.

(8) Using equation (3.28) calculate the supply airflow rate.

(9) Knowing the minimum fresh airflow rate to be supplied, calculate the fractions of fresh and recirculated air.

(10) Determine the mixture state, *M*, and identify this on the psychrometric chart.

(11) Knowing the temperature rise through the supply fan and duct system, identify the off-coil state, *W*, having the same moisture content, g_w, as the supply state, g_s, but with a temperature, t_w, that is lower than that of the supply air by the amount of the air temperature rise through the supply fan and duct.

(12) Join the states *M* and *W* by a straight line on the psychrometric chart.

(13) Identify the apparatus dew-point, *A*, on the chart so that the points *M*, *W* and *A* lie on a straight line and *A* is on the saturation curve.
(14) Determine the cooler coil contact factor by equation (3.27).
(15) Knowing the probable number of rows for the cooler coil likely to be used (see Section 3.6) in the air handling plant, review the practical value of the contact factor determined.
(16) If the contact factor has a practical value, retain the choice of *S*, made in (7) above. If it is not a practical value go back to step (7) and choose another value for t_s, half a degree different from the earlier choice, and repeat the procedure.

The procedure is simpler than it may appear and, with a little practice, is easy to use.

EXAMPLE 3.4

Given sensible and latent gains of 1401 W and 134 W, respectively, for a room conditioned at 22°C dry-bulb, and 50 per cent saturation when the outside state is 28°C. 19.5°C wet-bulb (sling), select a suitable supply air state for a simple all-air system. Assume that the minimum fresh air quantity is 24 l/s expressed at the supply state. The temperature rise through the supply fan and duct system is 1.5° and the rise through the extract fan and duct is 0.2°.

Answer (see Figure 3.20).

Identify the points *O* (28°C dry-bulb, 19.5° wet-bulb, 10.65 g/kg), *R* (22°C dry-bulb, 50 per cent saturation, 8.366 g/kg) and *R*' (22.2°C dry-bulb, 8.366 g/kg) on a psychrometric chart.

Sensible/total ratio = 1401/(1401 + 134) = 0.91

Identify this on the protractor and draw the design room ratio line through R.

Make a first choice of the supply air temperature, t_s, say 13°C. Identify the supply air state, *S*, on the psychrometric chart. By equation (3.28)

$$\dot{v}_{13} = \frac{1401}{(22-13)} \times \frac{(273+13)}{358}$$

$$= 124.4 \text{ l/s at } 13°\text{C}$$

Proportion of fresh air = 24/124.4 = 0.193.

Proportion of recirculated air = 100.4/124.4 = 0.807.

Mixture state (*M*):

$$t_m = 0.193 \times 28 + 0.807 \times 22.2 = 23.3°\text{C dry-bulb}$$

$$g_m = 0.193 \times 10.65 + 0.807 \times 8.366 = 8.807 \text{ g/kg}$$

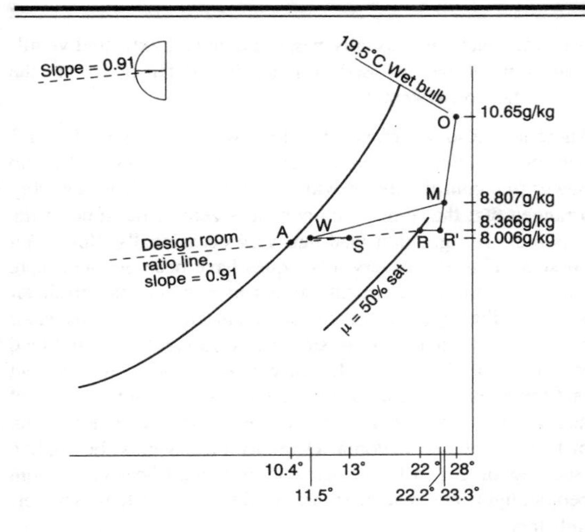

Figure 3.20 Psychometry for Example 3.4

Identify the off-coil state, *W*:

$t_w = 13° - 1.5° = 11.5°\text{C dry-bulb}$

Join *M* to *W* and identify *A* at 10.4°C on the saturation curve. By equation (3.27):

$$\text{contact factor} = (23.3 - 11.5)/(23.3 - 10.4) = 0.91$$

This is probably a practical contact factor for a cooler coil with six rows of tubes, 319 fins/m (8 fins/inch) and a face velocity of about 2.25 or 2.5 m/s. (see Section 3.6.)

Use equation (3.29) to determine the supply air moisture content:

$$g_s = 8.366 - \frac{134}{124.4} \times \frac{(273 + 13)}{856} = 8.006 \text{ g/kg}$$

This identifies the supply state, *S*, at 13°C dry-bulb and 8.006 g/kg on the psychrometric chart.

Since the contact factor is practical, accept this as the supply air state.

3.4.6 Maximum refrigeration load for a building

The components of the refrigeration load for a building may comprise:

sensible heat gain + re-heat + latent heat gain + fresh air load + supply fan power + supply duct gain + recirculation fan power

+ extract duct gain (usually negligible unless extracted ventilated light fittings are used) + fan coil unit fan power (in the case of fan coil systems).

There may also sometimes be an allowance of 1 per cent or 2 per cent to cover heat gain to chilled water piping and pump power (the contribution of which is very small). It is generally arranged that the re-heat component is zero at the time of the maximum refrigeration load but, very occasionally, this is not possible. If heat recovery techniques have been adopted there may be additional elements, aimed at reducing the fresh air load by utilizing some of the cooling capacity of the exhaust air before it is discharged to waste. In calculating the cooling load for the whole building, allowance must be made for the fact that the maximum simultaneous sum of the sensible and latent heat gains to the rooms or modules is not the same as the sum of their individual maxima. Diversity factors must be applied (see Section 3.1.7) to the sensible and latent heat gains from people, lights and machines, provided that the system used permits this.

The largest refrigeration load usually occurs at the time for which the fresh air load is greatest. This is commonly at about 1500 h suntime in July, in the UK, when the outside air enthalpy is at its maximum.

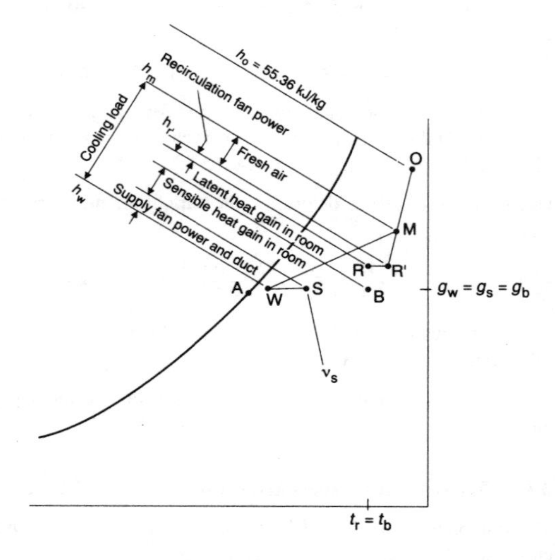

Figure 3.21 The components of the cooling load for an all-air system

Owing to their importance in the air conditioning design process, cooling loads are always checked and this should be done in a way that is as different as possible from the manner in which they were first calculated.

Figure 3.21 illustrates how this is done for a simple, all-air system. Five components of the cooling load are shown: the fresh air load, recirculated fan power, latent heat gain, sensible heat gain and the load due to the supply fan power and supply duct gain. The only one of these that cannot be checked in an independent manner is the fresh air load, which equals the enthalpy difference $(h_m - h_{r'})$ multiplied by the mass flow rate of air supplied, $\dot{v}_{ts}/v_s$. The relevant proportion of the extract fan power can be determined by equation (3.28), using only the fraction of the supply air that is recirculated. The supply fan power and the supply duct gain can also be independently determined by equation (3.28), using the full supply airflow rate, and the sensible and latent heat gains have been previously established by entirely different methods, prior to reaching this stage of the calculations. Hence the check on the refrigeration load, Q_{ref}, is as follows:

$$Q_{ref} = (\dot{v}_{ts}/v_s)(h_m - h_w) \quad (3.30)$$

$$= \text{Fresh air load} + \text{Recirculated fan power} + \text{Latent heat gain} + \text{Sensible heat gain} + \text{Supply fan power and duct gain}$$

$$= (\dot{v}_{ts}/v_s)(h_m - h_{r'}) + (OM/MR')(\dot{v}_{ts})(t_r' - t_r)358/(273 + t_s) + \text{Latent heat gain} + \text{sensible heat gain} + (\dot{v}_{ts})(t_s - t_w)358/(273 + t_s)$$

EXAMPLE 3.5

Calculate the cooling load for the case of Example 3.4 and check your answer.

Answer (see Figure 3.21)

Fresh airflow rate = 24 l/s at 13°C (19.3 per cent). Supply airflow rate

$$= \dot{v}_{13} = \frac{1401}{(22-13)} \times \frac{(273+13)}{358}$$

$$= 124.4 \text{ l/s at } 13°\text{C} = 100 \text{ per cent.}$$

Recirculated airflow rate = 100 – 19.3 = 80.7 per cent.

$$t_m = 0.193 \times 28 + 0.807 \times 22.2 = 23.3°\text{C}$$

$$g_m = 0.193 \times 10.65 + 0.807 \times 8.366 = 8.807 \text{ g/kg}$$

The outside air enthalpy, h_o, is 55.36 kJ/kg from a psychrometric chart and, by equation (3.23)

$h_{r'}$ = (1.007 × 23.3 – 0.026) = 0.008366 (2501 + 1.84 × 22.2)
= 43.59 kJ/kg

hence

h_m = 0.193 × 55.36 + 0.807 × 43.59 = 45.86 kJ/kg

For the off-coil state, W:

t_w = 13 – 1.5 = 11.5°C

By equation (3.29)

$$g_w = g_s = 8.366 - \frac{134}{124.4} \times \frac{(273 + 13)}{856} = 8.006 \text{ g/kg}$$

By equation (3.23)

h_w = (1.007 × 11.5 – 0.026) + 0.008 006 (2501 + 1.84 × 11.5)
= 31.75 kJ/kg

From the psychrometric chart or by interpolation in psychrometric tables, the specific volume at the supply state is

$\dot{v}_{13}$ = 0.8163 m³/kg.

$$\text{Cooling load} = \frac{0.1244}{0.8163}(45.86 - 31.75) = 2.150 \text{ kW}$$

Check:	watts	%
Sensible heat gain:	1401	66
Latent heat gain:	134	6
Supply fan and duct:		
(124.4 × 1.5 × 358)/(273 + 13) =	234	11
Recirculation fan:		
0.0807(124.4 x 0.2 × 358)/(273 + 13) =	25	1
Fresh air:		
(0.024/0.8163)(55.36 – 43.59) × 1000 =	346	16
Total cooling load	2140	100

This is within one half of one per cent of the other answer and represent a good check. If the check is more than two percent different from the original calculation a search should be carried out to determine the error.

3.5 AIR CONDITIONING SYSTEMS [11]

3.5.1 Unitary systems

Self-contained, room air conditioning units

Each unit comprises a direct-expansion air cooler coil (see Section 8), an air-cooled condenser, a hermetic compressor and fans to circulate the air over the cooler coil and the condenser coil. The whole assembly is contained within a sheet steel casing and units are usually mounted in a hole cut in an external wall, beneath a window sill. A simplified diagram is given in Figure 3.22. The air-cooled condenser rejects to outside the sensible and latent heat gains from the room, plus the heat corresponding to the power absorbed by the compressor and the heat liberated by the fans and driving motors.

Automatic control over room temperature is by cycling the compressor on–off but sometimes this is in sequence with a heater battery, which may be electrical or may use low temperature hot water (LTHW). Condensate that forms on the cooler coil drains into a collection tray and is piped to a slinger ring, mounted on the periphery of the condenser fan. The rotation of the fan and ring scatters the condensate over the fins of the condenser, where it evaporates, with a small increase in heat rejection but with extra corrosion. It is often arranged that some outside air is drawn into the unit cabinet, through the hole in the wall.

Such units have been used extensively, worldwide and are in a range of refrigeration capacities from 1.75 kW to 9 kW. Larger self-contained units are also available, up to about 60 kW of refrigeration but these cannot be mounted under window sills. They are usually located against an exterior wall, so that the condenser can project through the wall, for heat rejection to outside. They are used to air condition small shops and the like.

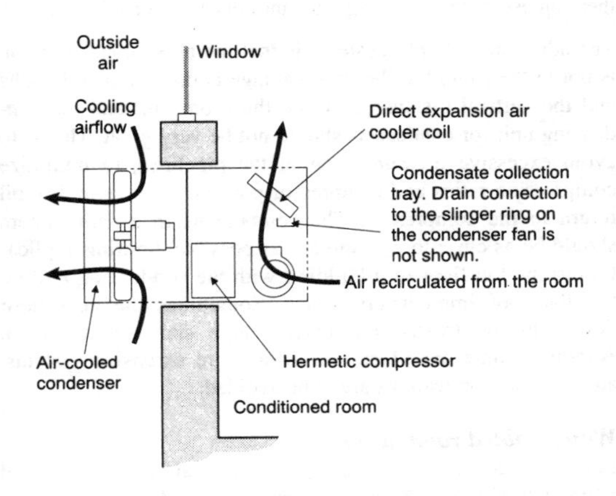

Figure 3.22 A diagram of an air-cooled, self-contained, room air conditioning unit. Refrigerant pipelines not shown

Stopping and starting units too often and operating them in cold weather will cause the compressor motor to burn out. If a filter is provided at the room recirculation grille this will reduce the airflow as its gets dirty and also lead to motor burn-out, unless it is regularly cleaned. Unit lives are between one and ten years, depending on the aggressive nature of the climate and the length and frequency of use.

Split-system air-cooled units

A condensing unit (compressor plus condenser) is located on the roof above the room or on a nearby exterior wall. The room unit consists of a direct expansion air cooler coil and a recirculation fan. The piping connections from the room unit to the condensing unit are comparatively small in diameter, comprising a suction line up to the compressor and a liquid line down from the condenser (see Chapter 8). Sometimes only the air-cooled condenser is on the roof and the compressor is in the room unit. The rising pipe to the condenser is then the hot gas line but the descending pipe is still the liquid line. Condensate drainage piping from the cooler coil must be provided.

Supply and extract, ducted mechanical ventilation is desirable to give the occupants the necessary amount of fresh air and to provide a measure of pressurization to discourage infiltration. When an auxiliary ducted supply of outside air is provided in this way the fresh air introduced increases the sensible and latent heat gains in the room and this must be dealt with: equations (3.5 and 3.17) are relevant. The symbol n in these equations then represents the air change rate introduced mechanically.

The advantage of split systems is that the noisy condenser fan is not in the room but the disadvantage is that the pipe lengths and the vertical distance between the room unit and the condensing unit (or condenser) should not be very great. This is to avoid excessive pressure drop in the pipelines, to minimize compression ratios and compressor powers, and to simplify oil return to the compressor. The components in a split system should be as close to one another as possible, limiting application to the top floor of a building with the condensing unit on the flat roof immediately above. Sometimes the next floor down can be treated but careful pipe sizing is essential. Although some manufacturers offer more extensive systems, adventurous applications are to be avoided.

Water-cooled room units

Each unit consists of a direct-expansion air cooler coil and recirculation fan, a hermetic compressor and a water-cooled condenser. A two-pipe system distributes cooling water from a remote cooling tower, desirably via a plate heat exchanger to ensure clean water flow to the room units.

There is no hole in the exterior wall of the room and there is no limitation on the lengths of piping, hence permitting an entire multi-room building to be treated.

It is customary to run the coils in the units wet and drain the condensate by a system of piping. Mechanical ventilation is desirable.

Water-cooled room units operate at lower condensing pressures than do air-cooled units and consequently less electrical power is needed to drive the compressor and the units are quieter. Changes in outside wet-bulb temperature are slower than changes in the dry-bulb, hence allowing water-cooled units to operate for longer periods when the outside air state is higher than the summer design state.

Water-cooled rooms units are available over roughly the same range of cooling capacity as self-contained, air-cooled room units but, in a larger form, they can be used for cooling duties up to about 250 kW of refrigeration.

Water loop air conditioning/heat pump units

Units are similar in form and content to the water-cooled units described above and, when operating as air conditioners, the condensers reject heat into a two-pipe system that distributes water from a cooling tower. The water flow rate through the units is critical and it is essential that a plate heat exchanger is interposed between the clean units and the dirty water in the cooling tower. The units should be fitted with bypass connections that allow the piping system to be flushed through, when commissioning, without passing dirt through the unit coils.

When working as air conditioners (Figure 3.23 (a)), hot gas flows from the compressor to the condenser and cold suction gas flows from the direct-expansion air cooler coil (the evaporator) to the compressor. A typical water temperature rise through the condenser is from 27°C to 38°C.

To work as a heat pump (Figure 3.23(b)), a reversing valve in the unit casing is operated to alter the direction of flow of the refrigerant gas. Hot gas flows from the compressor to the coil in the unit, which now acts as an air-cooled condenser, rejecting heat into the room in order to offset heat losses.

In warm weather all the units act as air conditioners and put heat into the water loop for rejection at the cooling tower. During mid-season, heat is transferred, through the water loop, from the side of the building where there is a net heat gain to the other side, where there is a net heat loss. Any surplus heat in the loop is dealt with by the cooling tower and any deficit of heat is made good by a boiler. In cold weather a boiler is

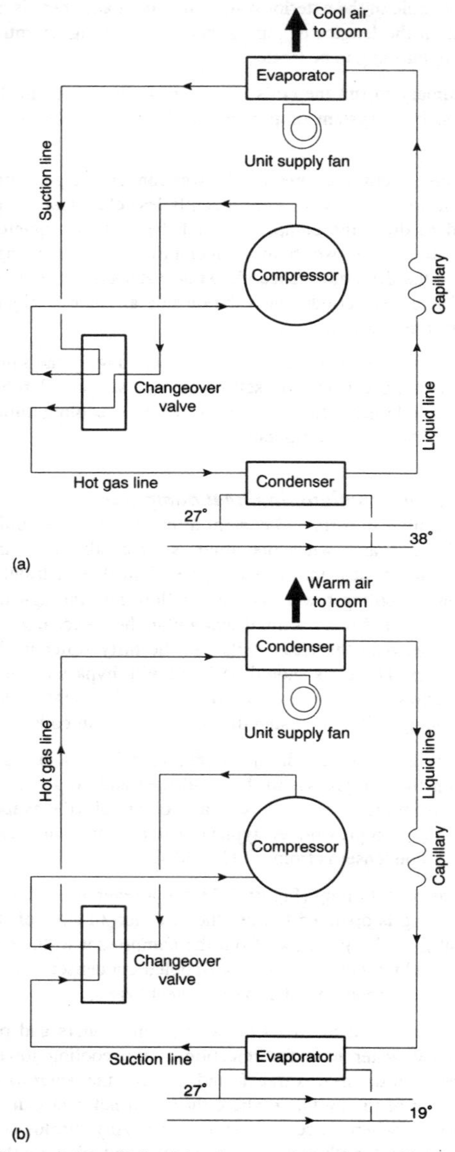

Figure 3.23 Principle of operation of water loop air conditioning/heat pump units: (a) Working as air conditioning units; (b) Working as heat pump units.

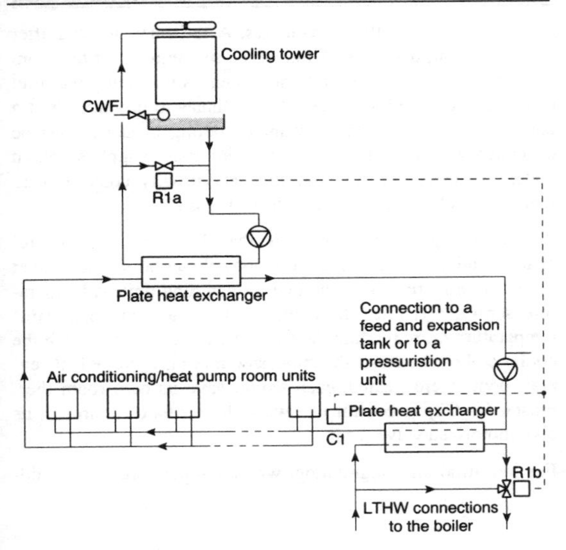

Figure 3.24 Schematic arrangement of piping for water loop air conditioning/heat pump units

required to supply most of the heat needed but part of the heat is supplied electrically by the motors driving the compressors in the units, in order to transfer the heat from the water loop to the coil in the unit, for rejection into the room. Figure 3.24 shows a schematic arrangement of the piping.

It must be remembered when selecting the units, that the presence of the necessary plate heat exchangers requires a small extra temperature difference, of a degree or two, in order to effect the heat transfer.

Condensate drainage and mechanical ventilation are essential.

3.5.2 All-air systems

Constant volume re-heat and sequence heat

For most of the time the heat gains in an air conditioned room are less than the design heat gains and, in winter, heat losses occur. Hence, with a system that delivers a constant airflow rate to the room the supply air temperature must be varied to suit the changes in the gains/losses. A re-heater, located after the cooler coil, does this. Although close control over the room temperature can be achieved, the system is inherently wasteful of energy because the cooler coil continues to produce air at a constant temperature and unwanted cooling capacity must be cancelled with re-heat: the refrigeration plant supplies chilled water to the cooler coil and runs at virtually constant load while the boilers provide the re-heat (Figure 3.25).

The advantage of the system is that, by using multiple re-heaters, independent temperature control can be provided over several rooms treated by the one air handling unit. A temperature sensor, C1, after the cooler coil in the plant, maintains temperature t_w, regulating the flow of chilled water through the cooler coil by means of the motorized mixing valve, R1. If several rooms were treated, each would have its own room thermostat (as C2) to control its own re-heater battery through its own motorized valve (as R2).

For industrial air conditioning, where temperature and humid-

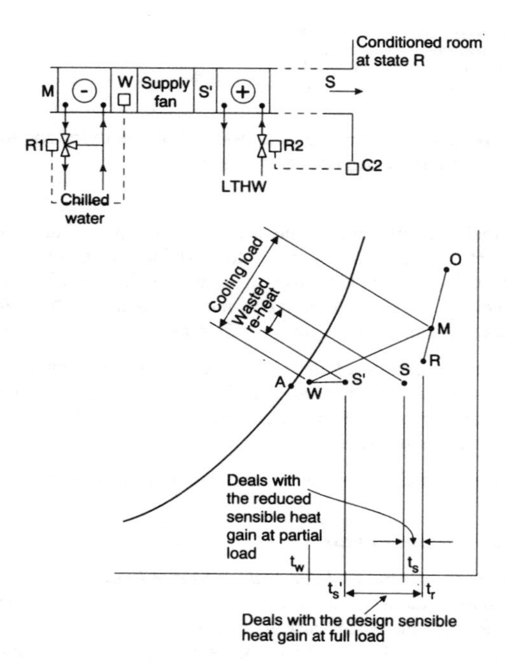

Figure 3.25 Re-heat at partial load

ity must be controlled independently in different rooms, such an arrangement would be acceptable, it then being necessary to provide dry steam injection in the supply duct to each room, immediately before the supply terminal.

For small commercial applications it is not acceptable and the alternative used is to operate the heater battery in sequence with the cooler coil. Figure 3.26 shows the psychrometry, 100 per cent fresh air being used for simplicity. As the sensible gains diminish, the room temperature falls and the thermostat C1a starts to open the by-pass port of the mixing valve, R1a, on the cooler coil. After the port is fully open, the control valve on the re-heater, R1b, begins to open.

As the chilled water flow rate through the coil is reduced at partial load (Section 3.6.5), the moisture content of the air leaving the coil increases. This gives a higher humidity in the treated room, under conditions of reduced sensible heat gain. It is therefore usual to fit a high limit humidistat, C1b, in the

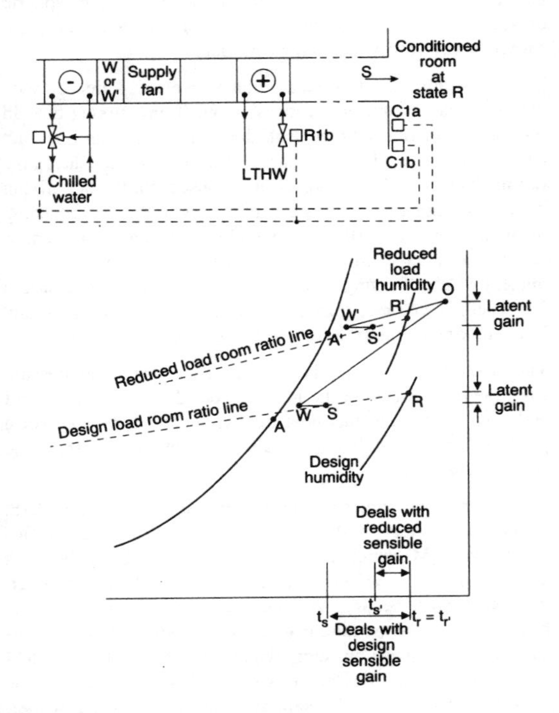

Figure 3.26 Cooling and heating in sequence

conditioned room. If the humidity rises, the control sequence between the heater battery and the cooler coil is interrupted and the humidistat starts to close the by-pass port of the mixing valve on the latter to increase its dehumidifying capacity. Meanwhile, until the humidity is corrected, room temperature is controlled by the heater battery, the system temporarily acting like a constant volume re-heat system.

Double duct systems

The air handling plant filters a mixture of recirculated and fresh air which it delivers into a pair of ducts, one containing a cooler coil and the other a heater battery (Figure 3.27). The hot and cold ducts feed mixing boxes, distributed throughout the building on a modular or a room basis. Each box mixes the hot and cold airstreams under thermostatic control for the module or room. Hot and cold ducted air distribution must be at high velocity because of the building space occupied, but it is desirable to use a low velocity duct system for extract because of the unstable nature of ducts containing air at sub-atmospheric pressures and because of potential problems with any air pressure reducing valves that may be needed.

When first introduced, serious problems were experienced with the high velocity ducted air distribution. If any mixing box, in response to thermostatic control, drew more air from one duct and less from the other, the change of airflow in each duct gave rise to changes in the static pressure. Consequently other mixing boxes had higher or lower static pressures in the ducts feeding them, upsetting the airflow supplied by the box and the thermostatic control achieved. The difficulty was solved by the development of a self-acting constant volume regulator to ensure a fixed supply volume, regardless of changes in duct pressure. Every mixing box must have a constant volume regulator.

The functions of a mixing box are: to reduce the static pressure from the high values in the hot and cold ducts, to mix the hot and cold airstreams thermostatically, to attenuate any noise produced and to supply a nominally constant airflow rate to the conditioned room.

The summer and winter psychrometry is shown in Figure 3.28. In summer, outside (O) air mixed with recirculated air (R') gives a state (M) which, after passing through the supply fan is at state M' and is warm enough to be the hot duct state, H, without further heating. As the season becomes cooler the heater battery warms the air to a temperature that is compensated against outside air temperature. The coil in the cold duct cools and dehumidifies air to a state W, which becomes the cold duct state, C, usually kept at a constant value. In winter, when it is cold enough, motorized mixing dampers in the air

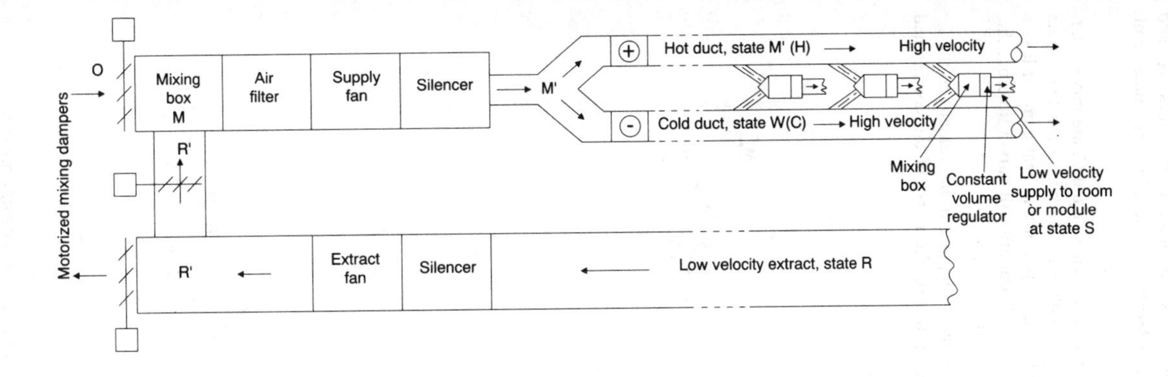

Figure 3.27 Duct distribution diagram for double duct mixing boxes

handling plant (Figure 3.27) can achieve the necessary cold duct state, C, (Figure 3.28(b)) without using the cooler coil, which is off.

The hot and cold airstreams, at states H and C, are mixed thermostatically to give the supply state, S, which has the correct dry-bulb temperature to offset the sensible heat gains or losses from the conditioned room. Humidity will vary a little but this does not matter for comfort.

After mixing, the airstream is supplied at low velocity to the air distribution terminals (grilles or diffusers) in the conditioned

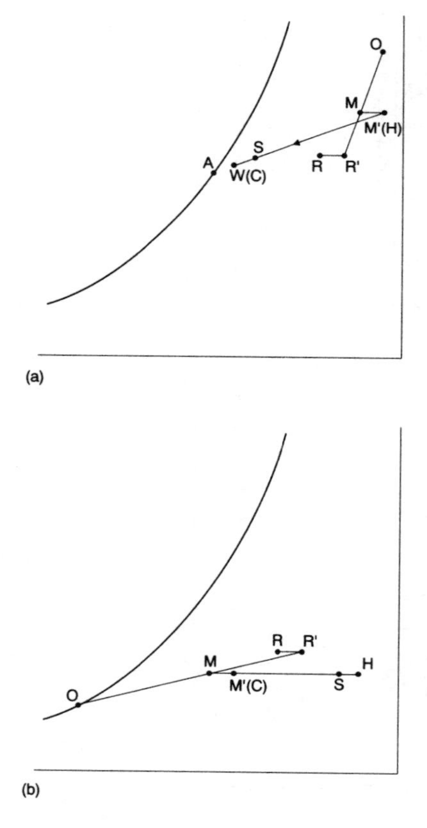

Figure 3.28 Double duct system psychrometry: (a) summer psychrometry; (b) winter psychometry

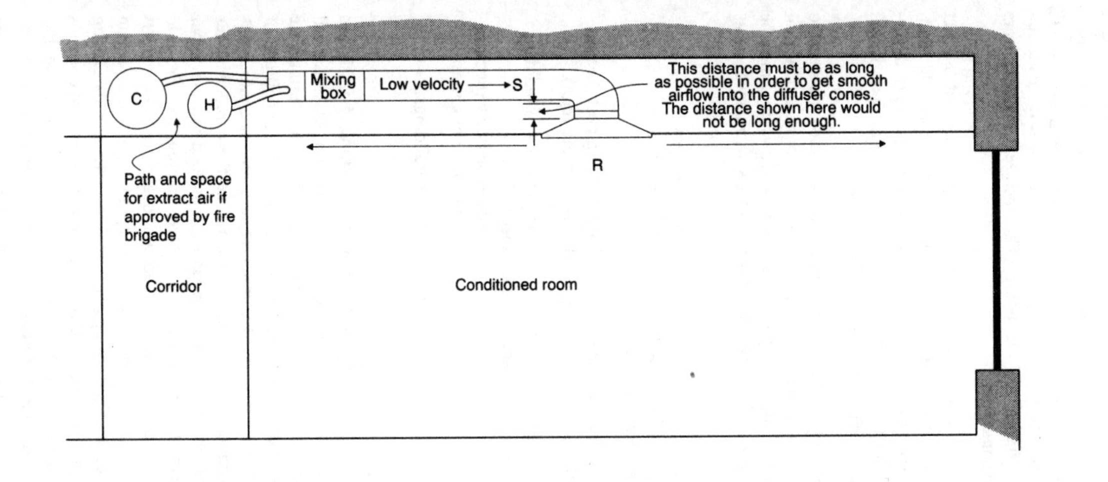

Figure 3.29 Double duct air distribution for a room. Accommodating the ducts requires a lot of building space and is a problem

space. Figure 3.29 illustrates this and it is evident that accommodating the ducts poses difficulties. It is essential that the longest possible straight duct should feed vertically into the ceiling diffuser (the length shown in the figure is not enough). This is to ensure smooth, quiet airflow into the diffuser cones. If insufficient space between the suspended ceiling and the soffit of the slab is allowed for this, the airflow will be disturbed and the diffuser will be noisy. The diffuser manufacturers should be consulted.

There are also cross-over problems. This is in spite of the fact that the hot duct main becomes smaller than the cold duct, nearer to the fan discharge, since a bigger temperature difference (say 35°C – 20°C) is available to deal with heat losses than to deal with heat gains (say 23°C – 14°C).

As a result of these restrictions on space, duct velocities tend to be higher than engineering prudence would suggest, fan total pressures are high, fan powers are high and systems may be noisy. Summarizing, the advantages and disadvantages are as follows:

Advantages	*Disadvantages*
Full cooling and full heating can always be available.	High capital cost.
100 per cent fresh air can be handled, if desired.	High running cost (because fan powers are high).
There are no dirty heat transfer surfaces in the conditioned room.	Occupies a lot of building space.
	Tends to be noisy (because of high duct velocities

The system is not suitable for modern office buildings.

Multizone units

See Figure 3.30. A supply fan blows a mixture of fresh and recirculated air over a pair of coils arranged with the heater coil above and the cooler coil beneath. These are often termed the hot and cold decks. On the downstream face of the coil the air flows through a motorized damper section, divided into zones of approximately equal size. For each zone the damper blades have common vertical spindles and are divided into an upper section, over the hot deck, and a lower section over the cold deck. The dampers for the hot and cold decks are mounted 90° out of phase, so that rotation of the spindles in one direction opens the upper blade group (say) while simultaneously closing the lower group. Thus a signal from a room thermostat operates the zone damper motor to mix an upper airstream that has passed through the hot deck with a lower airstream from the cold deck. The multizone unit operates like a double duct mix-

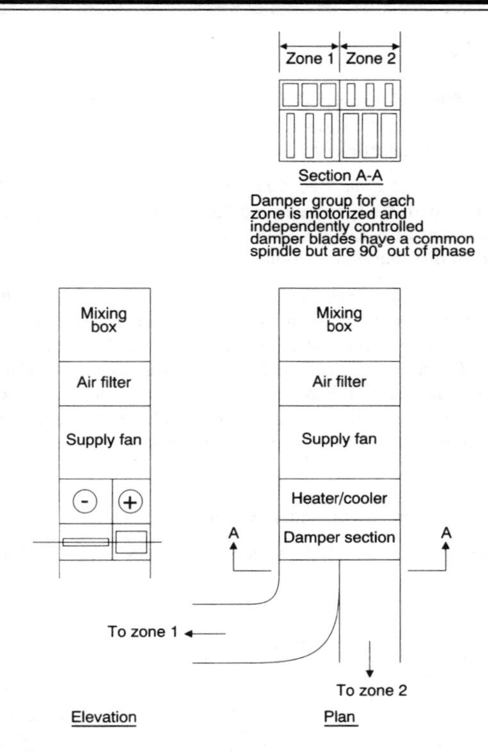

Figure 3.30 Multizone unit

ing box, except that mixing takes place in the plant room instead of in the conditioned room. Separate ducts from the plant feed the zones and air distribution is at low velocity. It follows that plant rooms must be located fairly close to conditioned areas.

Variable air volume (VAV)

Principle. The supply air temperature is kept at a nominally constant value and cooling capacity is varied, to match changes in the sensible heat gain, by altering the supply airflow rate, in accordance with equation (3.28). The attractive features of this are that no wasteful re-heat is needed to control room temperature and that, potentially, since fan power is proportional to volumetric airflow rate (see Chapter 6), running costs should be reduced.

Fan power, W_f, in watts, is expressed by

$$W_f = p_{tF}\ \dot{V}_t/\eta \tag{3.31}$$

where

p_{tF} = fan total pressure (Pa)
$\dot{V}_t$ = volumetric airflow rate (m^3/s)
η = total fan efficiency as a fraction.

Room air distribution (see Chapter 7). This is of critical importance with VAV systems and Figures 3.31 and 3.32 illustrate the cases of constant volume air distribution and variable volume air distribution in a room.

Cool air supplied at, say 13°C to offset sensible heat gains is denser than the warmer air in the room and would fall into the room and cause a draught if counter-effects did not prevail. The jet of supply air is delivered into the room above the occupied zone and, by entrainment with rising convection currents, its mass increases and its temperature approaches the room air temperature, because momentum flow is conserved:

$$m_1V_1 = m_2V_2$$

where m = mass (kg) and V = velocity (m/s).

Air flowing over the ceiling suffers a frictional pressure drop

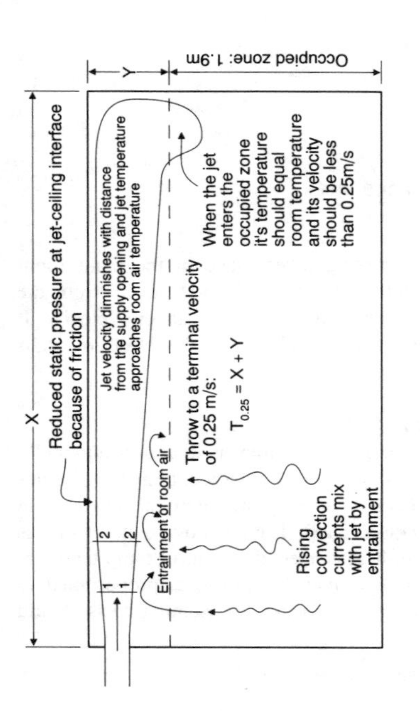

Figure 3.31 Typical constant volume air distribution. Momentum flow is conserved and, with reference to section 1–1 and 2–2: $m_1V_1 = m_2V_2$; as the mass, m, increases by entrainment, the jet velocity, *v*, reduces

and the static pressure at the jet–ceiling interface falls below the air pressure in the room under the jet. Hence, the static pressure in the room tends to press the jet onto the ceiling. This is called the Coanda effect. The effect diminishes with increasing distance from the supply opening because the jet velocity reduces as air is entrained from the room, causing the frictional loss to reduce, this being dependent on the square of the velocity.

Air is supplied to a room through sidewall grilles, linear slot ceiling diffusers, or circular ceiling diffusers (see Chapter 7), in order of the length of throw they give, circular ceiling diffusers having the least throw. Throw is defined by the following (see Figure 3.31):

$$T_{0.25} = X + Y \tag{3.32}$$

Dumping. As the airflow from a supply air is throttled the momentum reduces and the jet may not entrain enough room air to continue to increase its temperature. The anti-buoyancy of the cold jet may then overcome the Coanda effect and the air could leave the ceiling, enter the occupied part of the room to cause local discomfort as a draught. This is called dumping.

A pattern of rotational air movement may be set up in the far side of the room (Figure 3.32). When the VAV terminal later increases the airflow rate, the energy in the eddies may be sufficient to prevent the jet from re-attaching to the ceiling. The supply airflow must then increase to a bigger percentage than the one at which it dumped, so that it has enough energy to disperse the pattern of eddies and re-attach to the ceiling. The

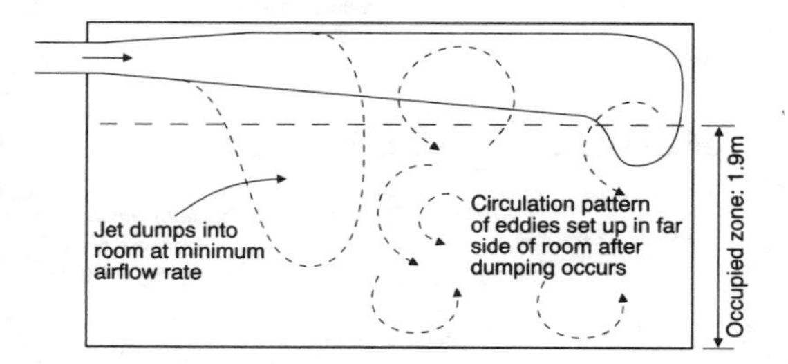

Figure 3.32 The anti-buoyancy of the cold jet causes it to dump into the occupied zone at minimum allowable airflow rate. Considerably more than this minimum rate may then be needed for the jet to have energy to break up the pattern of eddies that develops after dumping, and re-attach to the ceiling

extent to which dumping is possible with supply air terminals depends on the type of terminal used and its geometry.

Terminal geometry. The geometry of the supply air terminal influences the minimum airflow rate to which a variable air volume system can turn down

(i) *Fixed geometry supply terminals* (Figure 3.33(a)). A room thermostat, C1, sends a signal to a VAV regulator, R1, to throttle the supply airflow rate in response to diminished heat gains. The air velocity on the low pressure side of the VAV box also reduces as less air is supplied and hence the velocity over the ceiling falls off rapidly with a reduction in airflow. As a result, the air may leave the ceiling before the end of its intended throw and enter the occupied zone. This imposes a lower limit of about 40 per cent of the design airflow rate for this type of supply terminal.

(ii) *Variable geometry supply terminals* (Figure 3.33(b)). The branch duct contains a pressure regulating valve, R1, that maintains a constant static pressure on its downstream side and in the neck of the diffuser, before a movable plate. The static pressure is sensed by C1. A room thermo-

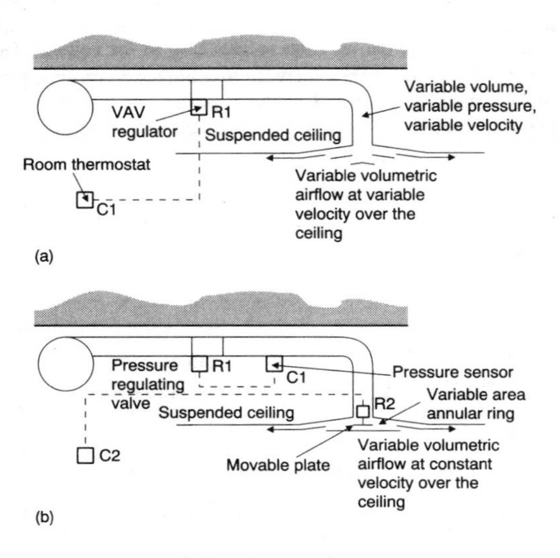

Figure 3.33 The principles of VAV terminal geometry: (a) constant geometry VAV terminal, (b) variable geometry VAV terinal

stat, C2, controls the position of the movable plate through a motor, R2. The movement of the plate provides a ring of varying area, through which air can flow to the diffuser. Since static pressure is constant on the upstream side of the plate the velocity through the annular ring is also constant, as the volumetric airflow rate varies. Constant air velocity exists over the ceiling and the Coanda effect prevails for a longer distance. Terminals can throttle to about 25 per cent of the design rate. Various proprietary devices are available that assist the Coanda effect by splitting the airflow into two parts, one constant and the other variable. The variable part of the jet is delivered next to the ceiling and the minimum, constant airflow is supplied beneath, helping to keep the variable jet on the ceiling.

Pressure dependence. As with the double duct system, when one VAV terminal throttles the flow of air, the static pressure of the duct for all the other VAV units alters, because the ducted airflow rate and frictional pressure drop also change. The performance of the unit becomes unstable if it has no in-built device that stabilizes the static pressure of the air fed to the throttling part of the terminal. This is illustrated in Figure 3.33(a) where the fixed geometry unit has no pressure stabilizer and the static pressure before the unit is unpredictable, depending on the performance of the other VAV terminals in the rest of the system. VAV terminals with this characteristic are termed pressure dependent. They should never be used in medium or high velocity duct systems because duct pressure variations will be large and the performance of the VAV terminals will vary enormously as a consequence. They can be used on small systems with low velocity duct distribution, because the variations in static pressure in the ducts will be comparatively small. There will, however, be some variation in the performance of individual terminals and this must be recognized and accepted.

If a VAV terminal has an in-built pressure regulator there will be a stable static pressure on the upstream side of the throttling section of the terminal and its performance will not be affected by static pressure changes in the ducting as the duties of other units change. Such VAV units are pressure independent.

Heating methods. A VAV system has no inherent ability to heat. The controlling thermostat sends a signal to the terminal to throttle the airflow upon fall in air temperature and eventually the unit delivers its minimum airflow rate, before dumping. It is possible to reverse the action of a thermostat but this would be of no value if the ducted air were simply heated in winter: adjoining rooms might be simultaneously suffering heat gains and heat losses and the warmed airflow would be unable to satisfy both.

Three methods are available for dealing with heat losses, without sacrificing control:

(i) *VAV with compensated perimeter heating.* Radiators, finned tube, or the like, are located around the perimeter and fed with LTHW, the flow temperature of which is compensated against outside air temperature. The VAV system deals with any heat gains. No account is taken of casual gains and hence the system is somewhat wasteful of thermal energy. Nevertheless, the method is the cheapest of the three and has proved commercially popular, with fairly effective results.

(ii) *VAV units with terminal re-heaters.* Each VAV unit is provided with its own re-heater battery which is commonly fed with compensated LTHW but, sometimes, it may be electrical. Upon fall in room temperature the VAV terminal first throttles the constant temperature cold airflow to its minimum rate. Upon further fall in air temperature the airflow rate is kept constant but the re-heater comes on to warm the air. The VAV terminal operates in sequence with the heater battery and control over the latter can be simple proportional or proportional plus integral (see Chapter 4). Occasionally, particularly when the heating is electric, the re-heater control may be two-position.

The system is thermally more efficient than the one using perimeter heating because cooling and heating are in sequence and re-heating is only applied when the airflow is at its minimum value. Casual gains are taken account of. One difficulty is that since the VAV terminals are usually above the suspended ceiling heat is provided in the wrong place: the best place is beneath windows where down-draughts, cold infiltrating air and so-called cold radiation can be dealt with. One manufacturer deals with this by locating the VAV units in the ceiling next to the windows and arranging to blow the air across the ceiling, back into the room when cooling, but downwards over the window when the re-heater is on.

(iii) *Double duct VAV.* Upon fall in room temperature cold air fed to the VAV box is throttled to its minimum value, no hot air being used. Upon further fall in room temperature warm air is drawn from the hot duct and mixed with air from the cold duct, the total supply airflow being kept constant at its minimum value. Eventually the terminal is delivering only hot air to the room.

The system suffers from all the objections of the constant volume double duct system plus any problems that may arise with the variable air volume system. It has not been popular and is not recommended.

Fan-assisted VAV. Figure 3.34 shows a notional fan-assisted VAV terminal. At full cooling duty the fan and the heater battery are both off. Medium velocity ducted air from the central air handling plant is fed to the VAV box at the back of the terminal, which is mounted above the suspended ceiling of the room being treated. Damper flat D1 is closed by the static pressure of the air ducted in from the VAV box and damper flap D2 is held open. Air bypasses the fan through D2 and then flows through the heater battery into the low velocity duct leading to the air distribution diffuser in the ceiling. Upon fall in room temperature, sensed by C1, the VAV box throttles towards its minimum stable airflow rate. On further fall in temperature the fan starts. The static pressure at fan suction opens damper D1 and the static pressure at fan discharge closes damper D2. The static pressure in the ceiling void is greater than in the box and air flows through D1 into the fan suction chamber of the unit. The fan delivers the minimum stable airflow rate, which consists of a mixture of air from the VAV box and air from the ceiling void. Upon further fall in room temperature the heater battery motorized valve, R1b, opens.

An alternative, sometimes used when the sensible heat gains are particularly high, is to fit a cooler coil behind damper D1 in order to cool the air drawn from the ceiling void. The fan runs continuously and the system is no longer VAV.

Noise from VAV terminals. Noise can travel in the air path along the supply duct and through the VAV terminal directly into the room. Noise may also break out through the duct walls and pass through the suspended ceiling into the room.

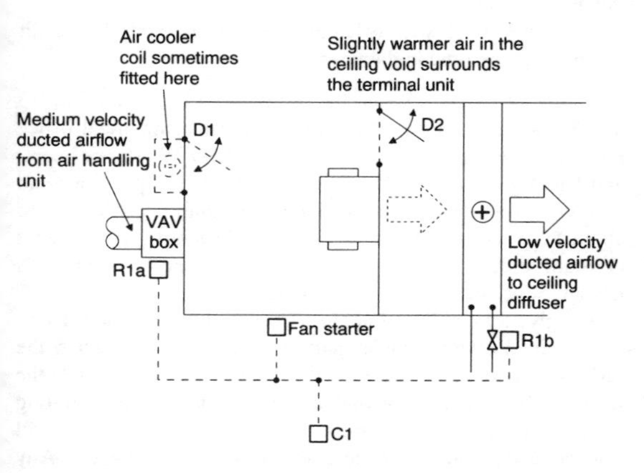

Figure 3.34 A notional fan-assisted VAV terminal unit

Turbulence within the VAV terminal can generate additional noise which will be emitted as acoustic radiation from its walls.

Terminals emit less noise as the volumetric airflow rate is throttled but more noise as the upstream static pressure rises. An increase in upstream pressure is the inevitable consequence of throttling and it is essential that fan capacity is controlled, as the system duty diminishes, to prevent this happening. If it is well engineered, fan capacity control can result in the system operating at less than its design noise rating for much of the time, since airflow rates are mostly less than the design rate.

Fresh air supply. A typical design supply airflow rate for a commercial office building is about 8 l/s m^2, referred to the treated floor area (see Example 3.4). Of this, a suitable minimum fresh air supply rate would be 1.4 l/s m^2, which is 17.5 per cent of the whole. As the system throttles to its minimum stable supply airflow rate of, say, 30 per cent, the fresh air rate falls to 0.42 l/s m^2, which is not enough. The fresh air supply rate must be increased as the system throttles to ensure an adequate average supply to the entire building. It should be possible to measure continuously the flow rate through the fresh air intake duct to the air handling unit and to increase this to at least the minimum, as necessary. Alternatively, since it is necessary to control fan capacity as the system throttles (see Section 6.9), the total rate could be monitored and the fresh air intake dampers modulated to give the correct quantity, in relation to the total airflow rate.

However, it is usual to vary the proportions of fresh and recirculated air handled by the central plant as the outside psychrometric state changes with the seasons. Figure 3.35 shows how this is done to economize in the energy for refrigeration. It follows that a VAV system is handling a large amount of fresh air for much of the time in the year.

Humidity variation. Using the results from Examples 3.2, 3.3 and 3.4 it can be shown that, for a typical commercial office module, if design sensible heat gains of 1247 W (100 per cent) and latent gains of 134 W change to 249 W (20 per cent) and 134 W, respectively, the room relative humidity will rise to about 59 per cent. This is if the dry-bulb temperature is kept constant at 22°C by throttling the airflow to 20 per cent of its design value. Allowing the room temperature to drop to 20°C would give a humidity of about 67 per cent at the reduced sensible gain. However, under partial load conditions, when the airflow rate over the face of the cooler coil is reduced, the dehumidification exceeds that achieved under design operating conditions (see Section 3.6.5). This means the humidity will not increase as much as the above calculations suggest. Any humidity change with a VAV system will be tolerable in a

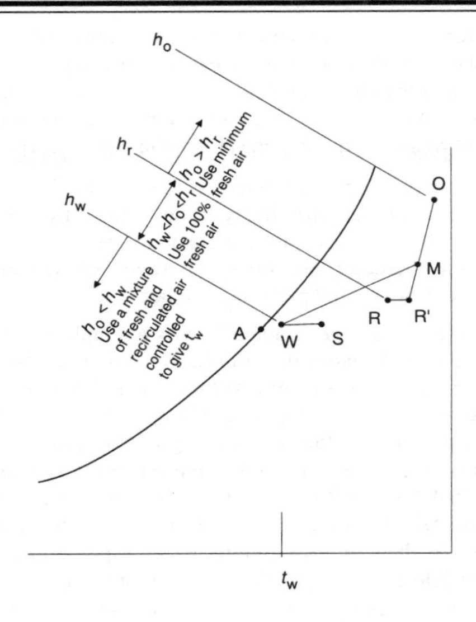

Figure 3.35 Varying the mixing proportions of fresh and recirculated air, as the seasons change, to economize on the energy used for refrigeration

temperate climate, since it is a comparatively unimportant part in human comfort. In a hot humid climate further thought should be given to possible humidity changes at partial load, before a VAV system is used.

The size of the air handling plant. The air handling plant must be large enough to supply the air necessary to deal with the maximum simultaneous sensible heat gains for the part of the building served by the plant. This means that the maximum simultaneous sensible heat gains must be calculated (see Section 3.1), taking account of diversity factors for people, lights and business machines and also allowing for the natural diversity of the solar heat gain as the sun moves round the building. Hence an air handling plant should deal with opposite faces of the building, which suffer maximum solar gains at different times of the day.

Duct sizing See Chapter 7. Ducts at the ends of the system, feeding air to individual VAV terminals, must be large enough to handle 100 percent of the design air flow rate, no diversity factors being allowed for the heat gains. On the other hand, having calculated the maximum simultaneous sensible heat gains for the part of the building served by a particular air handling plant, the total air quantity handled by the plant is known. This is less than the sum of the individual design air quantities for the VAV terminals because diversity factors have been allowed when determining the air handling plant duty.

Prudent engineering judgement is needed to decide on the appropriate air quantities likely to be handled by the ducts in between the plant and the remote terminal units. Sizing the ducts for these air quantities should be conservative with respect to velocity and pressure drop rate.

Fan capacity control. Figure 3.36 is a simple VAV system, showing only the index duct run (with the largest total pressure drop) to a single VAV unit. At the design airflow rate the fan runs at speed n_1 and its characteristic *p-v* curve intersects the system *p-v* curve at the point P_1 to give flow rate v_1. The system curve shown as a broken line represents the system without a terminal VAV unit present. The design pressure drop across the terminal unit is the vertical distance between the points P_1 and Q_1. As the terminal unit throttles in response to changes in the sensible heat gain in the treated room, the system curve rotates anti-clockwise about the origin and cuts the fan curve at the point P_2 in order to give an airflow rate v_2. The pressure drop across the terminal unit has risen considerably and equals the vertical distance between the points P_2 and Q_2.

If a pressure sensor, S_1, is placed immediately before the terminal unit an increase in static pressure can be used to reduce the fan speed from n_1 to n_2. The system curve will then intersect the fan curve for speed n_2 at the point P_3 and the pressure drop across the terminal unit is very much less. The unit operates more quietly and the fan power used is greatly reduced, since it is proportional to the product of the airflow rate and the fan total pressure (see equation (3.30)).

The intersection of the fan and system curves will be along a control line, shown chain dotted in the figure. If the sensor S_1 exercises simple proportional control over fan speed the pressure drop across the terminal unit will rise somewhat as the terminal throttles, but if proportional plus integral control is exercised (Chapter 4) this can be avoided and the pressure drop across the unit kept at its design value, regardless of the airflow rate.

In a real installation, involving many units, the picture is more complicated and cannot be easily analysed. Common practice has been to place the pressure sensor in a position shown by Z in the figure, which is between two-thirds and three-quarters of the way from the fan discharge to the index terminal. If direct digital control of pressure is used at each unit, it should be possible to scan the static pressures of all the units in the system, at

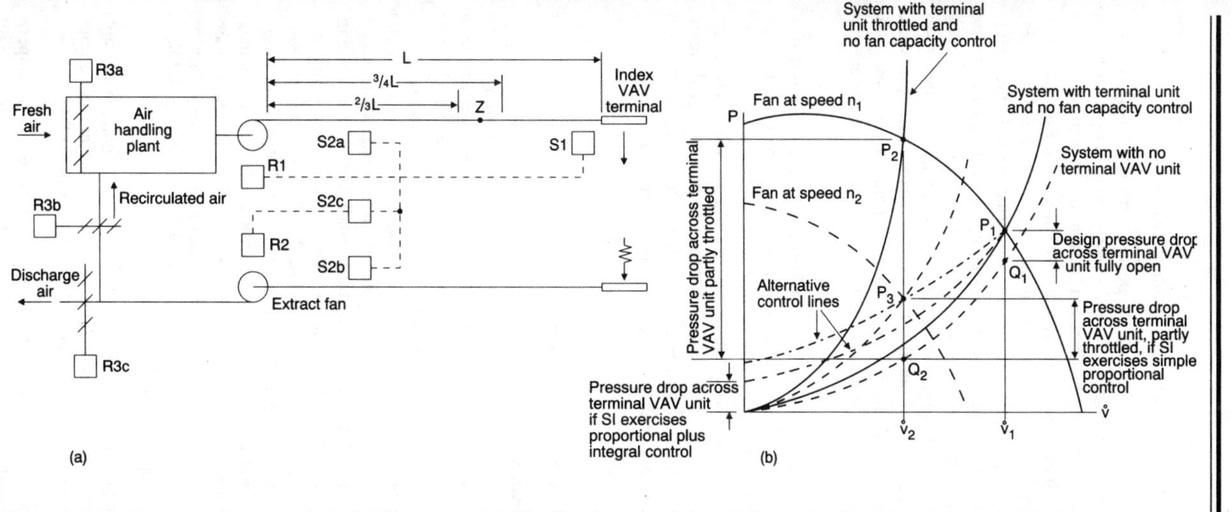

Figure 3.36 Fan capacity control of a VAV system. (a) Plant and duct schematic for a simple notional system. (b) Pressure-volume relationships

short intervals of time, and to use the signal from the unit having the lowest pressure to control fan speed.

The capacity of the extract fan must be turned down in harmony with that of the supply fan. If a pair of velocity pressure sensors (S2a and S2b) is located in sections of the main supply and extract ducting where representative values of the air quantities are flowing, the measured values can be compared (by S2c) and the capacity of the extract fan adjusted in small increments until it is handling the correct quantity of air, in relation to the supply air quantity. It is no good using the signal from the pressure sensor, S1, regulating the capacity of the supply fan, to control the extract fan directly: the two fans are different types, they have different pressure–volume characteristics, and they are in different systems, which also have different characteristics.

Even if the supply and extract fans are properly controlled in unison and the VAV terminal units are supplying the correct air quantities, there is a chance that static pressures in the rooms could vary slightly. This may be noticed when doors are opened or shut. Using variable volume extract terminals regulated in parallel with individual supply terminals might deal with this but would be a very expensive solution.

3.5.3 Air water systems

Fan coil systems

Fan coil units. Fan coil units are small air handling plants, usually located in the room being treated. They are simple in construction and are fed with chilled water and/or LTHW. Sometimes they are provided with electric re-heating. Air is recirculated from the room and blown over a cooler coil and fresh air may be supplied, depending on the application and the economics. Although intended for installation as free-standing, sheet metal, cased units, they are commonly located above suspended ceilings and compete with VAV systems. In the simplest form of system the units are fed with chilled water from a central water chiller and run wet with condensate in summer. There are no arrangements for supplying fresh air and condensate drainage must be provided. This may be satisfactory for dealing with heat gains but any needs for ventilation and heating are not considered.

Although fan coil units are available in quite large sizes, suitable to deal with small shops, bars, etc., they are more commonly used in commercial office buildings in the UK.

The cooling capacity of the unit depends on the flow temperature and flow rate of chilled water, the entering air dry-bulb and wet-bulb temperatures, the number of rows of finned tube in the cooler coil and the fan speed. Sensible cooling capacities range from 500 W to 5000 W and total capacities from 700 W to 6500 W. Chilled water flow rates lie between about 0.05 and 0.4 l/s with pressure drops through the cooler coil of 5 to 25 kPa.

The heat liberated by the fans in the units is deducted by the manufacturer before quoting sensible cooling capacity but the heat gain to the chilled water resulting from this must be included by the system designer when calculating the temperature rise of the chilled water flowing through the coil.

Single phase, permanent split-capacitor motors are used to drive the centrifugal fans in the units, with absorbed powers in the range from 20 W to 190 W. For a building of any size switchgear is necessary to give a random start for the fan coil units when the system is first switched on in the morning.

Units can generally operate at high, medium or low fan speed, absorbing about 140, 100 and 75 per cent power, with respective total cooling capacities of approximately 118, 100 and 83 per cent.

For most applications the units are selected to provide sensible cooling to offset sensible heat gains. To do this the chilled water flow temperature must be controlled at a value giving a cooler coil surface temperature above the room dew-point.

The general behaviour of cooler coils is dealt with in Section 3.6 and much of what is dealt with applies equally to the behaviour of fan coil units.

Automatic control of fan coil units. The possibilities are:

(i) On–off control for the fan. Temperature control may be acceptable but the variation in air movement and noise is usually not acceptable in the UK.

(ii) Two-position control over the chilled water flow rate by a solenoid valve. Temperature control can be good, if the air distribution in the room is good. The thermal inertia of the materials of the room construction damps the variation in room temperature. Noise and air movement are unchanged. A consequence of two-position control at the units, which must not be ignored, is that the chilled water flow rate in the piping circuit feeding the fan coil units will vary. This must not be allowed to affect the chilled water flow rate through the chiller. Hence the chiller must have its own primary, pumped, chilled water piping circuit and the fan coil units must have their own secondary, pumped, chilled water piping circuit. This is necessary anyway to control the flow

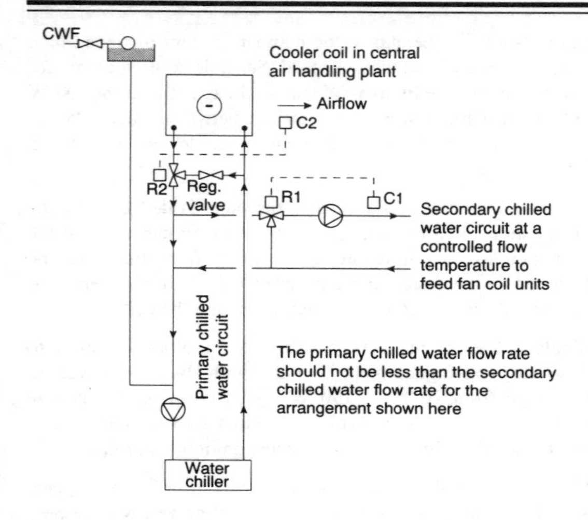

Figure 3.37 A simple arrangement of primary and secondary chilled water circuits for fan coil units that are to do sensible cooling only

temperature to the fan coil units. See Figures 3.37 and 3.38.

(iii) Modulating control using two-port motorized valves. This gives good control over temperature, particularly if the valve is characterized (see Chapter 4). Primary and secondary piping circuits are needed (Figure 3.37). With two-port valves, two-position or modulating, branches feeding remote units may become dead legs under partial load operation. To avoid this and keep branches alive, a three-port valve or a cross-connection with a bleed valve should be fitted at the end of branches. (Figure 3.38).
The build up of excessive pressure and noise at throttled valves under partial load, must be prevented. This is done by sensing the pressure difference across the flow and return branches in a suitable place and using the increase of pressure to throttle a valve at pump discharge (R2 in Figure 3.38(a)) or to open a valve in a bypass across the pump connections (R2 in Figure 3.38(b)). The latter uses more pumping energy because pump power is proportional to flow rate. A third possibility is to vary pump speed.

(iv) Modulating control using three-port motorized valves. This gives good control, provided that the three-port

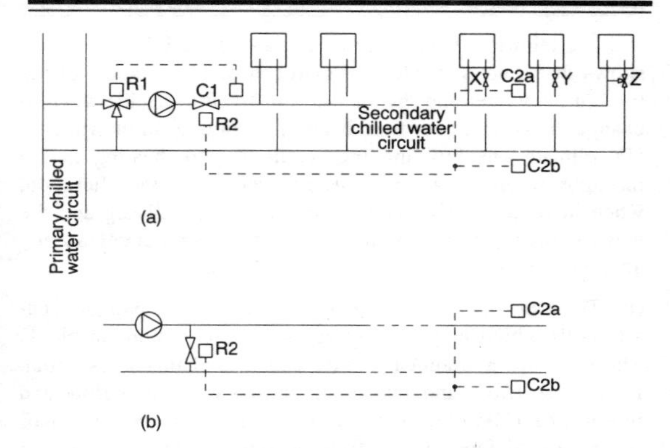

Figure 3.38 Arrangements for reducing the pressure drop across two-port valves used to control fan coil units, at partial load. (a) Three-port valve at Z keeps the chilled water lines alive. Two-port valves at X and Y control room temperature. Temperature sensor C1 controls the chilled water flow temperature by means of three-port valve R1. C2a and C2b are pressure sensors that throttle R2 upon increase in pressure difference. (b) This shows an alternative arrangement to (a). Pressure sensors C2a and C2b open R2 to divert secondary water as the pressure difference increases

valves are characterized. A disadvantage is that no diversity can be allowed for pipe and pump sizing because the flow rate is nominally constant at all loads.

Fresh air supply. It is usually possible to bring a small amount of fresh air through a hole cut in the wall behind the unit but this is uncontrollable and admits traffic noise and pollution. A ducted supply and extract mechanical ventilation system is recommended. This gives a controlled ventilation rate, discourages natural infiltration and offers the possibility of using the fan coil units to do sensible cooling only. The ducted system then contains a cooling coil in its air handling plant which does all the dehumidification required and the fan coil units are dry.

Two-pipe and four-pipe systems

(i) *The two-pipe system.* Each fan coil unit contains only one coil which is fed by two pipes, one flow and one return. In summer, chilled water is fed to the fan coil units and, in winter, LTHW is provided instead. This is called a changeover system and is a failure in the temperate climate of the UK. Figure 3.1 shows that a net sensible heat gain can occur for some of the time in the winter and this makes it difficult to decide when to change the system over from heating to cooling. In the variable UK climate one building face could require heating in the morning, when it was in shade, and cooling in the afternoon, when in sunlight. The thermal inertia of the building and the water in the piping system introduce a time lag that defeats any attempt to achieve effective changeover control.

(ii) *The four-pipe system.* Each unit contains a pair of separate coils which may or may not share a common fin block. The coils are hydraulically independent and this is essential. Four pipes (flow and return for chilled water and flow and return for LTHW) feed each fan coil unit (Figure 3.39). A pair of centrifugal fans, driven from a common motor, blows air recirculated from the room over the two coils.

A temperature sensor, C1, located in the return air path within the casing, regulated two motorized valves, R1a and R1b, in sequence, with a dead band in between so that both valves cannot be open together. Two-port and three-port valves may be used and controlled in sequence but it is unwise to use a pair of such valves in a common valve body, particularly with three-port valves. Heat flow through the common valve body from the LTHW to the chilled water and spoils performance. Such valves often have small clearances between the plugs and seatings with the risk of blockage by scale. Protecting the valves by fitting a strainer upstream is not practical: the strain-

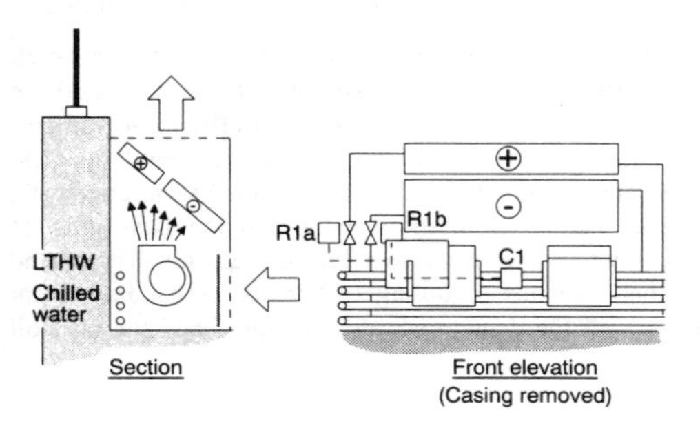

Figure 3.39 Four-pipe fan coil unit. (Two or more fans are used to give better air distribution over the coils. Motorized undulating valves, R1a and R1b, are operated in sequence and controlled from a recirculated air temperature sensor, C1

ers frequently foul up with dirt or scale and prevent water flow. Designing the piping system to vent air and avoiding the accumulation of scale by the provision of adequate dirt pockets is a better approach. If possible, control valves should be used that do not have small clearances between their plugs and seatings. Flushing out the system, prior to commissioning, is essential and, when doing this, the fan coil units should be bypassed to keep them clean, using special piping cross connections for this purpose.

The LTHW flow temperature is compensated against outside air temperature to improve the control of the heater coil.

Under no circumstances should a three-pipe system (one chilled water flow, one LTHW flow, and a common return) be contemplated. A four-pipe system using fan coil units with a single coil should also never be used. Both these systems introduce serious hydraulic and thermal problems.

Heating. The options appear to be: using a two-pipe changeover system, warming the auxiliary ducted supply air, or using a four-pipe system. The changeover system is a failure in the UK, as has been explained. Warming the ducted supply air is also a failure because of the heat loss from ducted air and the difficulty of balancing a low velocity air supply system to much better than at +/– 10 per cent. Furthermore, a common warmed air supply to terminal units introduces control difficulties, as discussed in Section 3.5.2 for variable air volume systems. The correct and only solution in the UK is to use a four-pipe system with two coils in the fan coil units.

Unit location. The best place to locate a fan coil unit is under the window where air circulation is up the window, over the ceiling towards the corridor wall, down the far wall and back for recirculation at the unit. The ducted supply and extract system distributes air from grilles in the far wall, or possibly, from ceiling diffusers (see Figure 3.40). Unfortunately this arrangement occupies lettable floor area and it has become common to put fan coil units above suspended ceilings. When this is done, consideration should be given to the air distribution across the ceiling and over the windows where, with single glazing, cold downdraughts can occur in winter and cause discomfort. It is also important to duct the supply air to the back of the fan coil unit (above the ceiling) to ensure that the ventilating air introduced does actually get into the room, as part of the fan coil air distribution. Failure to do this may mean that the air supplied by the duct system bypasses the room entirely through the ceiling void. Figure 3.41 shows one possible arrangement. The air supplied to the room by the fan coil unit above the suspended ceiling is a mixture of air recirculated from the room and air ducted from a central air handling unit. Air must be able to

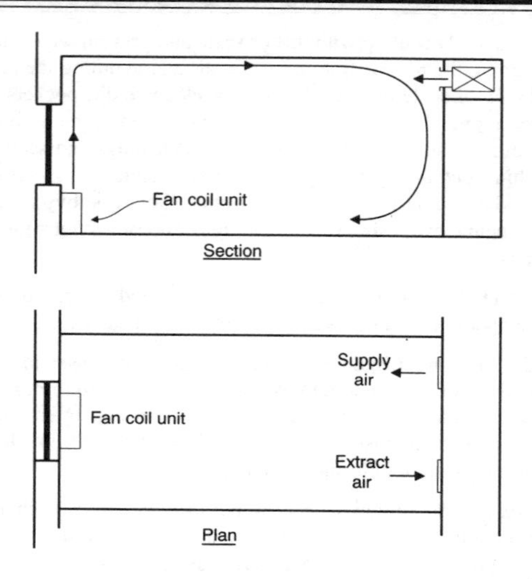

Figure 3.40 Conventional fan coil unit location

escape into the ceiling void from the room for recirculation by the fan coil unit.

Free cooling. Fan coil units must have a supply of secondary chilled water at all times of the year. The options are as follows.

(i) *Using the cooler coil in the air handling plant.* The refrigeration plant is switched off and both the primary and secondary chilled water pumps are kept running. Water flowing through the inside of the tubes of the cooler coil is then chilled by the air passing over the outside of the finned tubes. During winter weather, adequate frost protection is essential. Referring to Figure 3.37 it is seen that the three-port valve on the primary coil would have to have its bypass port closed to allow this method of free cooling. A higher secondary chilled water temperature is probably acceptable in winter and one or two degrees higher than the value used for the summer design performance is suggested. Investigation into the performance of the primary air cooler coil would establish what leaving water temperatures could be produced but it appears likely that chilled water at 12°C or 13°C is possible when the outside dry-bulb is less than about 6°C.

(ii) *Thermosiphon cooling with the refrigeration compressor*

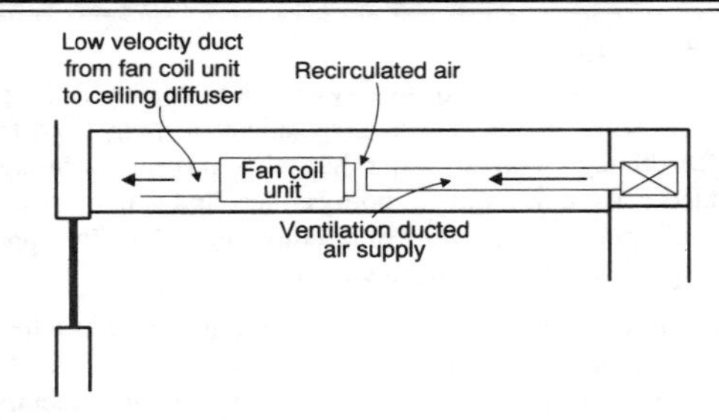

Figure 3.41 Fan coil unit above a suspended ceiling

off. This is possible with certain types of refrigeration plant, when suitably modified. The principle is that the compressor is switched off and a valve in a bypass across it is opened to allow the natural flow of refrigerant gas past the compressor. See Section 8.5.

(iii) *Using a cooling tower.* Cooling towers can give water at about 11°C for a significant proportion of the year. This may be directly, using suitable shut-off valves to permit water from the tower to be pumped into the secondary chilled water circuit. Any hydraulic imbalance must be considered and the water from the tower must be properly filtered. Close monitoring of the corrosion possibilities in the secondary circuit is necessary.

The cooling tower can be used indirectly, if a plate heat exchanger is inserted to separate the dirty cooling tower water from the clean water in the secondary circuit. This avoids the need for water filtration and special monitoring of the corrosion risk but gives a higher secondary chilled water flow temperature.

Noise. Noise largely originates from the fans used and can vary according to the type of fan, the type and arrangement of the bearings, the way in which the fans are mounted, the rigidity of the metal casing of the unit and the smoothness of the airflow through the discharge grills. Changes of fan speed alter the noise produced.

As with VAV terminal units, the manufacturer must provide details of the sound power level across the audio spectrum. It is then possible to establish the NC or NR value likely in the room, if the acoustic properties of the room and furnishings are known.

Chilled ceilings.

Background. The origin of this system is the embedded pipe coil in the soffit of the slab to give radiant heating from the ceiling. Although chilled water has also been fed through embedded coils in this way, in both ceilings and walls, the suspended aluminium pan, with pipe coils clipped to its upper side, has replaced the embedded version.

Ceiling types. Different types of ceiling are available with a fairly wide choice of options and applications, giving the architect a measure of freedom in ceiling design. One manufacturer offers a system that may be part of almost any standard metal suspended ceiling, or in anodized aluminium, or as painted metal. It is possible to apply ceiling paper or an adhesive foil or a fine grade plaster, after installation. A perforated finish is also available to provide an acoustic quality. Networks of plastic piping that may be incorporated in ceilings or walls are also possible and chilled beams, with or without fans, have ben used.

Auxiliary air supply. An auxiliary ducted air supply is essential because the ceiling cannot be allowed to run wet: all the latent heat gain must be dealt with by the supply of dehumidified air. To deal with typical latent heat gains of 134 W for a treated floor area of 14.4 m^2 (see Examples 3.2 and 3.3) the necessary supply airflow rate can be calculated by equation (3.29). If it is assumed that a practical off-coil state for the air leaving the cooler coil in the central air handling plant is 11°C dry-bulb, 10.5°C wet-bulb (sling) then the moisture content of the air supplied to the room would be 7.702 g/kg. If the room state is 22°C dry-bulb, 50 per cent saturation and 8.366 g/kg, then the supply airflow rate is given by

$$\dot{v}_{13} = \frac{134}{(8.366 - 7.702)} \times \frac{(273 + 13)}{856} = 67.43 \text{ l/s}$$

Over a treated floor area of 14.4 m^2 this represents a specific airflow rate of 4.68 l/s m^2, referred to the floor area.

Ceiling piping. Pipe coils in the ceiling are typically 15 mm diameter and, for cooling purposes, pipe spacings are commonly from 200 mm to a maximum of 450 mm. Spacings greater than 450 mm will not give sufficient cooling to be effective.

The installation is in four-pipe form: the outer two or three loops of piping, next to the window, carry LTHW at a flow temperature compensated against outside temperature. The whole of the remainder of the ceiling is provided with piping coils carrying chilled water at a flow temperature that is above the room dew-point. Control over this flow temperature should be proportional plus integral plus derivative (see Chapter 4). Hence primary and secondary chilled water circuits are required (see Figure 3.37).

Acoustic properties. An insulating blanket is placed above the piping coils to minimize the heat gain from the room above and this also provides an acoustic quality. The material used is glass fibre or the like and should be coated in a gel to prevent the erosion of fibres. Alternatively, the material may be packed in polyethylene bags to prevent erosion but with a small loss of some acoustic performance. A typical acoustic performance for a ceiling comprising 600 mm × 600 mm panels is given in Table 3.8.

Table 3.8 Acoustic properties of a typical suspended metal pan ceiling.

Octave band (Hz)	125	250	500	1000	2000	4000
Sabine absoprtion coefficient	0.2	0.4	0.65	0.6	0.5	0.4

The method of ceiling suspension and the panel dimensions affect the absorption coefficient. This is seen in Table 3.8 at 500 Hz with a wavelength of 670 mm, for which the table gives a peak absorption, related to the 600 mm panel dimension.

Air distribution. Ceiling diffusers can be used to supply the auxiliary air but a large amount of ceiling depth is required for this, and if sufficient depth is not provided, the airflow over the diffuser cones will be disturbed and will generate objectionable noise. A better arrangement that is quieter and cheaper in capital cost, is to use side wall grilles at high level on the corridor wall, for both supply and extract air. With this arrangement the depth of ceiling void needed can be as little as 200 mm, under favourable circumstances, if pipes do not have to cross over one another.

Sensible cooling capacity. The auxiliary airflow rate to deal with the latent heat gain has been shown to have a typical value of 4.68 l/s m². Assuming a room dry-bulb temperature of 22°C and a supply air temperature of 13°C, equation (3.28) can be used to show that its sensible cooling capacity is 52.7 W/m². If comfort is expressed in terms of a dry resultant temperature of 22°C this might be equated to a dry-bulb of 23°C and a mean radiant temperature of 21°C. In this case the sensible cooling capacity of the auxiliary air supply is 59.2 W/m². Typical mean panel surface temperatures are 16°C to 18°C and sensible cooling capacities are in the range 20 to 100 W/m² (Figure 3.42), depending on the mean temperature difference between the chilled water and the room, and on the piping arrangement used. Assuming a mean panel surface temperature of 17°C and an air temperature of 23°C the difference is 6°C and, from

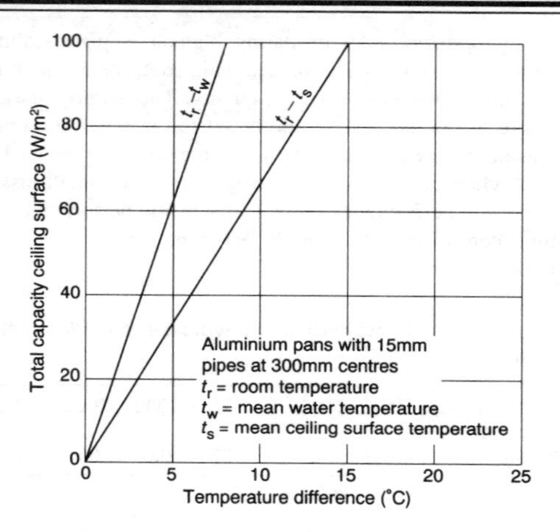

Figure 3.42 Typical total cooling/heating capacities of a ceiling

Figure 3.42, the ceiling cooling capacity is about 40 W/m². Arrangements must be made for free cooling in winter.

In terms of human comfort it is desirable that the mean radiant temperature of a room should be less than the dry-bulb when heat gains are occurring and this is what a chilled ceiling provides. When net heat losses occur the reverse is true and this also is provided by a warmed ceiling.

Advantages and disadvantages. Advantages of the system are:

(1) It is quiet if properly designed and installed because there are no terminal units with moving parts in the room.
(2) It is probably the cheapest air conditioning system to operate because there are no terminal unit fans and the auxiliary duct system is low velocity with a low fan total pressure.
(3) It takes up no lettable floor area.
(4) It provides an acoustic ceiling.
(5) There is no need to provide a special unit enclosure, as is sometimes necessary with fan coil or induction units.

Disadvantages of the system are:

(1) It may be inflexible in responding to partition rearrangement, but this depends on the module dimensions.

(2) It may be dearer in capital costs than some other systems if proper discounts are not allowed for the benefits of the acoustic ceiling and the absence of unit enclosures.

Chilled beams. Chilled beams have been used, principally in Scandinavia. A metal enclosure, in the shape of a beam, runs along the modular grid lines, at right angles to the windows, and partitions can be fixed beneath them. The enclosure contains pipe coils carrying chilled water/LTHW and there may also be a ducted air supply within the beam to deal with the latent gain. Local cooler coils, heater batteries and fans are also sometimes incorporated.

It is possible that this arrangement could give more flexibility for accommodating partition changes.

Perimeter induction systems

The induction unit. Primary air is delivered at high pressure (125 Pa to 850 Pa) to a plenum chamber, forming part of each terminal unit, and issues from the chamber through multiple nozzles as high velocity jets. Each jet entrains about three to five volumes of surrounding air which comes through a recirculation opening in the unit casing, from the treated room. A secondary cooler coil is fitted in the path of the induced airflow and control over room temperature is achieved by regulating the chilled water flow through a motorized valve (Figure 3.43). The secondary coil does sensible cooling only and maximum coil capacities, over a typical commercial range, are from 800 W to 1600 W, depending on the unit size, nozzle type, pressure and arrangement.

Primary air. Primary air is generally a mixture of fresh and recirculated air that is filtered, cooled and dehumidified, and re-heated to a temperature that is compensated against outside

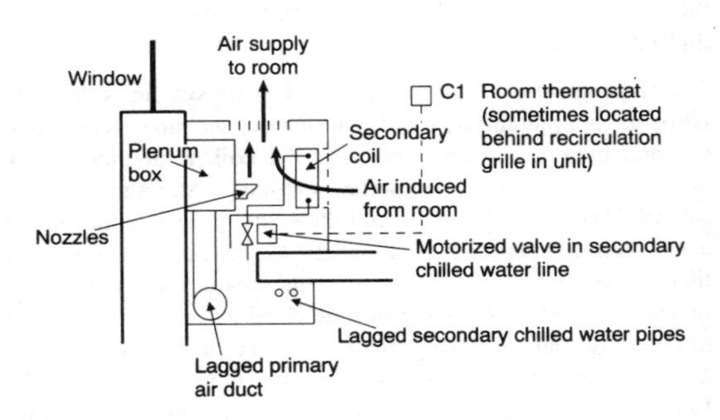

Figure 3.43 Two-pipe perimeter induction unit

air temperature at a central air handling plant. It is then fed to the terminal units through spirally-wound circular ducts at high velocity. The functions of the primary air are as follows:

(i) To provide enough fresh air (see Section 3.2).
(ii) To induce sufficient air over the secondary coil to deal with about 80 per cent of the sensible heat gains.
(iii) To deal with all the latent heat gains in the conditioned room (so that the secondary coil can run dry and a condensate drainage system is not needed).
(iv) To offset heat loss from the treated room.

The unit coils receive secondary water at a constant temperature of about 11°C or 12°C, related to the room dew-point.

Depending on the unit size and the nozzle pressure, primary air quantities are from 47.5 l/s to 65 l/s with sensible cooling capacities from 474 W to 649 W, based on a primary air temperature of 14°C and a room temperature of 22°C.

Two-pipe non-changeover system. When first introduced to Europe the induction system was used in a changeover form. In winter, the secondary coils received compensated LTHW and the primary air was supplied to the units at a constant temperature of about 14°C. There was adequate heating capacity but little cooling capacity. In summer, the action of the room thermostat was reversed, secondary chilled water at a controlled temperature (to avoid condensation) was delivered to the unit coils and primary air was supplied at a compensated temperature (as described above).

This changeover form of the system was designed for the severe extremes of the North American climate and proved unworkable in the varied climate of the UK, and much of Europe. It was replaced by the two-pipe non-changeover version, where the system operates in its summer mode throughout the year, and was used in many office buildings in the UK throughout the 1960s and 1970s.

Four-pipe system. The four-pipe induction system is the best. Units have a plenum box and two distinct secondary coils, one a heater battery and the other a cooler coil. Four pipes (flow and return for constant temperature chilled water and compensated LTHW) feed the coils and the primary air is unheated, at a constant temperature of about 14°C. A room or return air thermostat controls the capacity of the heater and cooler coils in sequence, either by a pair of motorized valves or by a pair of modulating dampers (Figure 3.44). The dampers are self-acting, using duct static pressure as the source of energy. The four-pipe unit gives good control over room temperature, summer and winter, without any changeover problems.

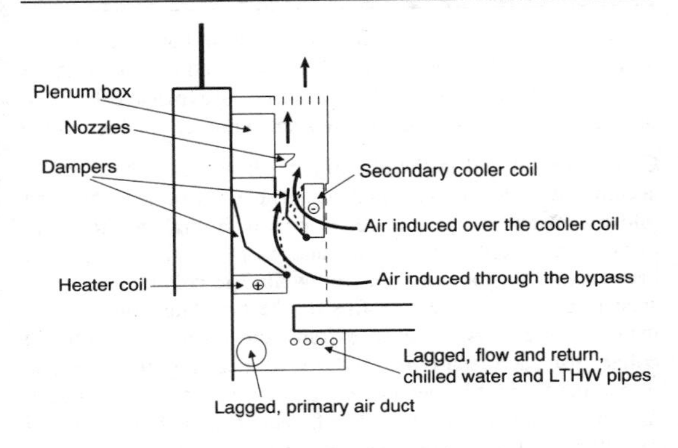

Figure 3.44 Four-pipe induction unit with air damper control. The dampers are shown in a position that gives partial induced flow over the cooler coil, partial flow through the bypass, and no airflow over the heater coil. The broken lines show the two dampers in the full heating position.

Single-coil, four-pipe units have been used but are not recommended because the chilled water and LTHW piping circuits are interconnected and hydraulic difficulties are a certainty.

Three-pipe systems (one chilled water flow, one LTHW flow and one common return) have been attempted in the past but have not been successful. They should never be used.

3.6 CHILLED WATER AIR COOLER COILS

3.6.1 Construction

Tubes run horizontally across a cooler coil face with vertical fins (for easy condensate disposal) and are generally made of copper, with diameters between 8 and 25 mm, the larger diameters being used for bigger cooling duties and for structural reasons (where the span of the tubes is long). For most commercial applications aluminium plate fins are used, some corrosion of the metals in the presence of slightly acid condensate being tolerated. Tinned copper tubes with tinned copper fins is a combination less likely to corrode but, electro-tinned fins and tubes are better. Organic varnishes have been used instead of tinning, for corrosion protective. The weak link in protection against corrosion is the steel frame in which the finned tubes are mounted to form a cooler coil. Black steel and, worse, galvanized steel, often corrode before the finned tubes. Stainless steel has been used for cooler coils operating in aggressive circumstances. This material is expensive because of welding and working difficulties and a poorer conductivity.

Collars are formed in the aluminium plates when the holes to accommodate the tubes are punched. The tubes are then threaded into the plates and a mandrel is pulled through the tubes to expand them onto the collars and form a reasonable joint between the fin root and the tube wall. The collars also act as spacers between successive fins. If the fin plates are too thin there is not enough metal to extrude and form a conventional collar. Instead, a 'star burst' collar is formed (Figure 3.45). The grip at the fin root is then not as tight and heat transfer between the fins and the tubes is less. The fins are from 0.42 mm to 0.15 mm thick and are commonly spaced at 316, 394 and 476 per metre length of tube (8, 10 and 12 fins per inch).

Fin plates may be smooth or corrugated. Alternatively, individual fin strip may be spirally wound onto individual tubes, a spiral groove being cut in the external tube wall to accommodate the fin root. The fin strip is stretched in the process of winding, resulting in a thinner outer edge but a crimped root. The grip on the tube is tight and heat transfer between the fin and the tube wall is good but manufacturing is expensive and the pressure drop in the airstream is greater than with plate fins.

The tubing arrangement is single or double serpentine (Figure 3.46), single being commoner. Rows are always piped for contra-flow, with respect to the airstream: the coldest water next to the coldest air. Odd numbers of rows are possible but an even number is the norm, to give the flow and return headers on the same side of the coil, facilitating pipe connections. Four or six rows are used for coils in the UK but six or eight are used abroad in hot and/or humid climates. Water velocities within the tubes are between 0.6 m/s (self-purging of air) and 2.4 m/s (above which erosion is likely, especially at the heels of return bends). Coils are made short and wide, rather than tall and narrow. This reduces the number of return bends, keeps down the cost of manufacture and minimizes the need for intermediate drain trays.

3.6.2 Condensate drainage

Condensate must drain freely down the fins. If this does not happen, condensate may be blown off the fins and carried down the duct, where a pool will form and perhaps drain out of the duct, through a seam or a joint, to cause an expensive nuisance. Furthermore, if space between the lower fins is filled with condensate, effective heat transfer is prevented and the

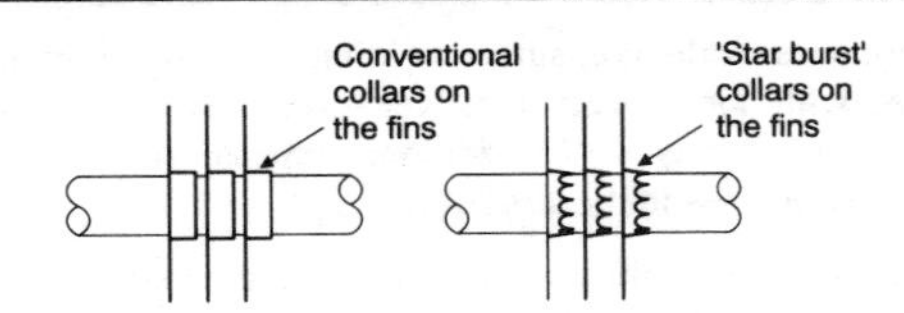

Figure 3.45 Conventional collars grip the tubes more tightly and give better heat transfer than do 'star burst' collars

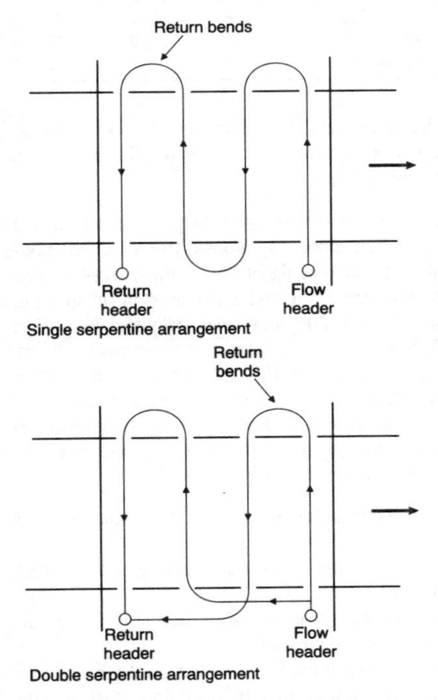

Figure 3.46 Tubing arrangements for cooler coils. Single serpentine is commonly used. Coils are always piped for contra-flow, with respect to the rows and the airflow, as shown: coldest water next to the coldest air

performance of the coil suffers. Fitting downstream eliminator plates (with the correct face velocity) prevents condensate carry-over but they are an additional expense and increase the total pressure loss in the airstream.

For sensible–total ratios not less than 0.65 the maximum allowable face velocities, without the use of downstream eliminator plates, are given in Table 3.9.

Table 3.9 Fin spacing and allowable face velocities without the need for downstream eliminator plates, when the sensible-total heat ratio of the cooling process is not less than 0.65

Fin spacing per metre of tube	316	394	476
Fin spacing per inch of tube	8	10	12
Maximum face velocity m/s	2.5	2.2	2.1

As a general rule, more than 316 fins/m (8/inch) should not be used and the face velocity should not exceed 2.5 m/s, when a cooler coil is to dehumidify.

One or more horizontal drain trays must be provided across the full depth and width of the coil. Opinions differ as to the maximum permissible vertical spacing of the multiple trays needed on tall coils. The maximum dimension depends on fin spacing, face velocity and the sensible–total heat ratio for the design cooling load. With sensible–total heat ratios less than 0.8, the maximum vertical dimension of the coil face or the maximum distance between drain trays is 900 mm, with 316 or 394 fins/metre. If the sensible–total heat ratio is 0.85 or more, 1200 mm may be allowable, provided that the face velocity and finning are as suggested above. See Figure 3.47.

(i) The drain trays must be over the full width and depth of the coil.
(ii) The drain connection in the tray must be at the lowest part of the tray.
(iii) There must be adequate access on each side of the coil for maintenance and cleaning.
(iv) A condensate trap must be provided for each separate drain tray and the trap must be deep enough to remain flooded and allow the flow of condensate from the tray, taking into account the negative or positive static pressure of the airstream flowing over the coil.
(v) Multiple traps should join a common vertical header.
(vi) The vertical header must terminate in an air gap before feeding a tundish.
(vii) The tundish must be connected to the drainage system by a condensate line with an adequate fall.

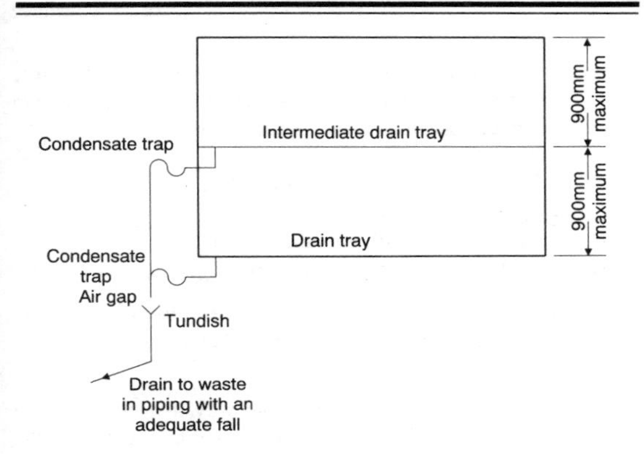

Figure 3.47 Maximum vertical dimensions for drain trays on cooler coils

The air gap is essential for two reason: to be able to verify visually that condensate is actually flowing and to prevent contamination from the building drainage system to the air conditioning plant.

Airflow over the face of a cooler coil is not uniform. The natural shape of the cross-section of an unrestrained jet of air is circular (see Section 3.5.2). Hence there is less airflow over the corners of a cooler coil than over the middle. This complicates calculations of heat transfer but also means that the velocity over the central part of the coil is higher than the face velocity, with an increased risk of condensate carryover. This is particularly true if blow-through coils are used instead of the more common draw-through arrangement (when the coil is on the suction side of the fan). The pattern of airflow leaving the discharge side of the fan is very disturbed and must be smoothed and reduced to the required coil face velocity before the air can be allowed to enter the cooler coil. Achieving this is often very difficult. Although there is an advantage that the temperature rise through the fan occurs before the coil, it is generally prudent to avoid the use of blow-through coils.

3.6.3 Sensible and latent heat transfer

This is a complicated topic. One approach [7], which is commonly used but which gives only an approximate solution, expresses the total heat transfer, Q_t, by the following equation:

$$Q_t = A_t U_t \frac{(t_{a2} - t_{w1}) - (t_{a1} - t_{w2})}{\ln[(t_{a2} - t_{w1})/(t_{a1} - t_{w2})]} \tag{3.33}$$

where

A_t = total external surface area of the fins and the tubes (m^2)
U_t = thermal transmittance coefficient enhanced to take account of latent as well as sensible heat transfer (W/m^2C)
t_{a2} = final air dry-bulb temperature (°C)
t_{w1} = initial chilled water temperature (°C)
t_{a1} = initial air dry-bulb temperature (°C)
t_{w2} = final chilled water temperature (°C).

See Figure 3.48. The temperature differences used in equation (3.33) can be evaluated by starting at either end of the coil — the same answer is obtained.

The U-value is calculated in the usual way, by summing the thermal resistances of: the water film within the tubes, the metal of the tube wall and fins, and the air film on the outside of the finned tubes. One way of enhancing the U-valve is to reduce the thermal resistance of the air film by dividing it by the sensible–total heat ratio for the psychrometric process. Equations are available [26, 27] for determining the various thermal resistances. However, there is usually doubt about coefficients for the air side of the finned tubes and the best source of information on this is the manufacturer. Although references 26 and 27 provide some information on the matter it is difficult to express a U-value for the whole coil if only part is wet with condensate. Determining the boundary between the wet and dry areas is not straightforward.

Sensible cooling is only possible if all the external surface of the coil is above the dew-point of the airstream. Primary and secondary chilled water circuits are necessary (Figure 3.17).

3.6.4 Cooler coil contact factor

The contact factor was defined in terms of the geometry of the cooler coil in Section 3.4.2. With the simplifying assumption

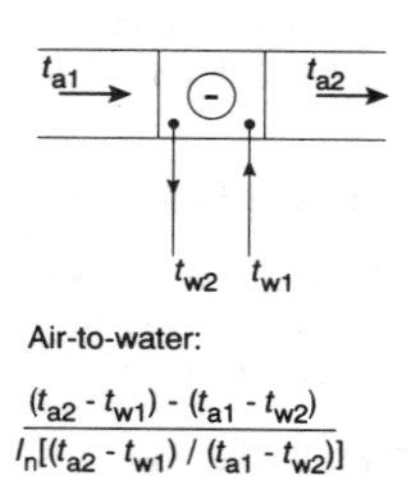

Figure 3.48 Log mean temperature difference (LMTD)

that the mass flow rates of air and chilled water remain substantially constant the contact factor can be expressed [7,28] in term of the coil construction by the following equation:

$$\beta = 1 - \exp[-A_r/A_f)(r/1.25\,R_a\,v_f)] \quad (3.4)$$

where

β = contact factor
A_r = total external surface area per row (m^2)
A_f = face area (m^2)
r = number of rows
R_a = thermal resistance of the air film (m^2 °C/W)
v_f = face velocity (m/s).

Some typical, approximate values are given in Table 3.10.

Table 3.10 Approximate contact factors for coils with 316 fins/m

Face velocity (m/s)	Rows	Contact factor
2.0	4	0.91
2.5	4	0.85
3.0	4	0.79
2.5	2	0.61
2.5	6	0.94
2.5	8	0.98

3.6.5 Performance at partial load

This is difficult to predict without complicated calculations but in simple terms, the possibilities are as follows.

(1) *Constant entering dry-bulb, varying entering wet-bulb.* Figure 3.49(a) illustrates what happens. Under design load conditions air at state O enters the cooler coil and leaves it at state W. The apparatus dew-point is A. If the entering dry-bulb temperature stays constant but the entering wet-bulb reduces, the on-coil state is O', the off-coil state is W' and the apparatus dew-point is A'. Since the wet-bulb onto the coil has reduced, the load on the coil has also reduced, wet-bulb lines being nearly parallel to lines of constant enthalpy. Hence the temperature rise of the chilled water is less and consequently the mean coil surface temperature is less. Thus A' is lower down the saturation curve than A. The contact factor is unchanged because this depends on the construction of the coil, if the flow rates of air and chilled water are unaltered. The partial load cannot be predicted without involved calculations.

(2) *Constant entering wet-bulb, varying entering dry-bulb.*

Referring to Figure 3.49(b) it is seen that the on-coil state alters from O to O'. Since the wet-bulb onto the coil is constant the cooling load is also substantially constant, the rise in the chilled water temperature is unchanged and the mean coil surface temperature does not alter. The apparatus dew-point stays in the same position at A. The contact factor has not altered so it is possible to identify the position of W' by geometry on the psychrometric chart.

(3) *Constant chilled water flow temperature, varying chilled water flow rate.* Figure 3.49(c) illustrates what happens when the performance of the coil is intentionally reduced by partially opening the bypass port of a three-port motorized control valve, R. Less chilled water flows through the coil and since the on-coil state, O, is unchanged the rise in water temperature is increased. The mean coil surface temperature also increases and the apparatus dew-point moves up the saturation curve from A to A'. The contact factor stays constant for a while and then the flow characteristics within the tubes change from turbulent to transitional and the position of W' breaks away from a predictable position on the broken line (the locus of the off-coil state). The contact factor cannot then be determined by geometry on the psychrometric chart. The probability is that the end of the broken line, after this happens, is never parallel to a line of constant moisture content, even when the bypass port is nearly fully open. This is because the chilled water flow temperature is constant and part of the coil always does some dehumidification. This is the most common way of controlling an air cooler coil.

(4) *Constant chilled water flow rate, varying chilled water flow temperature.* In Figure 3.49(d) the piping arrangement is more complicated. A pumped secondary piping circuit varies the chilled water flow temperature and keeps the flow rate constant, by means of R, a three-port motorized mixing valve. As the flow temperature of the chilled water is increased, under thermostatic control, the mean coil surface temperature increases and A slides up the saturation curve. The contact factor is geometrically predictable for a while but eventually the mean coil surface temperature approaches and exceeds the dew-point of the entering air. The point A' is then no longer on the saturation curve and its position cannot be determined without involved calculation. The last part of the coil performance is sensible cooling only.

Wild coils without any control over capacity are sometimes used. They overcool and waste some energy and running cost but do not significantly affect comfort in the conditioned space. A limit to the overcooling is provided by the control over the chilled water temperature at the refrigeration plant. The chilled

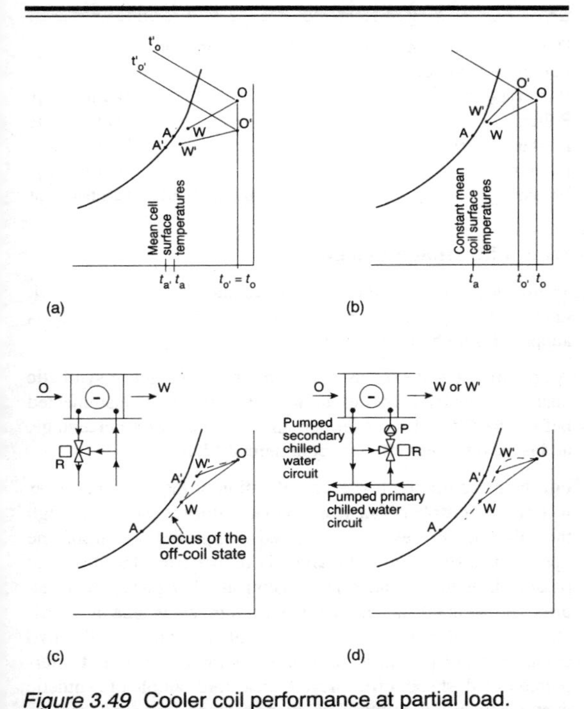

Figure 3.49 Cooler coil performance at partial load. (a) Constant entering dry-bulb, varying entering wet-bulb. (b) Constant entering wet-bulb, varying entering dry-bulb. (c) Constant chilled water flow temperature, varying chilled water flow rate. (d) Constant chilled water flow rate, varying chilled water flow temperature

water temperature might be allowed to rise in the winter, offsetting some of the overcooling. The saving in the capital cost of the controls must be weighed against the costs of the wasted energy.

3.7 AIR FILTERS

3.7.1 Particle sizes

The unit of size is the micron (one millionth of a metre) which is abbreviated as μ and particle sizes may be roughly categorized as follows:

Dusts (formed by natural or mechanical abrasion):	<1 μ
Fumes and smokes	1 μ
Bacteria	0.2 to 5.0 μ
Pollen	5.0 to 150 μ
Fungus spores	1.0 to 120 μ
Human hair	30 to 200 μ
Viruses	much smaller than bacteria

3.7.2 Filtration efficiency

There are two basic methods of testing efficiency [29, 30]: gravimetric and discoloration. A third method [31, 32] is adopted for high efficiency, absolute filters.

A gravimetric test expresses the ratio of the weight of synthetic dust collected by a filter to the weight of synthetic dust injected before the filter. This is termed an arrestance, as a percentage, and is used for less efficient commercial filters.

For the discoloration test a simplification of the procedure is as follows. Atmospheric air from the laboratory is passed through the filter under test. A measured sample of air from the upstream, dirty side of the filter is drawn from the duct and passed through a standardized chemical filter paper. A measured sample is taken from the downstream, clean side of the filter, in a similar way. The intensity of the stains on the two chemical filter papers is compared photometrically and interpreted as a dust spot efficiency. A measured sample of synthetic dust (a specified mixture of carbon black, lint and grit) is injected into the duct on the upstream side of the filter and the test is repeated. This is done several times and the average used to express the dust spot efficiency of the filter.

Absolute filters (termed high efficiency particulate, or HEPA filters in USA) are tested [31, 32] by injecting a solution of sodium chloride in water into the test duct before the filter. A measured sample of air is drawn from the clean side of the filter and passed through the flame of a bunsen burner. A bright yellow colour is produced and its intensity is an indication of the weight of sodium in the flame. Knowing the mass and strength of the solution injected, the sodium flame efficiency is established. This is usually expressed as a percentage penetration, the complement of the efficiency.

3.7.3 Classification of filter efficiency

Typical filter efficiencies are as shown in Table 3.11.

Table 3.11 Filtration efficiencies

Filter type	Face velocity (m/s)	Arrestance (%)	Atmospheric dust spot efficiency (%)
Automatic			
Viscous	1.9–2.5	80	–
Dry	2.5–3.6	70–80	–
Panel			
Cleanable viscous	1.9–2.7	65–85	–
Cleanable dry	1.7	70–75	–
Disposable	1.8–3.8	70–90	–
Bag			
Low efficiency	3.8	–	30–50
Medium efficiency	3.8	–	55–90
High efficiency	2.5	–	90–97
Absolute			
Low efficiency	1.4	–	95
Medium efficiency	1.3	–	99.7
High efficiency	0.45–1.3	–	99.99997
Electrostatic	2.5	–	90

Eurovent numbers [30] are also used as a classification

Table 3.12 Eurovent classification of arrestance and efficiency

Eurovent number	Average arrestance, A (%)	Average dust spot efficiency, E (%)
EU1	< 65	–
EU2	$65 \leqslant A < 80$	–
EU3	$80 \leqslant A < 90$	–
EU4	$90 \leqslant A$	–
EU5	–	$40 \leqslant E < 60$
EU6	–	$60 \leqslant E < 80$
EU7	–	$80 \leqslant E < 90$
EU8	–	$90 \leqslant E < 95$
EU9	–	$95 \leqslant E$

3.7.4 Filter types

Dry panel filters

Dry filters use a filtration medium consisting of continuous elements of glass strands in an open structure. The medium is packed with a graded density, in the direction of airflow, to give a fairly uniform collection of dust through the filter.

Polyester fibres are also used. Filters using these media often have an overspray giving a gel type coating to assist filtration. Fire-resistant treatment is also provided. The more efficient dry filters use pleated glass paper as the filtration medium. Foamed plastic has also been adopted, to offer a cleanable filter, but this may give a fire and health risk and a much lower arrestance. Foamed plastic is not recommended. Nominal panel dimensions are 600 mm × 300 mm, 500 mm or 600 mm and thicknesses are 25 mm or 50 mm. The framework of the cell is a comparatively flimsy construction, often in the form of rigid cardboard, with metal cleats at the corners. A retaining mesh of metal or glass fibre keeps the filtration material within the framework.

Filter panels are arranged as a group across the airstream, termed a battery, and it is important that the face velocities recommended by the manufacturers are not exceeded, otherwise the arrestance or efficiency obtained will be less than anticipated and the pressure drop will be more than expected. Arrestance or efficiency can be improved by fitting the filter panels obliquely across the duct section in order to provide more filtration surface.

Most panel filters are disposable although the life of a panel can sometimes be extended by using a vacuum cleaner over its dirty side as part of a maintenance schedule.

Viscous filters

Viscous filter panels use a durable fill of glass fibre, metal turnings, etc., that is coated with a suitable oil. In addition to the obvious necessary properties of being free of smell, not a fire hazard, not poisonous, and so on, the oil must have the right properties of viscosity and capillarity. Its viscosity must be such that particles of dust will penetrate the surface of the oil and its capillarity must be such that, once within the oil, the particles will move naturally through the filter to give dust retention in depth.

Dry panel filters generally have higher arrestances than viscous filters but viscous filters have greater dust-holding capacities.

Automatic filters

Automatic dry filters are widely used in commercial applications. The filtration material is woven to form a roll. A clean roll is located in a box at the top of the filter housing and stretches across the section of the airstream to form a roll of dirty material in a box at the bottom of the housing. The two rolls are rotated by a geared-down electric motor under the control of a time switch. Sometimes the rolls can be located at the sides of the filter housing. Maintenance is comparatively straightforward but arrestance is usually less than with panel filters.

It is also possible to have automatic viscous filters. These may be somewhat similar to the roll form, described above, with a series of hinged, oiled plates travelling across the airstream instead of a roll of filtration material. Another version uses fixed vertical plates and pumps oil over them. They are not used in commercial applications in the UK.

Bag filters

These comprise modules of filtration material in bag form to offer a large surface area across the airstream. When the fan is not running the bags hang limply but flow out horizontally when in normal operation with the fan working. Depending on the filtration material they have high efficiencies and are often used as an intermediate filter in clean room applications.

Absolute filters

These are made of dense glass paper, formed in deep pleats. Interleaves of corrugated aluminium are sometimes fitted between the paper pleats to channel the dirty airstream through the depth of the filter. Paper interleaves have also been used and some manufacturers construct the paper pleats to have horizontal cords on them, to form paper corrugations parallel to the airflow and channel it through the filter. The frames of the absolute cells must be of the highest quality, to prevent flanking leakage of dirty air, and metal or hardwood is used for the purpose. For the same reason, the assembly of filter cells to form an absolute filter battery must not permit flanking leakage.

The pressure drop across absolute filters is in the range of 100 Pa to 300 Pa, when clean. Since the filters are expensive it is usual to allow the pressure drop to build up, when dirty, to between three and six times the clean loss, depending on the type of filter. Fan capacity must be increased as the filter gets dirty, in order to continue delivering the required airflow rate. This is often done by increasing the fan speed, manually at intervals dictated by the observed pressure drop across the filter or, sometimes, automatically.

The typical, nominal dimensions of absolute filter cells are:

600 mm wide × 600 mm high × 150 mm or 300 mm deep. Face dimensions of 600 mm × 1200 mm are also used.

Figure 3.50 shows a typical use of different types of filter for a clean area application.

Electric (electrostatic) filters

These comprise two parts: an ionizing unit with about 12 kV applied across the electrodes and a collection unit consisting of multiple vertical plates with an applied voltage of about 6 kV. Negative electrodes and plates are earthed. Molecules of air passing through the ionizing unit are ionized and dust particles in the airstream acquire a charge by collision with the ionized molecules. The downstream plates collect and retain particles of opposite charge, 80 per cent going to the negative and 20 per cent to the positive plates. The plates are usually coated with an oil or gel to aid dust retention, and this must be replaced at intervals. One version of the filter retains the dust on the plates in a thickening layer, the dust particles agglomerating to form large flakes. As the space between the dirty plates reduces the air velocity increases, the large flakes are blown off and easily collected by a conventional filter, downstream.

With any type of electric filter an insect screen is necessary upstream and a pre-filter should also always be used. The electric filter does not fail safe, except for the filtration provided by the pre-filter.

Electric filters are very efficient (up to about 90 per cent atmospheric dust spot efficiency) but cannot compete with absolute filters for the very highest standards. One advantage they have over absolute filters is that the pressure drop across them, including that of the pre-filter, is much less than that of an absolute filter and hence the running cost, in terms of fan power, is less. Very little electric current is used.

Various techniques have been adopted in the past to reduce maintenance costs, with automatic washing etc. but these have generally proved complicated. The agglomerator type appears to require less maintenance in this respect.

Electric filters are expensive and have lost a good deal of popularity in recent years.

Some ozone is generated in the ionizing unit and the amount increases if the air velocity falls. While the small quantities of ozone normally produced by an electric filter are quite acceptable large quantities are not and hence this type of filter should not be used with a variable air volume system.

Activated carbon filters [7, 30]

These are the only effective way of removing smells. The carbon used is produced in a way that provides an enormous surface area within the carbon particles. There is an attraction (termed adsorption) between gaseous molecules and surfaces, the strength of which depends on the molecular weight of the gas and its boiling point, the higher the boiling point the greater

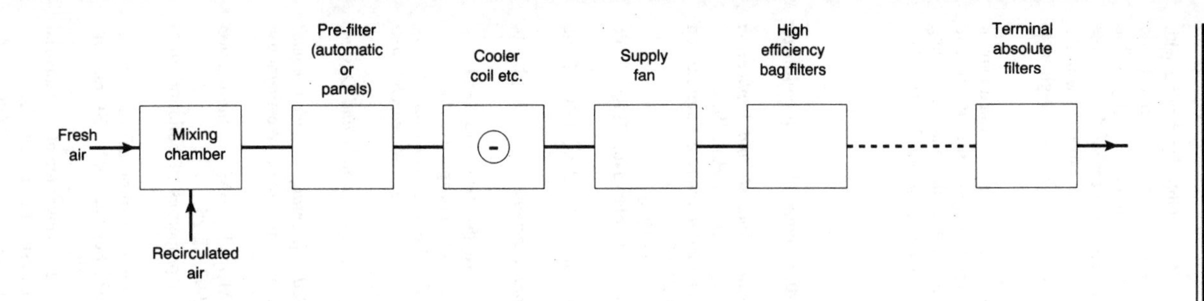

Figure 3.50 Filter application for clean areas. The pre-filter is to protect the plant. The bag filter is positioned after the supply fan, on its high pressure side so that any leakage is clean air passing to outside. The absolute terminal filters could cover the entire ceiling of the clean area. Heater batteries, humidifiers, silencers and the extract fan are not shown

the adsorption. This makes the filter suitable for dealing with organic gases related to smells such as body odour (butyric acid) but less suitable for gases such as ammonia. It is possible to modify the adsorption characteristics of a carbon filter by the addition of other chemicals. After the filter is saturated with the adsorbed gas it can be reactivated by raising it to a high temperature (about 600°C).

Proper filtration of the air before it enters the activated carbon filter is essential. One manufacturer suggests pre-filtration to EU6 or EU8. Pressure drops through the filter vary according to the airflow rate but a typical value is about 76 Pa, when clean and handling 750 l/s.

References

1. CIBSE Guide A8, Summertime temperatures in buildings, 1986. Table A8.3(e).
2. P. O. Fanger, *Thermal Comfort Analysis and Application in Environmental Engineering*, McGraw-Hill, 1972.
3. G. H. Green, The effect of indoor relative humidity on colds, *ASHRAE TRANS*, 1979, **85**, Part 1.
4. CIBSE Guide A4, Air infiltration 1986.
5. CIBSE Guide A2, Weather and solar data, 1986. Table A2.23.
6. CIBSE Guide A2, Weather and solar data, 1986, Table A2.27.
7. W. P. Jones, *Air Conditioning Engineering*, 4th edition, Edward Arnold, 1994.
8. *Glass and Thermal Safety*, Pilkington Bros. Ltd, 1980.
9. CIBSE Guide A9, Estimation of plant capacity, 1986.
10. R. H. L. Jones, Solar radiation through windows – theory and equations, *Building Services Research and Technology*, 1980 **1**, (2), 83–91.
11. W. P. Jones *Air Conditioning Applications and Design*, Edward Arnold, 1993.
12. *Thermolume Water-cooled Lighting Environmental Systems*, Applications Manual 61-3000, Westinghouse Electric Corporation, Dec. 1970.
13. CIBSE Applications Manual AM7, Information Technology in Buildings 1992, 22-25.
14. CIBSE Guide A3, Thermal properties of building structures, 1986.
15. CIBSE Guide A5, Thermal response of buildings, 1986.
16. ASHRAE Handbook, 1993. Fundamentals, SI Edition, Tables 30-33.
17. American Conference of Government Industrial Hygienists, Committee on Threshold Limits, 1971.
18. CIBSE Guide B2, Ventilation and Air Conditioning

requirements, 1986.
19. ASHRAE Standard 62 - 1989, Ventilation for acceptable indoor air quality.
20. DIN 1946, Part 2, Air conditioning health requirements (VDI ventilation rules), Jan 1983.
21. CIBSE Guide A2, Weather and solar data, 1986, Table A2.11.
22. Traffic noise and overheating in offices, *BRE Digest* **162** Feb. 1974.
23. Wyon *et al.*, European Concerted Action: Indoor air quality and its impact on man, Commission of the European Communities, Joint Research Centre, Environmental Institute 1992.
24. Jaakala *et al*, European Concerted Action: Indoor air quality and its impact on man, Commission of the European Communities, Joint Research Centre, Environmental Institute 1992.
25. ISO International Standard 7730 1994 08 15 Annex A.
26. ASHRAE Handbook Fundamentals, 1993, SI Edition, Chapter 12.
27. Forced circulation air-cooing and air-heating coils, AIR Standard 410-81 Air Conditioning and Refrigeration Institute, Arlington, Virginia, 1987.
28. W. H. Carrier, R. E. Cherne, W. A. Grant and W. H. Roberts, *Modern Air Conditioning Heating and Ventilation*, 3rd edition, Pitman, 1959.
29. BS 6540: 1985, Part 1, Methods of Test for Air Filters Used in Air Conditioning and General Ventilation, British Standards Institution.
30. Eurovent 4/5, Method of testing air filters used in general ventilation, 2nd edition, HEVAC Association, 1980.
31. BS3928: 1969, Method for sodium flame test for air filters (other than for air supply to IC engines and compressors), British Standards Institution.
32. Eurovent 4/4, Sodium chloride aerosol test for filters using flame photometry technique, HEVAC Association, 1980.
33. ASHRAE Handbook, 1991, Applications, SI Edition, chapter 40.

4 Automatic controls

4.1 DEFINITIONS

There are many terms used in the field of automatic controls and BS 1523 Part 1 — Glossary of Terms Relating To The Performance of Measuring Instruments is a useful reference. However, it does not include all of the terminology used in the building services control field. Some of the more common terms used in this industry are listed below:

Controlled variable (controlled condition) The quantity or physical property measured and controlled, e.g. room air temperature.

Controlled medium The substance which has a physical property that is under control e.g. air in the room.

Manipulated variable The physical property or quantity regulated by the control system in order to achieve a change of capacity which will match the change of load, e.g. the flow rate of hot water through a heater battery.

Control agent The substance whose physical property or quantity is regulated by the control system, e.g. the water fed to a heater battery.

Desired value The value of the controlled variable which it is desired that the control system will maintain.

Set-point (set-value) The value on the scale at which the controller indicator is set.

Control point The value of the controlled variable which the controller is trying to maintain under steady-state conditions.

Deviation The difference between the set-point and the value of the controlled variable at any instant.

Offset A sustained deviation between the value of the controlled variable corresponding to the set-point and the control point.

Lag The delay in the effect of a changed condition at one point in the system on some other condition to which it is related.

Primary element — measuring unit The part of the controller which responds to the value of the controlled variable (detecting element) and gives a measured value of the condition (measuring element).

Final control element This is the mechanism which directly acts to change the value of the manipulated variable in response to a signal initiated at the primary element, e.g. motorized valve.

Automatic controller A device which compares a signal from the measuring element with the set-point and initiates corrective action to reduce any deviation, e.g. a room thermostat.

Differential or differential gap (applies to two position control) The smallest range of values through which the controlled variable must pass for the final control element to move from the first to second position of its two possible positions.

Proportional band (throttling range) The range of values of the controlled variable which corresponds to the movement of the final control element between its extreme positions.

Cycling (hunting) A persistent periodic change in the controlled variable from one value to another.

4.2 CONTROL SYSTEMS

An understanding of the basic function of the control loop and the standard modes of control is essential if a stable control over the plant and systems used in buildings is to be achieved and a satisfactory environment obtained.

For the control system to perform correctly the plant to be controlled must be capable of being controlled. The selection of plant is invariably made to meet the full load condition, but the worst control problems occur during partial load, at which the plant operates for a large percentage of the time. This fact is often overlooked.

4.3 THE CONTROL LOOP

To achieve automatic control in any given HVAC system six basic functions must be achieved by the control loop. These are listed below, together with the component in the control system, which performs the related function:

Function	*Performed by*
Measure change in controlled variable	Sensing, measuring element of controller
Translate the change into a useable force or energy	Controller mechanism
Transmit force or energy to the point of corrective action	Connecting members of control system, namely linkages for mechanical; wiring for electric/electronic; piping for pneumatic
Use force or energy to position control element and effect a corrective change in the controlled condition	Controlled device, e.g. final actuator
Detect completion of corrective change	Sensing, measuring element of controller
Terminate corrective action to prevent over correction	Controller mechanism, also final control element and connecting members of control system

4.4 CLOSED-LOOP CONTROL — FEEDBACK

Closed-loop control may be illustrated by a simple ventilation system where the room temperature (controlled variable) is maintained at a given set-point by means of a heater battery which is supplied with hot water (manipulated variable). A change in the controlled variable measured by a room detector initiates action to adjust the heater battery output by adjusting the manipulated variable. The change in room air temperature is fed back via the room detector which in turn stops or reduces the corrective action.

4.5 OPEN-LOOP CONTROL

In an open-loop system there is no feedback from the controlled medium. The manipulated variable is adjusted in some pre-arranged manner. This may be illustrated by using the example above but instead of the detector being in the room it is placed in the outside air. The heater battery output is then adjusted as a function of the outside air temperature and there is no feedback from the room temperature. Such a system ignores load changes due to casual heat gains in the room.

4.6 TYPES OF CONTROL SYSTEMS

There are five types of system in common use and these are briefly described below.

4.6.1 Self-acting

With this form of system, the pressure, force or displacement produced by the primary sensing element, is used directly as the source of power for the final control element. A common use of this type is for the control of HWS systems. A bulb, located in the HWS flow, is filled with a temperature sensitive liquid and connected to a valve actuator by means of a capillary tube. The change in volume of the liquid fill exerts a force on the actuator diaphragm, which, being directly connected to the valve spindle, positions it in relation to the temperature sensed. This type of system gives simple proportional control.

4.6.2 Pneumatic

Compressed air is piped to each controller and the controller changes the air pressure in a manner proportional to the value of the controlled variable being sensed. This is achieved by bleeding some air to waste. The output is then transmitted to the final control element which is caused to move by the output pressure change. This system is inherently modulating, giving proportional control but can be modified to give all the modes of control described later.

4.6.3 Electrical

The primary control element causes the application of a voltage by the controller to the final control element, in order to provide the necessary force to give a corrective action.

Two-position and proportional modes of control can be obtained with this system.

4.6.4 Electronic

This system is similar to the electrical type but the controller mechanism is far more sophisticated, requiring only small voltage or current changes from the primary control element. These signals are amplified to values suitable for actuating the final control element. All modes of control are available.

4.6.5 Direct digital control (DDC)

This method of control (Figure 4.1) is sometimes referred to as

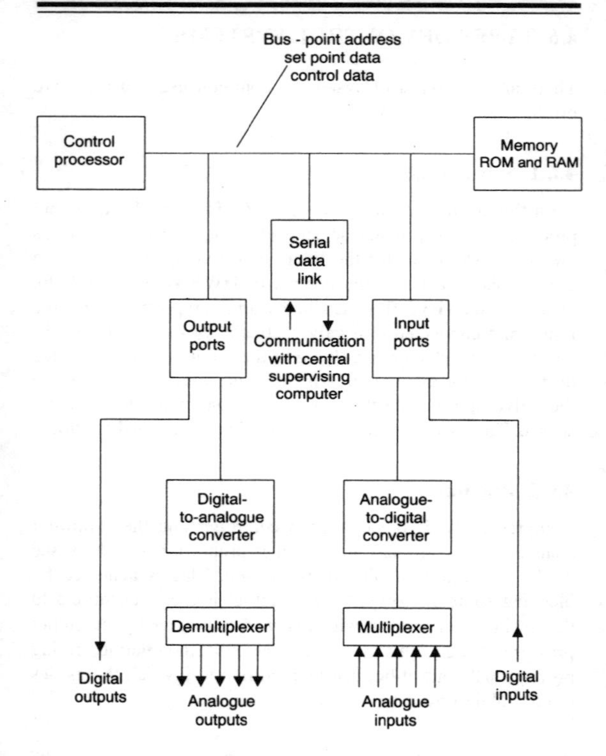

Figure 4.1 Direct digital control — simplified block diagram of a digital controller

'intelligent control' because a digital computer was initially used as the controller/processor. Today the computer has been replaced by the microprocessor which has two separate memories. The main operating programs (e.g. control algorithms) are held in the read only memory (ROM) while the temporary data (e.g. control loop set-points) are stored in the random access memory (RAM). As data stored in the RAM would be lost in the event of power failure battery back-up is normally provided to prevent this occurring.

The incoming and outgoing signals are connected to a multiplexer and demultiplexer, respectively. These devices enable simultaneous transmission of several messages along a single channel of communication. Input and output analogue signals are converted before and after being processed. Analogue-to -

digital and digital-to-analogue converters are used for this process.

The inputs referred to above are derived from sensors/transducers which measure the controlled variable, or from digital devices such as alarm thermostats or relay contacts. The outputs are in the form of analogue signals (e.g. 4–20 mA or 0–10V) or digital signals which are transmitted to the actuators driving control valves/dampers or to relays switching motor starters on/off.

Access to the data stored within the processor is achieved by a built-in keypad display unit, or a plug-in hand-held module.

In addition, a number of units may be connected to a communications network and supervised by a central computer. Direct digital control is now extensively used in the HVAC industry as it has the flexibility of pneumatics and the accuracy and ease of installation of standalone electronic controllers.

4.7 MODES OF CONTROL

4.7.1 Two-position control (on-off)

There are only two values that the manipulated variable can take, maximum or minimum.

There are also two values of the controlled variable which determine the position of the final control element. Between these values is a zone called the 'differential gap' or simply 'differential' in which the controller cannot initiate an action of the final control element. As the controlled variable reaches the higher of the two values, the final control element assumes the position which corresponds to the demands of the controller, and remains there until the controlled variable drops back to the lower value. The final control element then travels to the other position as rapidly as possible and remains there until the controlled variable again reaches the upper limit.

An example, using two-position control for an HWS calorifier is shown in Figure 4.2. The valve is open until the temperature reaches 60°C when the thermostat closes the valve. The valve remains closed until the water temperature drops to 54°C when the thermostat opens the valve. The 6°C gap is called the differential. The actual temperature of the water varies over a greater range than the differential (Figure 4.3).

Undershoot and overshoot are due to lag. The thermostat takes time to react to the change in temperature and the valve takes time to change its position. In the case of overshoot, the actual water temperature is above the set-point when the thermostat

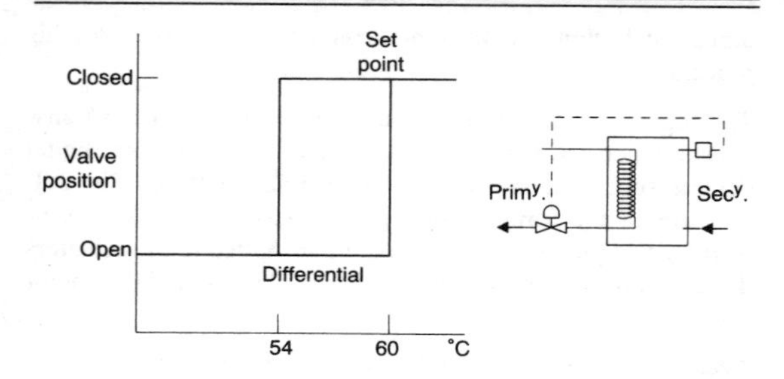

Figure 4.2 An example of two-position control

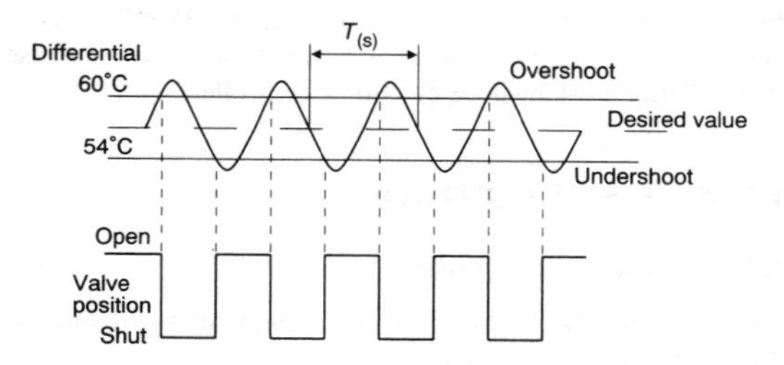

Figure 4.3 An illustration of overshoot and undershoot in two-position control

reacts and heat is still transferred while the valve is closing.

The main problem is that too much heat is supplied for too long. As it is not possible to alter the amount of heat input per unit time, the period of the cycle must be altered. Instead of 30 minutes on and 30 minutes off, the period can be artificially reduced to about 5 minutes. This is done by adding a small heater adjacent to the sensing element which emits heat to the element only when the thermostat is calling for heating, thus shortening the time taken for the thermostat to reach its set-point. This type of control is called timed two position and greatly reduces the swing resulting from the system lag.

4.7.2 Proportional control

If the output signal from the controller is directly proportional to the deviation, then the control mode is termed simple proportional. It is important to note that there is one, and only one, position of the final element for each value of the controlled variable within the throttling range of the controller. Outside

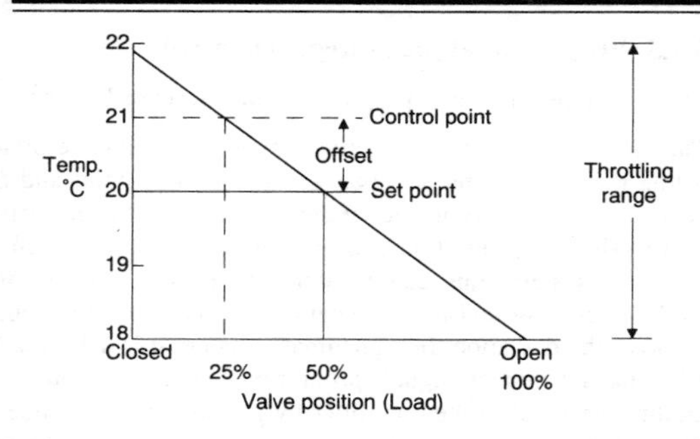

Figure 4.4 Proportional control

this the final control element is either at its maximum or minimum position. For example, consider a room being heated by a fan coil unit which is controlled by means of a valve, room temperature detector and proportional controller. From the graph of room temperature versus valve position (Figure 4.4) it will be seen that when the room temperature is 21°C the valve is 25 per cent open. If the heat input from the fan coil unit balances the heat loss from the room, an equilibrium condition has been reached. The room temperature will stay constant at 21°C even though the set-point of the controller is 20°C. The value of the space temperature is called the control point. A sustained deviation between the desired value and the control point is known as offset and is inherent in simple proportional control. If the load change causing the offset remained constant for a long period of time, one way to eliminate the offset would be to change the set-point by a corresponding amount. In the above example, the set-point would be moved to 19°C.

It can be seen from Figure 4.4 that the throttling range of the controller is 4°C. Since most controllers have adjustable throttling ranges, it seems sensible to suggest that offset could be reduced by setting a smaller range. Unfortunately this is not the case in practice because if the range is too small the system starts to cycle. In the extreme situation, a zero throttling range would give two-position control. So the selection of a proportional band is a compromise between minimizing offset and avoiding cycling.

The object of the more sophisticated modes of proportional control is to eliminate offset automatically and to do this as quickly as possible.

4.7.3 Proportional plus integral control

This is also referred to as proportional plus reset or P plus I.

The integral part of the corrective action is to eliminate offset. While there is a difference between the control point and the desired value, the controller indicates a signal for corrective action which is proportional to the size of the deviation. In an electrical system, this can be done by pulsing the actuator while a deviation exists, the length of the pulse being proportional to the deviation. In a pneumatic system it is achieved by allowing a feedback signal, proportional to the deviation, to modify the input to the controller by giving it a false value of the controlled variable. This is equivalent to altering the set-point. The rate that the reset action takes depends upon the characteristics of the controlled system. If it is too fast, the system cycles and if it is too slow the response is sluggish.

4.7.4 Derivative control

Derivative control action is where the corrective action produced by the controller is proportional to the rate of change of the controlled variable. Since the control action is not proportional to the deviation from the desired value, this action must be combined with either proportional and/or integral control action.

In Figure 4.5 the individual response is shown for all three modes of control due to a load change causing a deviation in the controlled variable. Also shown is the response when all three control actions are combined. This combination is referred to as PID or three-term control.

Today three-term control is used extensively as it is provided as standard in many of the microprocessor-based controllers which are now available.

4.7.5 Summary of modes of control

Each mode of control is applicable to systems having certain combinations of basic characteristics (Table 4.1). The simplest mode of control that will do the job satisfactorily is the best one to use, both for economy and for best results. Frequently, the application of too complicated a control mode will result in poor control, whereas, conversely, the use of too simple a control mode will often make control difficult, if not impossible.

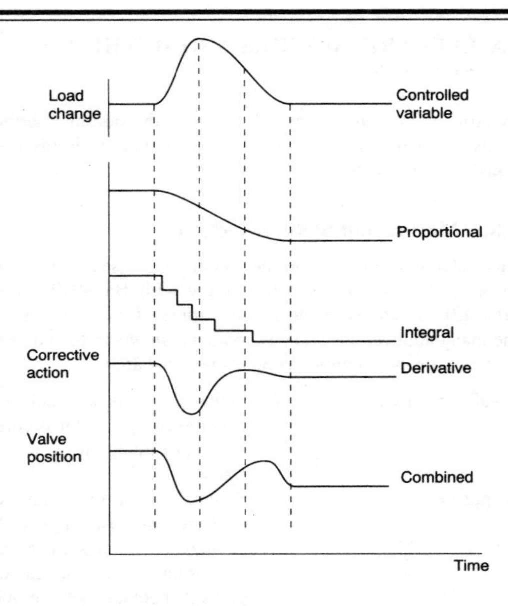

Figure 4.5 Individual and combined response of proportional, integral and derivative control action due to a load charge (deviation in controlled variable)

Table 4.1 Applications of the various forms of control

Mode	Application	Disadvantages
Two-position	Systems with large capacitance, minimum transfer lag, and the extreme positions giving inputs just above and below requirements for normal operation. Slow system response	Cycling, offset load changes must be slow
Proportional	Systems with large capacitance, small lag and slow response	Offset
Proportional plus integral	Systems with small capacitance, fast response and large load changes	Higher cost, takes longer to set up correctly

4.8 ELECTRIC MOTORS AND METHODS OF STARTING

Electric motors are extensively used in the building services industry. In this section the selection of motors is discussed together with the methods of starting.

4.8.1 Motors and motor selection

The induction motor is the most commonly used motor in the industry. Motors are built to comply with BS 4999 [1] and 5000 [2] and are available with a variety of enclosures to suit the many applications and environments in which they operate. Some of the more common enclosures used are:

Totally enclosed fan ventilated	Suitable for dusty and dirty conditions. The motor is cooled by a fan forcing air over motor frame.
Dripproof	Suitable for a clean and dry environment. The motor has ventilated openings in the end shields and screens are fitted to prevent contact with moving parts.
Weatherproof/hoseproof	Suitable for outside duty and can withstand extremes of atmospheric conditions, including corrosion due to chemicals.
Dust-tight	Can operate in a dusty environment, e.g. flour mills. Special seals are provided to prevent the ingress of dust.
Flameproof	Designed to operate in hazardous areas which are classified according to the potential risk of explosion. These motors carry certification by the British Approvals Service for Electrical Equipment in Flammable Atmospheres (BASEEFA).

Three-phase motors are self-starting but single-phase motors require an additional winding for starting. The start winding is connected to the supply initially and taken out of circuit, once the motor is up to speed. Single-phase motors, which are sometimes referred to as FHP (fractional horse power) motors are generally used at loads below 0.5 kW. British Standard BS 2048 [3] defines the dimensions and standard frame sizes of small motors. The speed of the induction motor is dependent on the design of the windings and the frequency of the electrical supply but it is basically a fixed speed motor. When the motor is fully loaded there is a marginal fall in speed of 3–4 per cent. The fixed speed is referred to as the synchronous speed. In Table 4.2 typical synchronous and full load speeds for a 50 Hz supply are listed.

Table 4.2 Induction motor speeds

Winding	Synchronous speed (rpm)	Full-load speed (rpm)
2 pole	3000	2900
4 pole	1500	1440
6 pole	1000	960
8 pole	750	720
10 pole	600	580
12 pole	500	480

Variations of speed are possible by the insertion of resistances into the rotor circuit, but to achieve this the rotor winding of the motor must be brought out to slip rings.

Where two fixed speeds are required, it is possible to obtain motors with windings that either vary the number of poles or have dual windings to give two different speeds. Pole-change motors have a 2:1 ratio, e.g. 1500/750 RPM but in the case of a dual wound motor, the variation can be greater, e.g. 3000/1000 RPM.

Where variable speeds are required then a special type of coupling may be used between the motor and the load. Alternatively, an inverter can be connected between the electrical supply and the motor in order to vary the frequency supplied to the motor. This will be discussed in more detail in Section 4.8.4. When using any method of speed variation which substantially reduces the motor, special care must be taken to ensure that the motor is not overheated at lower speeds due to a loss of cooling effect.

4.8.2 Motor selection

To enable a supplier/manufacturer to propose the correct motor for a given application certain information is required. A summary of this information is as follows:

Rating

- Output in kW and full-load speed
- Standard specification BS 4999 [1], BS 5000 [2], BS 4683 [4] Lloyds [5]

- Overseas — national or international standards may be applicable, e.g. IEC [6]
- Motor speed

Electrical supply
- Frequency, number of phases, availability of neutral wire
- Voltage. Variations of voltage, i.e. long cable runs between supply and motor
- Limitation in maximum connected load
- Supply authority regulations on maximum kVA or power factor

Mechanical conditions
- Type of mounting
- Direction of mounting, horizontal or vertical — shaft up or shaft down
- Mechanical power transmission arrangements
- Transmitted vibration
- Special stresses or end thrust on shaft
- Obstruction to free ventilation

Atmospheric conditions
- Ambient air temperature
- Altitude (if in excess of 1000 m above sea level)
- Barometric pressure
- Atmospheric pollution — dust, chemicals, etc.

Starting conditions
- Frequency of starting
- Reversal requirements
- Methods of starting and type of starter
- High static friction (motor/transmission/load, friction)
- High inertia load
- Acceleration
- Limitation on starting current

Load conditions
- Continuous, intermittent, fluctuating
- Speed variation — continuously variable or in step
- Prolonged shutdown

Others
- Terminal box left or right from driving end
- Bearings — ball, roller or silent sleeve type
- Coupling — vee pulley, flexible or fixed coupling, variable speed coupling, gear box

4.8.3 Energy efficient motors

Some manufacturers offer a type of motor which has an improved running efficiency of 1 or 2 per cent over the range of half to full load. These motors are marginally higher in price but with continuous running a good payback is possible.

The efficiency is gained by improvements in the mechanical details to reduce both windage and heating losses. The latter is achieved by reducing the rotor resistance which has a fundamental effect on motor starting torque and current. The torque falls and the starting current rises, both in the order of 20 per cent depending on the motor frame and manufacturer.

When using these motors it is necessary to make the following checks:

(a) Will the reduced torque accelerate the load with the particular method of starting selected? Centrifugal fans may be the worst affected.
(b) Will the increased starting current either exceed the local supply authorities limit or cause a more expensive starting method to be employed?
(c) Will the increased starting current necessitate fuse or switchgear re-selection. The longer starting time, due to reduced starting torque combined with higher starting currents, may cause tripping of a normal overload?

4.8.4 Methods of starting

The object of the motor starter is to fulfil a number of important functions which are summarized below:

- To connect the motor safely to the supply
- To disconnect the motor safely from the supply
- To provide the motor with protection from abnormal overloading
- To prevent the motor from re-starting after a supply failure
- To limit the current taken by the motor when starting
- To control the torque of the motor during the starting period
- To reverse the direction of rotation of the motor
- To control the speed of the motor
- To apply braking torque to the motor

The first three functions listed above are the most important and are mandatory requirements of the Institution of Electrical Engineers Wiring Regulations. The other functions are necessary, singly or in combination as required by the type of machine, supply limitation and its application.

There are four commonly used methods of starting three-phase motors, and these are:

- Direct on-line
- Star-delta

- Auto-transformer
- Soft-start

There are other variations on these methods but they are outside the scope of this book.

All forms of starter incorporate an overload trip to protect the motor. The overload unit may be of the thermal or electronic type, the latter being more sensitive. In addition the electrical supply to the starter is normally fused to provide short-circuit protection in the event of the motor cabling or windings failing.

Direct on-line starting (DOL)

This is the simplest and most common form of starting. In this method (Figure 4.6) the starter contactor connects the three mains supply lines to the motor terminals. The initial current drawn by the motor is between six and ten times full-load current and this sudden current surge can cause the supply voltage to dip which, in turn, may result in flickering lights and other disturbances to a range of equipment. Because of this the supply authorities normally limit the size of motor that may be started DOL to a maximum of 15 kW. Where there is no limitation on the supply, motors up to 120 kW may be started DOL. It should be noted, however, that the torque may rise to 200 per cent of full-load torque. The combined effect of this high torque and acceleration may then cause damage to belt or shaft drives. This must be considered during the selection process.

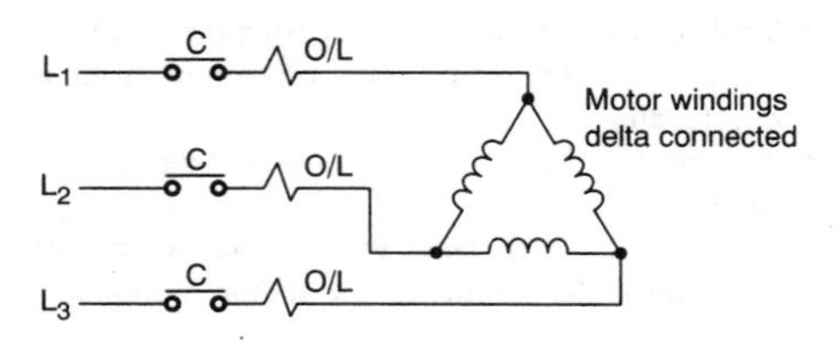

Figure 4.6 Direct on-line (DOL) starter

Star–delta starting (SDS)

Where direct on-line starting is unsuitable the initial current and starting torque may be reduced by applying a smaller voltage to the motor windings. The normal method of achieving this is to use star–delta starting (Figure 4.7). The three-phase windings of the motor are first connected in star while the motor is stationary. In this arrangement the voltage across each of the motor phase windings is $1/\sqrt{3}$ (i.e. 58 per cent of line voltage).

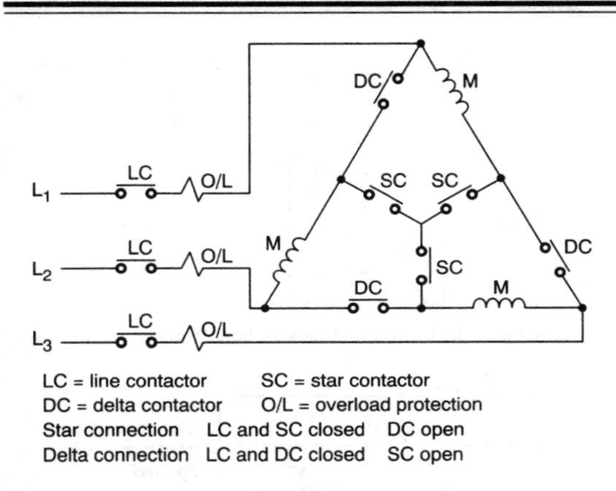

Figure 4.7 Star–delta starter

Consequently, the starting current is reduced to about twice full-load current with an accompanying reduction in starting torque to about 50 per cent full-load torque. Once the motor has achieved a speed of approximately 85 per cent of its rated speed, the starter connects the windings in a delta format, applying full voltage to each winding and allowing the motor to accelerate to full load speed as in direct on-line starting.

For star–delta starting all six ends of the motor windings must be brought out to the terminal box.

Auto-transformer starting (ATS)

This is an expensive method of starting as an auto transformer is used to provide the reduced voltage for starting (Figure 4.8). It has the advantage that by selecting an appropriate ratio of transformation, the starting torque and current can, within limits, be selected to suit the application. This method of starting may be used with three- or six-terminal motors.

The sequence of starting is as follows:

Stage 1
The mains supply is connected to one end of the auto transformer winding, the other ends of the windings are connected in star. The motor windings are connected to the transformer tappings.

Stage 2
The star point of the transformer is opened, leaving part of the auto-transformer winding in series with the motor.

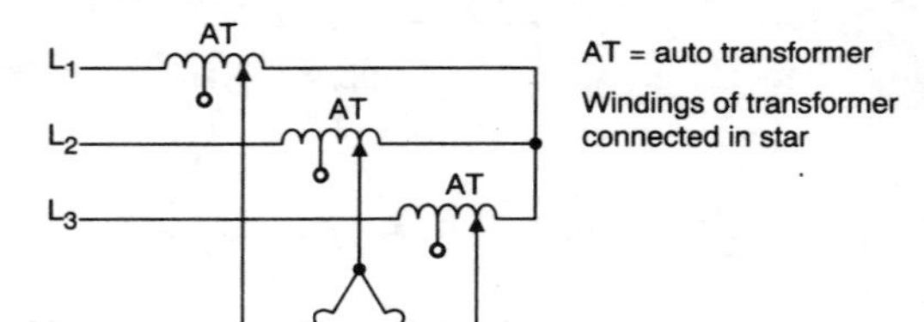

Stage 1 Motor supplied via auto transformer

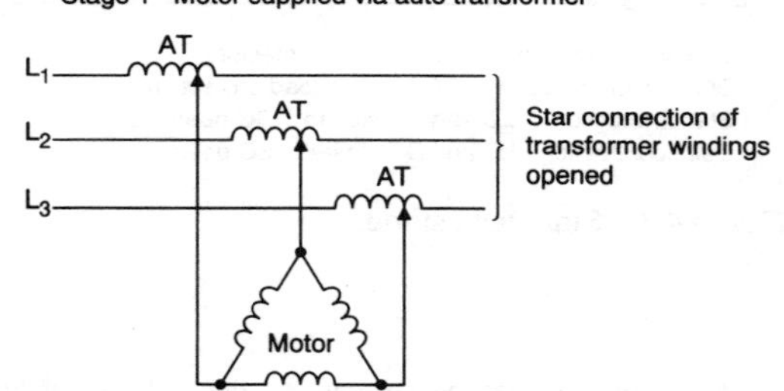

Stage 2 Auto transformer and motor in series

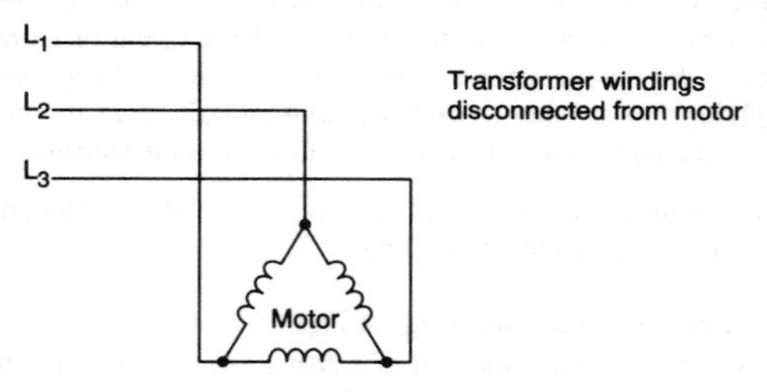

Stage 3 Motor connected directly to supply

Figure 4.8 Auto-transformer starter

Stage 3
The motor windings are connected directly to the supply and the transformer is disconnected leaving the motor in the normal running connection. The starting torque and current are dependent on the voltage tapping selected.

Soft start
In this method of starting, a smoothly increasing voltage is applied to the motor, by means of a pair of thyristors in each phase controlled by a microprocessor (Figure 4.9). The

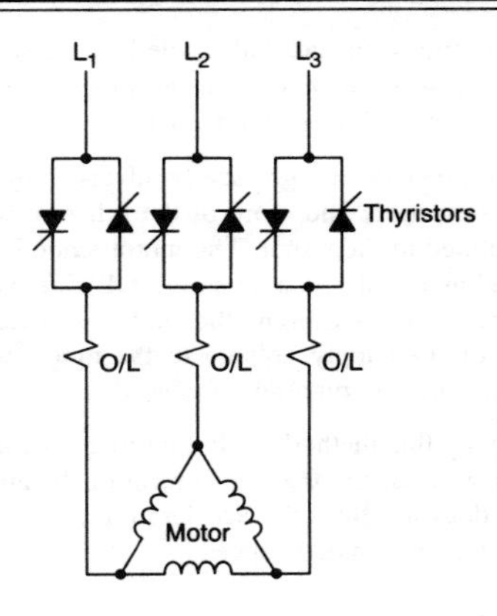

Figure 4.9 Thyristor starter (soft start)

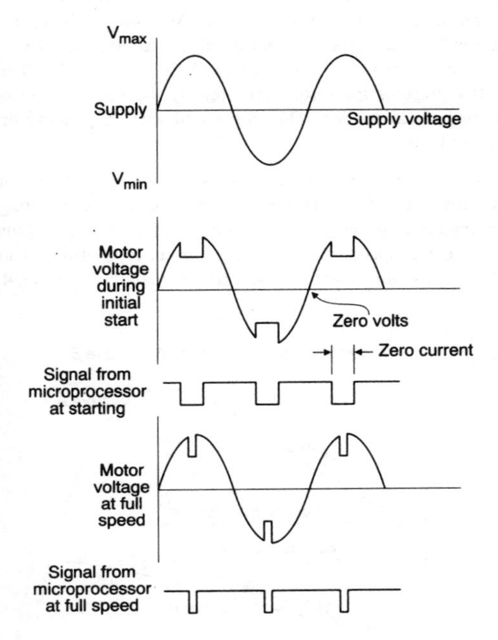

Figure 4.10 Soft starting – traces of motor voltage and thyristor firing for initial starting and full speed

thyristors are triggered each half cycle by a pulse generated by the microprocessor. Switch off takes place when the supply current passes through zero (Figure 4.10).

As the microprocessor progressively advances the firing angle of the thyristor, more and more of the full sine wave, of each phase, is applied to the motor. The motor smoothly accelerates to full speed as the starter applies the full sine wave (voltage) to the motor. In some designs the starter provides continuous adjustment of the supply voltage to the motor with a consequent improvement in running efficiency.

When applying this method of starting to centrifugal fans, care must be taken to ensure that the control of the applied voltage on starting does not limit the accelerating torque to the extent that the fan fails to run up to speed.

4.8.5 Starting problems

When a motor is energized it instantly develops a torque which is referred to as the locked rotor torque. Assuming a direct-on-line starter has been used, the motor will run up to speed developing a torque which follows curve A in Figure 4.11. The curve B represents the torque required by the motor to overcome mechanical friction. This curve will vary, depending on the type of load.

The difference between these two torque curves, i.e. motor torque curve A – Load Torque Curve B, is the accelerating torque available to accelerate the motor rotor and shaft, plus the load, to full speed. To enable the correct type of motor to be selected it is important to know precise details of the load.

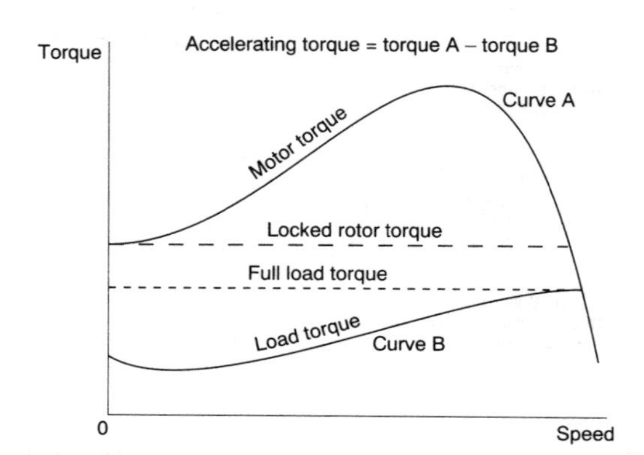

Figure 4.11 Motor torque and load torque curves

This is particularly important in the case of high inertia loads, such as some fans, where the starting load increases as the square of the speed.

When deciding the most suitable method of starting, the nature of the load and the starting current limitations are the important factors. A general summary of starting methods if given in Table 4.3.

4.8.6 Speed control

As previously stated the induction motor is basically a constant speed machine. However, there are applications where variations in speed are desirable as part of an overall control system, e.g. static pressure control on variable air volume systems.

One of the most cost effective methods of achieving variable speed is by means of an inverter (frequency converter). This device may be considered as a frequency generator which, when connected to a standard induction motor, will vary its speed according to the frequency generated.

Its principle of operation is as follows:

The 50 Hz alternating current supply is converted into direct current by means of a rectifier. The direct current is fed into an inverter, via a smoothing circuit, and is converted into an alternating current with a variable frequency. The inverter output voltage is also varied in proportion to the frequency.

In addition to infinitely variable speed control, savings in energy occur as the energy consumption of the motor is related to its load (torque) and speed. Other benefits include:

- Smooth starting — an inverter may be used in place of a 'soft starter'
- No inrush of current on starting
- Smooth motor acceleration
- May be remotely controlled
- Easily installed
- Low maintenance
- One inverter may control a number of motors

When applying an inverter to a given load it is important to consult with the motor manufacturer. Some increase in motor frame size may be necessary because of the reduced cooling of the motor at low speeds. There is also some marginal increase in motor losses at reduced speed due to the output voltage of the inverter not being a purely sinusoidal curve.

Table 4.3 Summary of starting methods

Method of starting	Starting current (% FLC)	Starting torque (% FLT)	Limitation on starting	Type of load	Comment
Direct-on-line	600/800	100/150	None	Low inertia quick run-up pumps, small fans	Least expensive Most reliable
Star–delta	200/300	30/50	None	High inertia slow run-up centrifugal fans	Requies six-terminal motor
Auto-transformer	150/400*	30/80	5 starts per hour	Centrifugal chillers, Hermetic machines	*Depends on tapping. Expensive Used on large motors with 3 terminals
Thyristor soft start	150/400	100/300	None	Where smooth run up is required pumps, centrifugal fans, chiller	These starters can achieve energy savings at less than full load

References

1. British Standard 4999: Specification of General Requirements for Rotating Electrical Machines.
2. British Standard 5000: Specification for Rotating Electrical Machines of Particular Types or for Particular Applications.
3. British Standard 2048: Specification for Dimensions of Fractional Horse Power Motors.
4. British Standard 4683: Specification of Electrical Apparatus for Explosive Atmospheres.
5. Lloyds Register of Shipping — relevant to motors for marine applications.
6. International Electrotechnical Commission (IEC).

5 Terminal capacity regulation

Terminal heat exchangers such as radiators and coils are usually oversized for the maximum thermal output required. This is because they are manufactured only in a limited range of sizes and normal design procedures incorporate safety margins in establishing room thermal requirements.

Except under preheat boost or pull-down cooling conditions the full thermal output of a terminal is rarely required and for most of the time the temperature control will require less than 50 per cent of the maximum thermal output available (see Figure 5.1).

Two basic types of control are available:

5.1 TWO-POSITION CONTROL

Two-position or on/off control, in which the terminal is switched at full output intermittently as required by a thermostat. This type of control is low cost and simple but gives satisfactory results for many applications such as control of low thermal capacity air recirculating room terminals. Although the heat energy input to the room is intermittent with this control, the damping effect of the room thermal capacity results in imperceptible swings in room temperature. Two-position control of heat input to hot water storage is quite satisfactory.

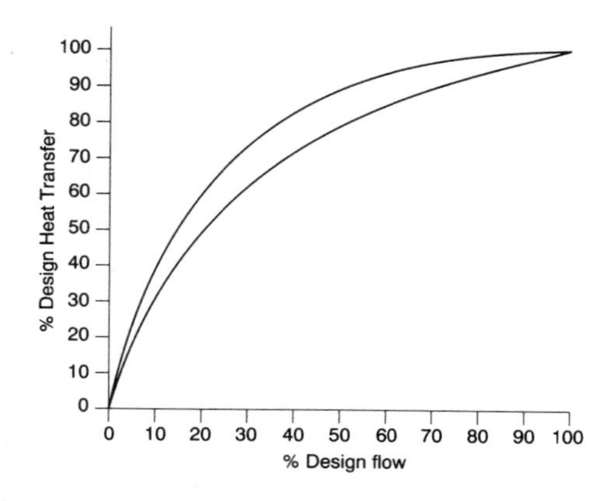

Figure 5.1 Typical range of heat transfer/flow characteristics for room terminals with constant inlet air temperature and air flow

5.2 MODULATING CONTROL

Modulating control in which the terminal output is regulated to match the room requirements. This more sophisticated and expensive control avoids room temperature cycling under most circumstances and is suitable for air reheating as well as air recirculating terminals.

Straight proportional modulating control gives a small offset in the controlled temperature dependent on the load, for comfort room temperature control this is usually acceptable but for applications where precise temperature control is required a proportion control plus integral reset will remove the offset (see Section 4.7).

5.3 VARIABLE WATER TEMPERATURE

Changing the water temperature to a terminal or group of terminals is a common method of control. For a group of terminals it is only satisfactory if all the terminals are in the same room and controlled from room temperature. Resetting the flow temperature against outside air temperature is now almost standard practice for space heating systems and has the merits of reducing mains losses, increasing boiler efficiency and reducing the throttling of terminal control valves. Flow temperature resetting is unsatisfactory as the sole means of room control and must be backed up by individual terminal control (see Figure 5.2).

5.4 VARIABLE WATER FLOW

This is the most common form of control, in which the terminal output is reduced by reducing the water flow. This is usually achieved by a three-way control valve diverting flow water

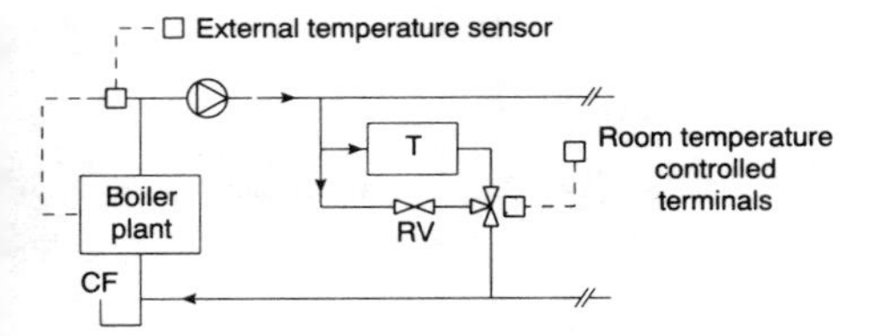

Figure 5.2 Boiler plant flow temperature controller reset by outside air temperature

directly to the return connection, bypassing the terminal and giving a nominally constant flow water distribution system. The return water temperature approaches the flow temperature as the thermal load reduces, increasing mains losses at part load.

Alternatively variable flow can be achieved by a two-way control valve throttling the water flow from the distribution pipework system; this makes the flow rate in the distribution system reduce and the water temperature drop or rise increase as the thermal load reduces, reducing mains losses at part load.

Variable water flow capacity control of heating coils causes temperature stratification of air leaving the coil at part load (see Figure 5.3). For some applications, such as frost protection or preheating before humidification, stratification is often the cause of poor humidification performance with water precipitation downstream of the humidifier (see Figure 5.4). It also makes sensing the mean air temperature after the coil inaccurate even with averaging capillary sensing elements. There is also a risk of dead water in the remote part of the coil freezing.

These problems are avoided by a separate circulating pump to the coil giving it constant flow and variable temperature, the coil circuit being fed from the distribution via an injection circuit (see Figure 5.5).

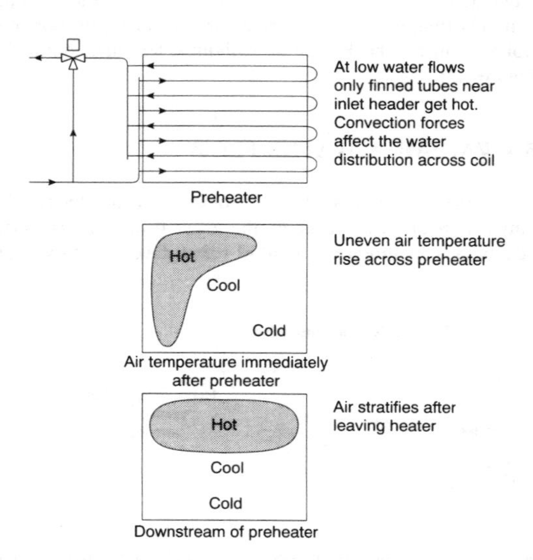

Figure 5.3 Air heater stratification

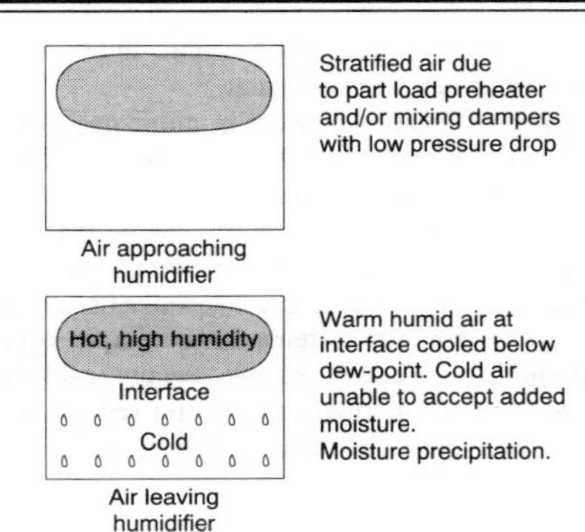

Figure 5.4

5.5 AIR SIDE FACE AND BYPASS DAMPER

The alternative to water control is air side control in which a system of dampers is arranged to throttle the air passing the heating or cooling coil and bypass the balance. With this control the water flow through the coil remains constant (see Figure 3.44).

The main features of face and bypass damper control are:

(1) Instant response to load change.
(2) Air side control takes up more space than water control making units larger.
(3) Possible problems of downstream air stratification although these can be reduced by arranging the dampers to have a high enough pressure drop to give turbulent mixing downstream.
(4) Some heat transfer still occurs when the coil damper is fully closed due to damper leakage and heat conduction

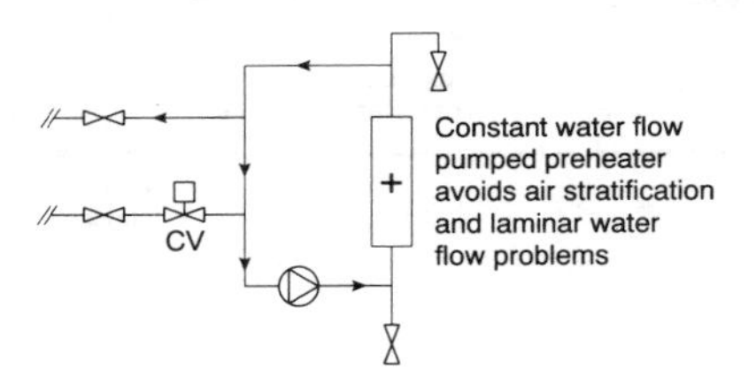

Figure 5.5 Pumped preheat coil with injection circuit

through the damper blades and duct connection between the damper and heater.

(5) This form of control is used in multizone air handling units with a hot deck (duct) and a cold deck (duct) from which each zone duct is fed with a mix of warm air and cold air via control dampers to achieve the air supply temperature required. Dual duct systems use a similar principle except that the mix takes place at the terminal box. These air mix systems have a rapid response to load change but are expensive, occupy a lot of space, are wasteful in thermal energy and require more fan energy than most other equivalent systems.

5.6 VARIABLE AIR FLOW (VAV)

VAV is a system in which the terminal cooling output is initially regulated by an air throttling damper changing the air flow rate. Two basic types are available:

(1) The pressure-dependent terminal in which a separate pressure reducing damper gives a nominally constant pressure upstream of a group of room supply terminals each of which has a throttling damper controlled by room temperature (see Figure 5.6).
(2) The pressure-independent terminal in which the terminal throttling damper is regulated by an air velocity controller to give the required air flow irrespective of increases in upstream pressure. The room temperature controller changes the set point of the velocity controller (see Figure 5.7).

In both types the air flow is not permitted to fall below a lower limit which is determined either by the fresh air requirement or by the ability of the air outlet terminals to mix the supply air

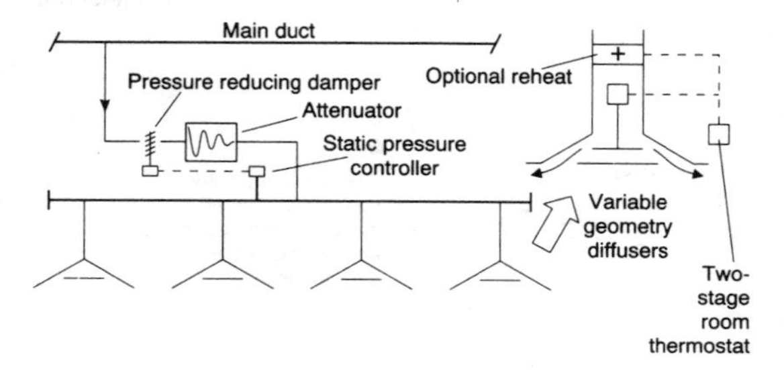

Figure 5.6 Pressure dependant VAV

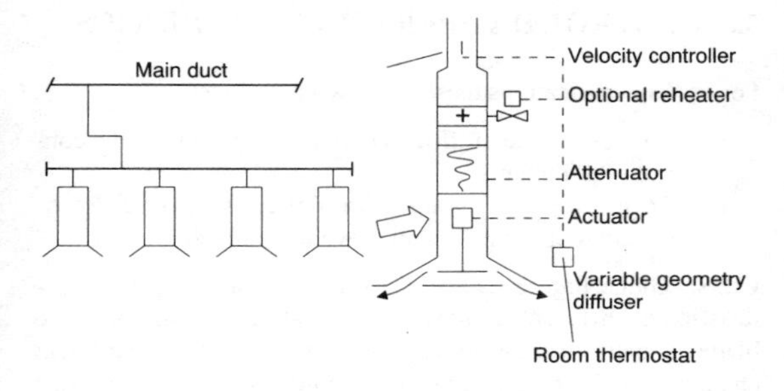

Figure 5.7 Pressure independant VAV

with the room air and effectively distribute the air evenly over the treated area.

At constant minimum air flow further reduction of cooling load and subsequent heating is achieved by reheating as a second stage in the control sequence.

5.7 CONSTANT VOLUME REHEAT

In these systems air at a suitable moisture content and temperature is distributed by a single duct system to a number of zones each having its own reheater regulated by a room temperature controller (see Figure 5.8). These systems are discussed in detail in Section 3.5.2 and the type of control is described in Section 5.4.

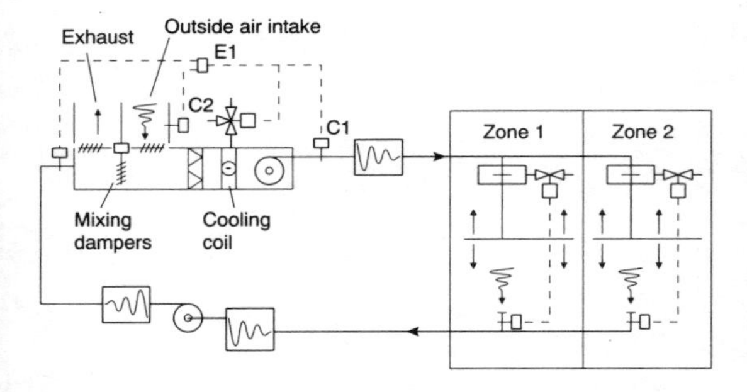

Figure 5.8 Constant volume system with reheat zones

5.8 CONTROL DAMPER CHARACTERISTICS

For mixing applications these have two functions:

(1) To regulate the air flow rate in accordance with the control requirements.
(2) To ensure that enough turbulence is created downstream to mix the air and prevent stratification.

CIBSE Guide Figure B11.33 [1] gives typical damper characteristics at different authorities suggesting that for opposed blade dampers an authority of about 5 per cent gives near linear characteristics. In many practical applications the damper size is predetermined by the size of the equipment and it is important that the actuators have the facility to adjust the angle of blade rotation over which the control is effective. Controllers are available which correct the dampers non-linearity giving a linear air flow response to a linear control signal.

Reference

1. CIBSE Guide Book B Figure 11.33 Installed damper characteristics.

6 Fans and their characteristics

6.1 DEFINITIONS

The terms in common use in fan engineering are as follows.

(1) Velocity pressure (p_v).

$$p_v = 0.5\ \rho v^2 \tag{6.1}$$

where ρ = air density (kg/m³) and v = mean air velocity (m/s). If a standard air density of 1.2 kg/m³ is adopted the equation becomes

$$p_v = 0.6\ v^2 \tag{6.2}$$

which is the form commonly used in the UK.

(2) Fan total pressure (p_{tF})

This is the total pressure rise through the fan, defined by

$$p_{tF} = p_{to} - p_{ti} \tag{6.3}$$

where p_{to} = total pressure at fan outlet (Pa or kPa) and p_{ti} = total pressure at fan inlet (Pa or kPa).

(3) Fan static pressure (p_{sF})

This is defined by

$$p_{sP} = p_{so} - p_{ti} \tag{6.4}$$

where p_{so} = static pressure at fan outlet (Pa or kPa)

Fan static pressure is measured in preference to fan total pressure, when testing fans, because the air distribution at a fan outlet is very disturbed, making it difficult to measure the velocity pressure at fan outlet.

In accordance with Bernoulli's theorem, total, static and velocity pressures are related by:

$$p_t = p_s + p_v \tag{6.5}$$

where p_t = total pressure (Pa or kPa) and p_s = static pressure (Pa or kPa) It follows from equation (6.5) that there is an alternative expression for fan total pressure:

$$p_{tF} = p_{sF} + p_{vo} \tag{6.6}$$

where

p_{vo} = velocity pressure at fan outlet (Pa or kPa)

The convention is adopted that the velocity pressure at fan outlet is based on the notional, mean, outlet velocity defined by

$$v_o = \frac{\text{volumetric airflow rate (m}^3\text{/s)}}{\text{area across the flanges at fan outlet (m}^2\text{)}} \tag{6.7}$$

A significantly large amount of kinetic energy is locked in the eddies that form the turbulent airflow at fan outlet but, if a sufficiently long length of straight duct is provided, a useful proportion of this can be recovered and converted into static pressure that can be used to offset frictional and other losses downstream in the duct system [1, 2].

$$L/D_e = 2.5 + 0.2(v_o - 12.5) \qquad (6.8)$$

subject to a minimum of 2.5 equivalent diameters

where
L = effective straight duct length to secure a smooth velocity profile over the duct section (m)
D_e = equivalent diameter of the fan outlet area over the flanges (m)
v_o = mean air velocity in the outlet area over the flanges (m/s).

(4) Fan and motor power
The rate at which the fan impeller delivers energy to the airstream is termed the air power and is defined by:

$$w_a = Q\,p_{tF} \qquad (6.9)$$

where
w_a = air power (W or kW)
Q = volumetric airflow rate (l/s or m^3/s)
p_{tF} = fan total pressure (Pa or kPa).

The rate at which power is supplied to the fan shaft is termed the fan power and is defined by:

$$w_f = Q\,p_{tF}/\eta \qquad (6.10)$$

where w_f = fan power (W or kW) and η = total fan efficiency as a fraction.

The power of the motor needed to drive the fan shaft must be greater than the fan power for several reasons:

(i) Account must be taken of the drive efficiency.
(ii) On start-up the impeller must be accelerated to full running speed within a reasonably short period of time typically about 18 seconds.
(iii) Margins [3] on the volumetric airflow rate (5 to 10 per cent) and the calculated fan total pressure (10 to 15 per cent) must be allowed to cover the gap between the drawing board and the site, and to allow for minor changes, air leakage and the unforeseen. Excessive margins are bad practice since they result in the fan operating at a lower efficiency.
(iv) The fan power should be calculated using the margins suggested above.
(v) To determine the motor power, a further margin of 25 per cent (for centrifugal fans with backward-curved impeller blades) or of 35 per cent (for centrifugal fans

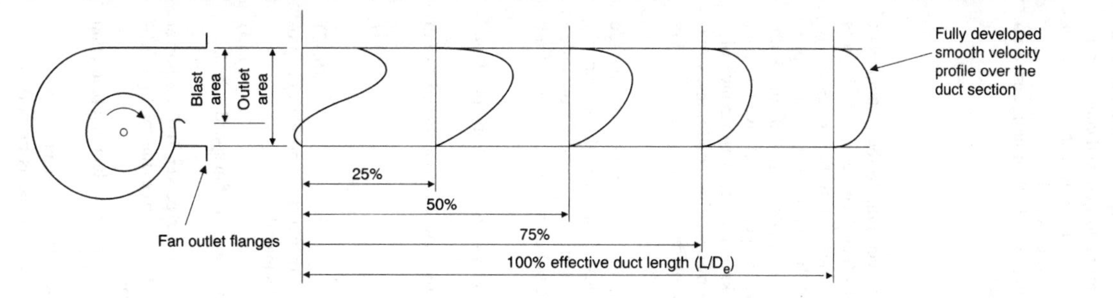

Figure 6.1 Duct length in equivalent fan outlet diameters to achieve a smooth velocity profile and to recover a conversion of wasted velocity pressure to useful static pressure. See equation (6.8). The minimum effective straight length is 2.5 equivalent fan outlet diameters

with forward-curved impeller blades) should be made. This covers the risk of overloading during commissioning, provides adequate starting torque, and caters for the possibility that the fan may have to run at a higher speed in order to achieve the design duty.

(vi) The calculated motor power would then be rounded up to the next commercial size.

6.2 FAN LAWS

A series of fans is manufactured according to the principles of dynamical similarity. This means that if any one size of fan is tested the performance of any other size in the series can be determined, without further test, by applying such principles. Fan performances are quoted for standard air (20° dry-bulb, 101.325 kPa barometric pressure, 50 per cent relative humidity and 1.2 kg/m³ air density) but knowing the performance of a given fan, under a specified set of operating conditions, variations in the performance can be predicted according to the fan laws.

The laws apply for a given point of rating on the characteristic curve that expresses the relationship between the volumetric airflow rate and the fan total pressure (or fan static pressure). The laws of most interest to the building services engineer are then as follows:

(i) For a particular fan, given system of duct and plant, and constant air density.

1. The volumetric airflow rate is proportional to fan speed:

$$Q_2 = Q_1(n_2/n_1) \qquad (6.11)$$

Hence

$$n_2 = n_1(Q_2/Q_1) \qquad (6.12)$$

2. The fan total pressure (or fan static pressure) is proportional to fan speed squared:

$$p_{tF2} = p_{tF1}(n_2/n_1)^2 \qquad (6.13)$$

3. The fan power is proportional to fan speed cubed:

$$w_{f2} = w_{f1}(n_2/n_1)^3 \qquad (6.14)$$

4. The relationship between air power and fan power means that, if the third law is true, the total fan efficiency must be constant.

(ii) For a particular fan, given system of duct and plant, and constant fan speed.

1. The volumetric airflow rate is constant.
2. The fan total pressure developed is proportional to air density:

$$p_{tF2} = p_{tF1}(\rho_2/\rho_1) \qquad (6.15)$$

3. The fan power is proportional to air density:
   $w_{f2} = w_{f1} (\rho_2/\rho_1)$ (6.16)
4. The total fan efficiency is constant

The fan laws cannot be used by a manufacturer to establish the fan performance, in the first instance. This can only be done by test or by the principles of dynamical similarity, once the test results are known.

6.3 CENTRIFUGAL FANS

6.3.1 Forward curved and backward curved impellers

Forward curved impellers

Figure 6.2 (a) is a diagram of the impeller of a centrifugal fan with forward curved blades, showing the relative air velocities vectors involved. The air velocity leaving the impeller blade tip tangentially to the curvature of the blade, is v_t. The peripheral velocity of the impeller is v_p. The resultant of these, v_r, is the absolute air velocity vector and represents the airflow rate handled by the fan for a given speed of impeller rotation.

The impeller is in the form of a runner, like a water wheel, open on both sides. It has between 32 and 66 blades, each with a short chord, of the order of 60 mm. (see Figures 6.2 (c) and (d)). Total efficiencies are in the range 60 per cent to 75 per cent and, because the blades are shallow, the maximum practical fan static pressure that can be developed is about 750 Pa, although some makers claim as much as 1 kPa.

Backward curved impellers

Figures 6.2 (a) and (b) show that, in order to achieve the same absolute air velocity vector, v_r, the peripheral speed vector, v_p, must be a good deal greater for a backward curved impeller than for a forward curved impeller. It follows that a fan with a backward curved impeller must run faster than one with a forward curved impeller, in order to deliver the same volumetric airflow rate. Hence backward curved fans tend to be larger and dearer than forward curved fans. Another consequence of the higher running speed is that backward curved impellers are noisier than forward curved.

The backward curved impeller has comparatively few blades, 14 to 24, and these are deep, extending over the full impeller diameter from its perimeter to the driving shaft, and the blades are attached to a backplate. Because the blades are deeper, backward curved impellers can develop much higher pressures than forward curved.

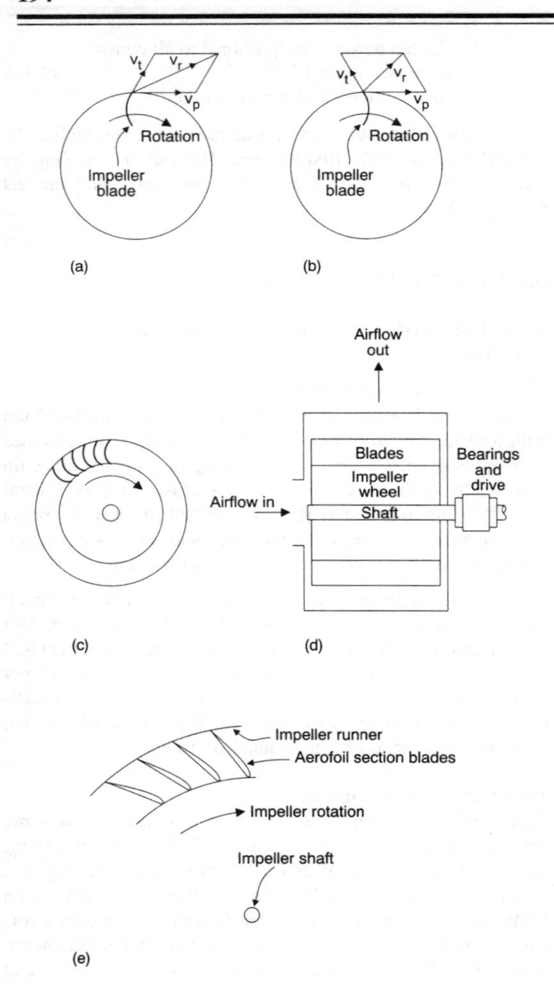

Figure 6.2 Various types of centrifugal fan impeller. (a) Forward curved impeller blades. (b) Backward curved impeller blades. (c) Forward curved impeller wheel. (d) Forward curved impeller wheel in section in a fan casing. (e) Backward curved impeller with aerofoil section blades

Backward curved impellers having blades with an aerofoil section (Figure 6.2(e)), instead of simple curved sheet metal (Figure 6.2(b)), are used with medium and high velocity systems of ductwork and have a higher total efficiency than any other fan, with a maximum of 89 per cent.

Characteristic curves

Three characteristic curves are used to express the performance of fans, in terms of volumetric flow rate versus pressure, total efficiency, and fan power.

Theoretically, the pressure developed by a backward curved impeller is a straight line that slopes downwards, from left to right (Figure 6.3), but with a forward curved impeller the slope is upwards. In practice, there are losses which change the shape of the characteristic. Molecules of air within the space between adjoining impeller blades rotate in a direction opposite to that of the impeller. This rotation helps airflow along one of the adjoining blade surfaces but retards it along the other blade surface. The loss becomes less significant as airflow increases. Secondly, air passing through the fan casing and impeller suffers frictional losses which are approximately proportional to the square of the airflow rate. Thirdly, there is a shock loss at entry: as the air enters the impeller it must change direction through 90°, as well as being rotated. The extent of this loss depends on the angle of entry to the impeller and any swirl in the entering airstream: with smooth, uniform, airflow into the

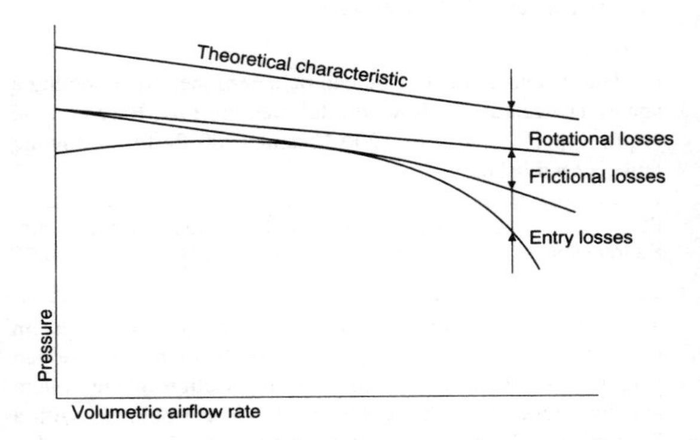

Figure 6.3 Characteristic curve build-up for a centrifugal fan with a backward curved impeller. If the impeller blades are forward curved the theoretical characteristic slopes upward and the actual curve is flatter with a point of inflexion in it; see Figure 6.4

inlet of the fan the entry loss diminishes to virtually zero at an optimum airflow rate, the value of which depends on the fan design. Poor ducted entry conditions, sagging flexible connections and pre-rotation in the entering air, will cause the fan performance to be much reduced.

The form of the impeller blades greatly affects the shape of the characteristic curves. Figures 6.4 (a) and (b) show typical curves for forward and backward curved fan impellers. The broken lines in Figure 6.4 (a) show how the pressure, efficiency and fan power are read against volumetric airflow rate.

There is also a system characteristic curve for the plant and ducting to which the fan is coupled, based on the simplifying assumption that the total pressure loss in the plant and duct is proportional to the square of the volumetric airflow rate. The curve that results is a parabola going through the origin. Where it cuts the pressure-volume curve for the fan defines the point of rating on the fan curve and shows the performance achieved in terms of pressure, airflow rate, total fan efficiency and fan power, for the particular fan speed.

EXAMPLE 6.1

A backward curved fan running at 600 rpm and having the characteristic performance shown in Figure 6.5 is connected to a system of plant and ductwork with a total pressure loss of 1000 Pa when handling 1200 l/s. (i) Determine the actual performance and the speed at which the fan should run if the design duty is required. (ii) Establish the required motor power, assuming a vee-belt drive efficiency of 98 per cent.

Answers

(i) The system characteristic can be determined by assuming a square law relating airflow to total pressure loss. Based on the required performance of 1200 l/s and 1000 Pa the following table is established:

l/s	200	400	600	800	1000	1200	1400
Pa total loss:	28	111	250	444	694	1000	1361

The characteristic curve is plotted on the coordinate system that expresses the pressure–volume performance of the fan (Figure 6.5), and it is seen that the intersection of the system and fan pressure-volume curves occurs at the point P_1 with a duty of 1000 l/s and a fan total pressure of 0.7 kPa. The efficiency is shown by the point E_1 and is 70 per cent. The first fan law, equation (6.12), is applied and the fan speed to achieve a flow rate of 1200 l is established:

$$n_2 = 600\ (1200/1000) = 720 \text{ rpm}$$

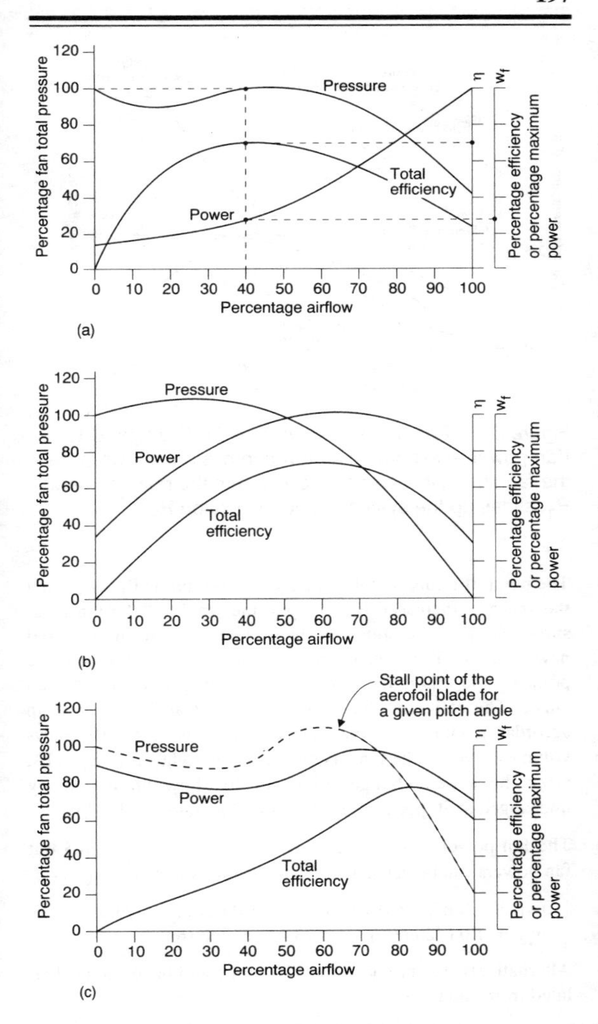

Figure 6.4 Characteristic performance curves for three types of fan. (a) Forward curved impeller centrifugal. (b) Backward curved impeller (centrifugal). (c) Axial flow fan

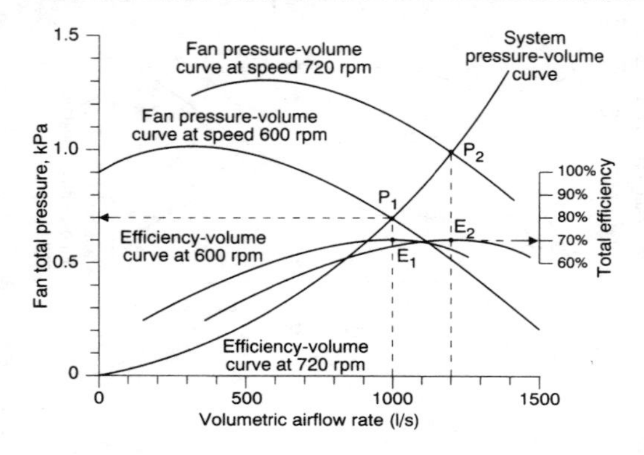

Figure 6.5 Fan and system curves for Example 6.1. Fan power – volumetric airflow curve is not shown. When the fan speed is increased to 720 rpm the point of rating, P_1, slides up the system curve to position P_2.

The point of rating on the fan curve at 600 rpm is P_1 and, when the speed is increased to 720 rpm, the whole of the fan pressure–volume curve slides up the system curve to the required new position. The intersection with the system curve is at the point P_2 and this is the same point of rating as P_1, on the fan curve. The efficiency has stayed constant at 70 per cent, in accordance with the fan laws. Hence the whole of the efficiency-volume curve for the fan has moved to the right, from E_1 to E_2, after the speed change, in order to show an efficiency unchanged at 70 per cent, for the new flow rate of 1200 l/s.

The fan power–volume curve is not shown in Figure 6.5 but fan powers can be calculated from equations (6.10) and (6.14):

$$w_{f1} = 1.0\ \text{m}^3/\text{s} \times 0.7\ \text{kPa}/0.7 = 1.0\ \text{kW at 600 rpm}$$
$$w_{f2} = 1.0\ (720/600)^3 = 1.73\ \text{kW at 720 rpm}$$

Alternatively, the fan power at 720 rpm could have been calculated from equation (6.10):

$$w_{f2} = 1.2\ \text{m}^3/\text{s} \times 1.0\ \text{kPa}/0.7 = 1.71\ \text{kW}$$

The slight discrepancy is due to small errors in reading the coordinates of the points P_1 and E_1, in Figure 6.5.

(b) Take margins of 7.5 per cent on the volume handled, 15 per cent on the fan total pressure and 25 per cent on the fan power (for a backward curved fan). Making use of equation (6.10) and incorporating the margins mentioned, the motor power is:

$[(1.2 \times 1.075) \times (1.0 \times 1.15)/(0.7 \times 0.98)] \times 1.25 = 2.70$ kW

This is then rounded up to the next commercial motor size, which could be 3kW.

Speed and temperature limitations

All fans used should be statically and dynamically balanced. Slight imperfections in the manufacture will still exist and the out of balance forces exerted will be magnified as the running speed increases. Furthermore, the rivets and welds used for joints and the strength of materials used for all the rotating parts, have limits on the stress they can safely accept without failure. These considerations give a critical speed for a fan. Taking account also of wear, corrosion and age[1], a fan should not be run at more than 55 per cent of the critical speed that the manufacturers quote.

There are also limits to the temperature at which a fan can safely operate. Temperature affects the yield stress of steel and its modulus of elasticity, both of which fall with increasing temperature. Although few fans used in ordinary ventilation and air conditioning applications handle air at temperatures exceeding about 60°C there are exceptions, for example when a fan is used for smoke extract. The manufacturers should be consulted for such applications.

Drives, bearings and handings

Most centrifugal fans used in constant volume systems are driven by vee-belts and pulleys from a four-pole, squirrel cage, electric motor, running at a synchronous speed of 1500 rpm and an actual speed of between 1415 and 1440 rpm. This gives scope to the manufacturer for expressing the performance of a fan against various speeds, and offers the facility of easily modifying fan speed on site, during commissioning, in order to get the intended duty, in accordance with the fan laws.

Space-saver vee-belt drives should never be used: because the fan and motor shafts are too close, the arc of contact on the smaller, driving pulley is less and the drive efficiency reduced.

Bearings for fans are usually roller or ball but sleave bearings are sometimes used. Different bearing arrangements are adopted but a pair of external bearings, with an overhung pulley, is practical and common.

Fans are made with eight possible directions of discharge, termed handings. To describe the handing, the fan is viewed from the driving motor side and if the air is being discharged in a counter-clockwise direction it is termed LG. If the direction of discharge is clockwise it is termed RD. A number after the two letters, in angular increments of 45°, completes the description of the handing. (see Figure 6.6(a)). The fan should be chosen with a handing that best suits the direction of airflow in terms of freedom from turbulence and quiet operation. Figure 6.6(b) gives some examples of good and bad arrangements.

6.3.2 Open paddle blade fans

The impeller is open and has no back plate. The construction of the casing is heavy and the blades on the impeller are commonly six or eight in number and also of heavy construction. Fans of this type are built to handle fairly dense concentrations of dust (sawdust, grinding wheel dust, wood shavings, etc.) and hence are designed to withstand excessive wear and abrasion. Impeller blades can be replaced and centrifugal forces have no bending effect because the blades are radial. The blades may be

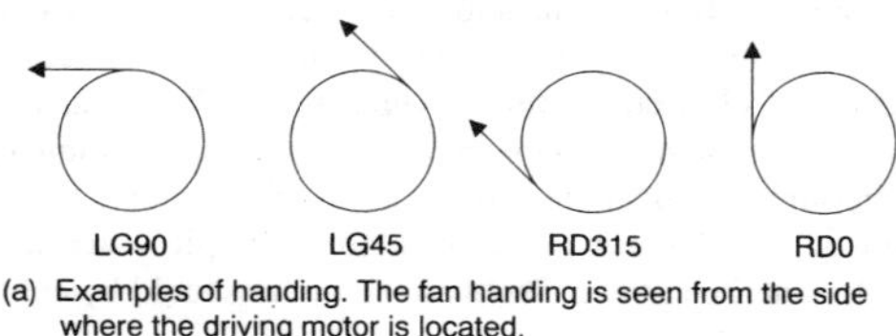

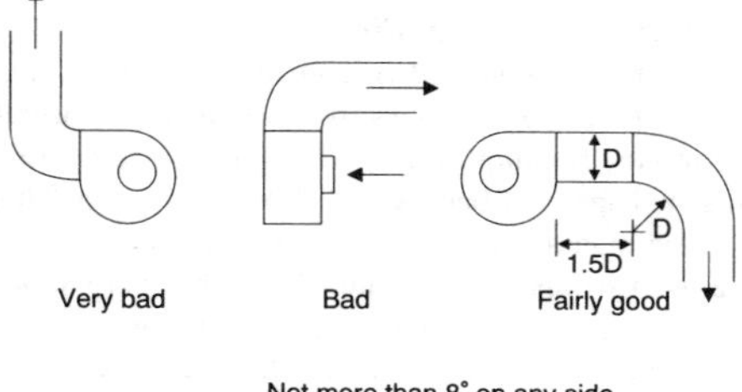

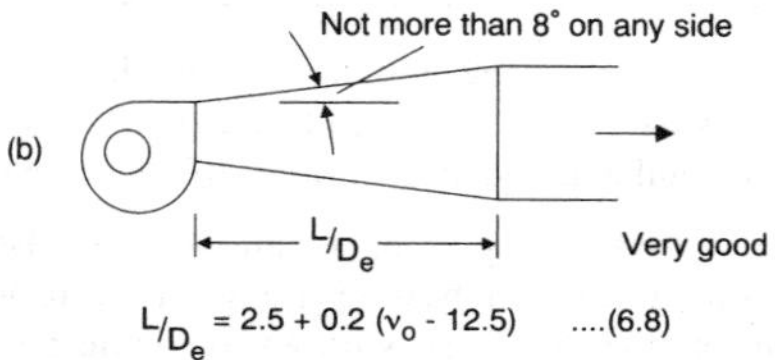

$$L/D_e = 2.5 + 0.2\,(v_o - 12.5) \qquad ....(6.8)$$

Where D_e is the equivalent diameter over the flanged outlet of the fan discharge, and v_o is the mean outlet air velocity over this area

Figure 6.6 Fan handings and some good and bad duct connections at fan outlet. L/D_e is subject to a minimum length of 2.5 equivalent diameters

paddles, fixed by radial tie rods to the fan shaft, or they may be of strengthened rectangular section, extending through the full radial depth of the impeller to the fan shaft. The absence of a back plate means that thermal expansion is not usually a problem and paddle blade fans can operate at temperatures up to 350°C.

The impellers have a pressure–volume characteristic in the form of a shallow curve, sloping downwards from top left to bottom right, and can develop up to about 3.5 kPa fan static pressure, when running at about 1440 rpm.

The best total efficiencies are less than 65 per cent and the fans are noisy but this is not a problem (except as a nuisance to neighbouring premises) because the applications are industrial. There is also a shrouded radial fan of similar characteristics which is slightly more efficient but cannot handle such dense concentrations of dust as the open paddle blade.

6.4 MIXED FLOW FANS

The impeller comprises a conically shaped shaft on which aerofoil section blades are mounted. Air enters the fan at the small end of the impeller cone and the compression of the airstream is both radial and axial (hence 'mixed flow') and is directed straight through to the outlet. Efficiency is about the same as a conventional axial flow fan but the mixed flow fan can develop higher pressures. The noise produced is less than that of the axial flow, for the same duty.

6.5 TANGENTIAL FLOW FANS (CROSS-FLOW FANS)

A forward curved impeller is used. The impeller is very wide and air enters the periphery of the impeller through a long slot, passes into the impeller blades and leaves through another long slot at the far side of the impeller, opposite the inlet slot. There is no airflow flow through the two conventional inlet eyes of the forward curved impeller and these are covered with end plates. The airstream is carried between each successive pair of adjoining blades on the runner of the impeller, as a series of rapidly rotating vortices, from the inlet slot to the outlet slot. Efficiency is less than 50 per cent and pressure development is small. The air discharge velocity is high and the fan laws are only approximately obeyed. Installation in ductwork is not practicable and the application is for room units (fan coil etc.) where the wide shape of the airjet leaving the fan is suited to airflow over a similarly wide air cooler coil.

6.6 AXIAL FLOW FANS

The impeller consists of a large diameter boss on which are mounted about eight to twelve aerofoil section blades. The motor is close-coupled to the fan, although an externally mounted motor, with drive belts passing through holes in the casing to a pulley on the fan shaft within, is possible. Clearance between the blade tips and the inside of the casing is critical, to avoid wasteful local air circulation and a consequent reduction in performance. The pressure generated by the aerofoil blades depends on the ratio of the lift produced by the airflow to the frictional drag it generates. This ratio is related to the angle between the entering airstream and the chord of the aerofoil blade. The lift–drag ratio for a simple aerofoil increases as the angle approaches about 15°, beyond which the airflow ceases to follow the contours of the blade and becomes turbulent. The lift vanishes and the blade is stalled. Other factors, that can keep the airstream from leaving the contour of the aerofoil, are influential and one finds that axial flow fans can have blade pitch angles up to about 30°, although the blade pitch angle cannot be usefully increased beyond the stall point.

The fan duty also depends on the fan speed and with the use of constant speed induction motors, close-coupled to the fan, a coarse control over the duty is obtained by using fans with different numbers of pairs of poles, giving synchronous speeds of 3000 rpm (two poles), 1500 rpm (four poles), 1000 rpm (six poles), etc., and actual speeds five or ten per cent less than this. A fine control over the fan duty can then be obtained by using various blade pitch angles. In the case of axial flow fans used for variable air volume systems, a continuous, pneumatically or hydraulically actuated, or motorized control over the pitch angle of the blades provides a very good control over fan capacity. See Section 6.9.

The static pressure developed can be improved by fitting fixed angle guide vanes downstream to reclaim some of the energy locked in the rotational components of the airstream leaving the fan. Mounting two axial flow fans in series will double the pressure developed, mounting three fans in series will treble it, etc., for the same volumetric airflow rate. There is virtually no limit to the fan total pressure that axial flow fans can develop, although the manufacturers should be referred to for pressures exceeding about 5 kPa, because of the air temperature rise accompanying the increase of pressure and for other reasons.

The noise produced increases as the fan total pressure goes up and also as the volume handled rises (see Figure 6.7).

The characteristic behaviour of an axial flow fan is shown in Figure 6.4(c). It has a non-overloading power curve and a peak

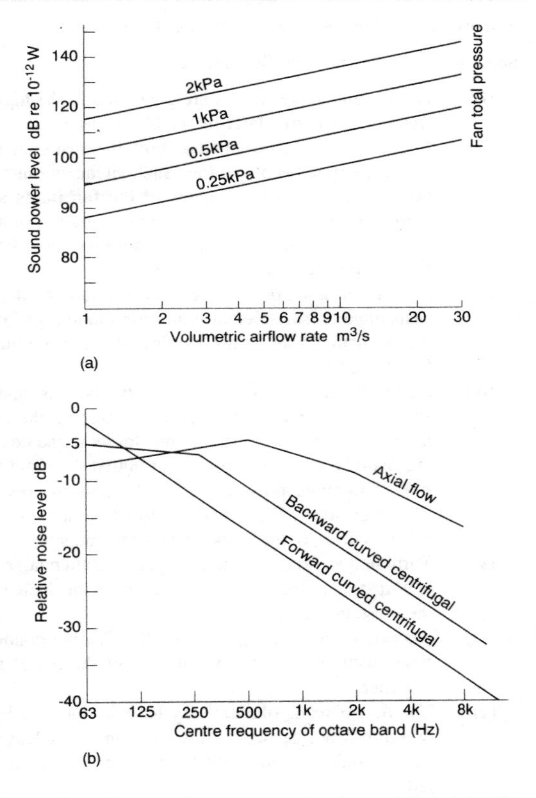

Figure 6.7 Fan performance and noise control: (a) relationship between fan performance and noise; (b) relative noise production of various fans

efficiency on a steep part of the pressure-volume curve (as also does the backward curved centrifugal). This latter feature is an advantage because it means that the volume handled does not change very much if the resistance of the duct and plant system, to which it is connected, increases a little. It is not usual to show efficiency as a separate scale. Loops of constant efficiency are shown on the pressure–volume coordinate system, instead. The maximum total fan efficiency is about 87 per cent, slightly less than the maximum achieved by backward curved fans with aerofoil impeller blades.

The standard range of operating temperatures is from about –20°C to +40°C. The manufacturers should be consulted if it is proposed to use fans outside this range of temperatures.

Some general notes on axial flow fans are:

(i) High sound power levels are generated in the higher frequencies, particularly by the blade tips.

(ii) Silencers must be provided on both the upstream and downstream sides of the fan. Such silencers must be bolted directly to the flanges of the fan inlets and outlets, in order to prevent noise breaking out of any flexible couplings that might otherwise have been fitted at the flanged fan openings.

(iii) The entire assembly of silencers and fan should be suspended on suitable anti-vibration mountings, paying particular care to the loading of the mountings. (see Figure 6.8).

(iv) Noise also radiates outwards from the fan casing and should be dealt with by carefully wrapping the casing with a sound barrier matting. Joints in the covering material must be lapped, not butted, to avoid the risk of flanking noise radiating through any poorly made butt joint. The covering should be done by the manufacturer, in the works, rather than on site.

(v) Turbulent airflow leaving the upstream attenuator, or any upstream duct fitting or piece of plant, increases the noise generated by the fan.

(vi) Distortion and misalignment of the flexible connections disturbs the airflow onto the blade tips of the impeller.

(vii) The downstream silencer may have an increased air pressure drop, caused by the rotation of air leaving the fan, unless fixed downstream guide vanes are fitted.

(viii) It may be necessary to fit additional silencers, possibly with axial pods, if enough attenuation cannot be provided by the silencers fitted directly to the fan (see Figure 6.8).

(ix) Fans should not be selected close to their stall point (see Figure 6.4(c)).

(x) Avoid using axial flow fans in occupied spaces, or above suspended ceilings in occupied spaces, or behind flimsy walls adjoining occupied spaces, unless the fan casing is efficiently covered with sound barrier matting and properly silenced.

(xi) Avoid duct fittings or transformation pieces close to axial flow fans.

(xii) Axial flow fans can be installed with vertical airflow, provided that proper attention is given to any lubrication difficulties that may arise.

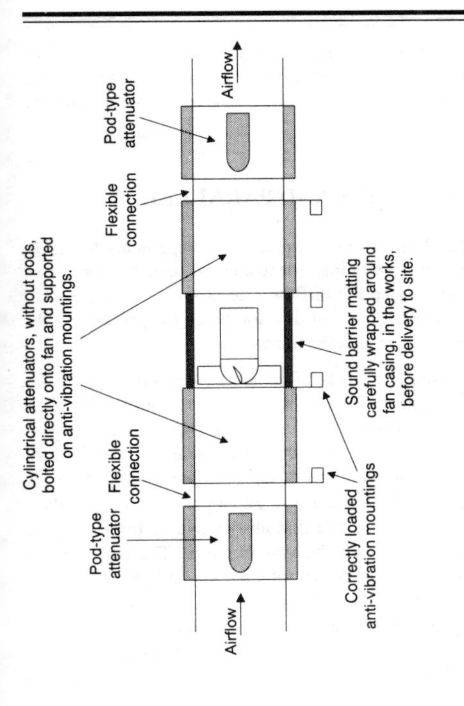

Figure 6.8 Silencers and anti-vibration mountings for an axial flow fan

6.7 PROPELLER FANS

The impeller comprises three or four sheets of warped metal plate of constant thickness. Airflow is mixed, part being radial and part axial. To avoid wasteful local recirculation at the blade tips the fan is mounted in an orifice or, if installed in a duct, it is fitted in a rectangular diaphragm plate across the duct section.

For free inlet and outlet conditions the volumetric airflow rate can be considerable but these fans do not develop much pressure (maximum fan static pressure is about 15 or 20 Pa) and they are not suitable for handling air against the resistance of ductwork and plant.

The pressure–volume characteristic of the propeller fan is a fairly shallow curve, falling downwards from left to right, as the volume handled increases. Efficiency is poor with a maximum of less than 40 per cent. The fan power–volume characteristic rises as the volume falls, with an increase in system resistance. This is because of the larger proportion of the fan power wasted in maintaining the local circulation of air at the blade tips. Fan power is consequently a maximum at zero airflow and the fan motor is certain to burn out if airflow is prevented by an increased system resistance or by closing dampers.

6.8 FANS IN SERIES AND PARALLEL

Figure 6.9 illustrates series and parallel arrangements for backward curved centrifugal fans. For two equal fans the rule is that they will handle twice the airflow rate at the same pressure if connected in parallel, but will develop twice the pressure at the same airflow rate if connected in series.

If forward curved centrifugal fans are connected in parallel a small complication may arise. Referring to Figure 6.4(a) it is seen that, in the left-hand part of the pressure–volume curve, where there is a dip, two or three airflow rates are possible for a given pressure. In some cases, the multiple options possible when adding the airflow rates at a given pressure, can produce an S-shaped combined curve that allows two or three possible points of intersection with the system curve. The performance oscillates between the points of intersection and produces some vibration. The solution is to increase the resistance of the system curve by partly closing a main damper. This makes the system curve rotate in an anti-clockwise direction and takes the intersection of the fan and system curves away from the unstable position.

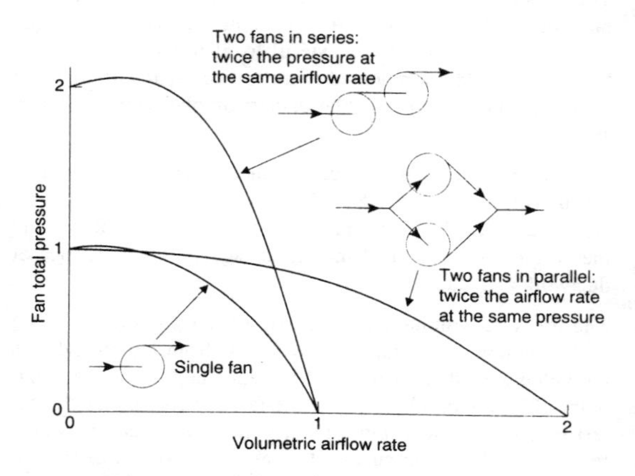

Figure 6.9 Fans in series and parallel

6.9 FAN CAPACITY CONTROL

There are five methods of altering fan capacity: dampering, variable inlet guide vanes (centrifugal only), varying the blade pitch angle (axial flow only), varying the impeller width (centrifugal only) and varying the fan speed.

6.9.1 System dampering

If a main damper is closed it increases the resistance of the system to airflow.

EXAMPLE 6.2

Determine the power wasted across a partly closed main duct damper that reduces the airflow rate from 1.0 m^3/s to 0.5 m^3/s.

Answer

Refer to Figure 6.10(a). When the main duct damper is fully open the fan and system pressure–volume curves intersect at the point P_1. The fan total pressure is 690 Pa and the volumetric airflow rate is 1.0 m^3/s. The total fan efficiency, given by the point E_1, is 68 per cent. By equation (6.10), the fan power is calculated as $1.0 \times 0.69/0.68 = 1.01$ kW.

When the damper is partly closed the two curves intersect at the point P_2. The duty is 0.5 m^3/s at 900 P_a and the total efficiency, given by the point E_2, is 52 per cent. The fan power is calculated as $0.5 \times 0.9/0.52 = 0.87$ kW. Part of this power is used to overcome the total pressure drop in the whole of the system and plant but the power wasted across the partly closed damper is large and is proportional to the total pressure difference between the points P_2 and P_3 (which has coordinates of 0.5 m^3/s and 175 P_a). The wasted fan power across the partly closed damper is calculated as $0.5 \times (0.900 - 0.175)/0.52 = 0.70$ kW. This method of fan capacity control is very wasteful.

6.9.2 Variable inlet guide vanes

These comprise a set of radial blades centred on the fan shaft and mounted in a ring that is located at the entry to the fan inlet eye. The blade spindles pass through holes in the ring and are attached to a linkage which is motorized, to give movement of the blades in unison, from fully open to fully closed. By adjusting the angle of the blades, through movement of the linkage, the airstream entering the fan inlet can be given a swirl. If the swirl is in the same direction as the impeller rotation the airflow is assisted, but if the imparted swirl is in the other direction the airflow is opposed and the fan capacity reduced.

The effect on the pressure volume characteristic curve is to

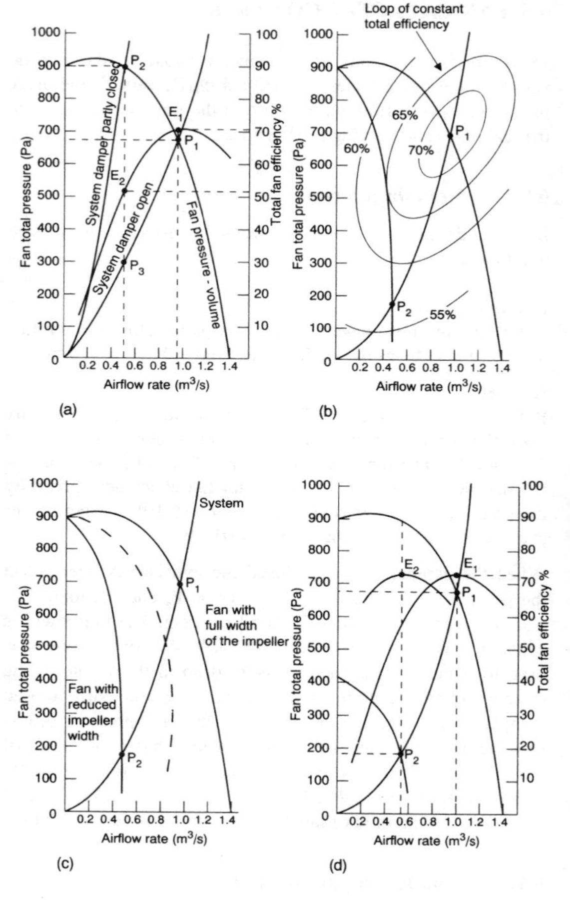

Figure 6.10 Methods of fan capacity control: (a) system dampering; (b) variable inlet guide vanes or variable blade pitch angle; (c) variable disc throttle; (d) variable fan speed

rotate it in a clockwise direction about the origin. This is shown in Figure 6.10(b). To reduce the airflow rate from 1.0 m^3/s to 0.5 m^3/s the inlet guide vanes are partly closed, the fan pressure–volume curve rotates and the point of intersection with the system curve moves from P_1 to P_2.

Loops of constant efficiency are shown on the pressure–

volume diagram. Although the values shown on the efficiency loops are notional, they are typical and suggest that the method of capacity control is much more efficient than using a system damper. In practice, the method works poorly because of mechanical problems with the linkage for moving the vanes and because of their poor response — the vanes have to close through 45° before there is any effect on the airflow.

6.9.3 Variable blade pitch angle

The impeller blades on the hub are motorized and continuously adjustable, while the fan is running. Figure 6.10(b) illustrates the behaviour of the method, which is similar to that of variable inlet guide vanes. The control over fan capacity is excellent and there are no mechanical problems.

6.9.4 Variable impeller width ('disc throttle control') [4]

A disc on the impeller shaft, within the runner, is automatically moved in an axial direction to change the width of the impeller (see Figure 6.11). The effect is similar to that shown in Figure 6.9. If the disc is at its extreme position, the full width of the impeller is available for airflow and corresponds to position 2 on the airflow coordinate for two fans in parallel in Figure 6.9. As the disc is pulled towards the fan inlet opening less impeller width is available for airflow, the part of the impeller between

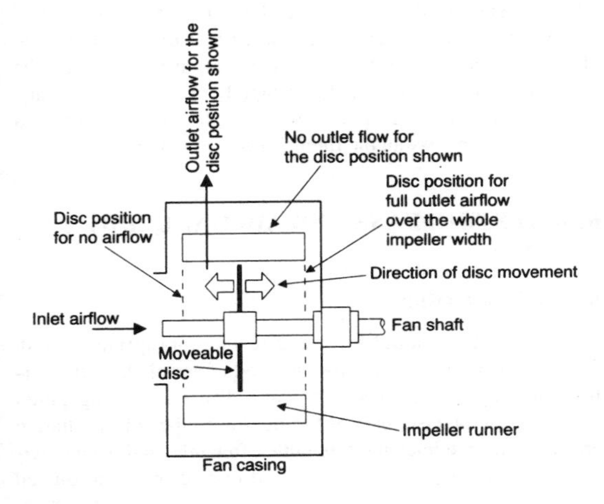

Figure 6.11 A diagram of disc throttle control

the disc and the side of the fan casing not being used. For example, if the disc were in the mid-way position this might correspond to position 1 on the airflow coordinate in Figure 6.9. Hence the fan pressure-volume curve moves progressively towards the position O, as shown in Figure 6.10(c), by the point of intersection sliding down the system curve from P_1 to P_2 and beyond, to the origin.

6.9.5 Fan speed variation

If the speed of the fan is varied the point of rating on the fan pressure–volume curve slides up and down the system curve, in accordance with the fan laws (Section 6.2), the total efficiency of the fan staying constant. This is shown in Figure 6.10(d). As the speed of the fan is reduced the point of intersection with the fan curve moves down the system curve from P_1 to P_2. Since P_1 and P_2 are the same point of rating on the fan curve, the total efficiency stays constant. The whole of the efficiency curve moves to the left and the point E_1 occupies a new position at E_2, to keep the efficiency at the same value.

The best method of varying fan speed is to use an inverter that changes the normal supply frequency to a different value, the speeds of the driving motor and the fan, changing accordingly. This method is efficient, the fan power falling as the motor speed is reduced. Some other, electro-magnetic methods are not efficient, the fan power actually rising as the motor speed falls.

If a vee-belt drive is used its efficiency diminishes as the motor speed is reduced [4]. A belt drive efficiency of 96 per cent at 100 per cent speed can reduce to 86 per cent at 50 per cent speed. On the other hand, if a toothed belt drive is used, the efficiency can stay constant at 98 per cent until the speed falls to 50 per cent, thereafter dropping rapidly to 60 per cent, with a speed reduction of 25 per cent. This should be taken into account when assessing the running costs of systems.

6.10 TESTING FANS AND AIR HANDLING UNITS

6.10.1 Fan testing

Fans in the UK should be tested to the appropriate British Standards [5]. This prescribes a method of establishing the performance by using a test rig, some discretion being given regarding the precise form of the rig. Four types of installation are covered: free inlet and free outlet, free inlet and ducted outlet, ducted inlet and free outlet, and ducted inlet and ducted outlet.

The way a fan is installed on site is never the way it was tested: it is always very different. Hence if manufacturers' catalogue data is based on test to the British Standard, it is unreasonable to expect the exact catalogue performance. This underlines the need to take the utmost care in providing good inlet and outlet conditions on site, so that the best performance can be obtained.

Since most installations used packaged air handling units, which bear no resemblance to the test rig used by the British Standard, it is evident that the performance of such packaged units will be uncertain, unless they also have been tested to an appropriate standard.

6.10.2 Testing air handling units

Packaged air handling units should be tested to the proper British Standard [6]. In place of a simple fan, this uses a fan mounted in a cabinet that is connected to a cooler coil cabinet. The cabinet combination is tested as a whole, to BS 848 [5], just as if it were a simple fan, and pressure–volume curves obtained. The cooler coil cabinet may be fitted on the upstream or downstream side of the fan cabinet, to correspond to the form of air handling unit, draw-through or blow-through.

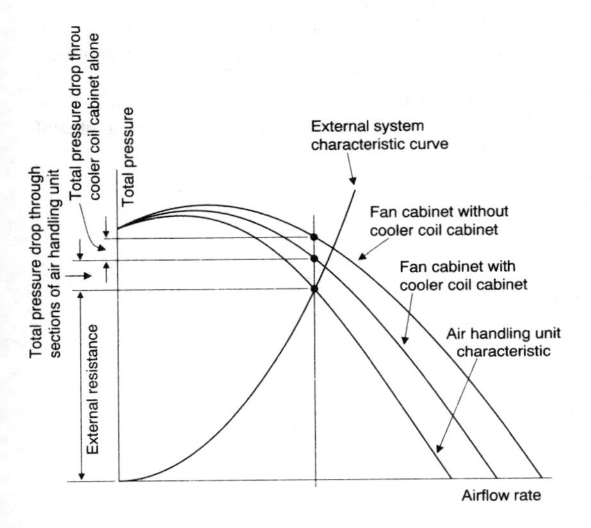

Figure 6.12 Establishing the pressure-volume characteristic to BS 6583, for an air handling unit

The manufacturer should establish the pressure drop through the cooler coil cabinet using another standard [7]. For a given airflow rate the cooler coil cabinet pressure drop is added to the pressure obtained for the combination test to BS 848 and a new pressure-volume curve obtained (see Figure 6.12). Knowing the tested pressure drops for the other cabinets (mixing box, filter chamber, etc.) that the manufacturer proposes to use to form the complete air handling unit, their total pressure drop can be deducted from the curve obtained for the fan cabinet alone and a new pressure-volume curve obtained for the air handling unit, in the form of external resistance against airflow rate.

References

1. Keith Blackman *Centrifugal Fan Guide*, 1980, Keith Blackman Ltd.
2. W. P., Jones, *Air Conditioning Engineering*, 4th edition, Edward Arnold, 1994.
3. TM8, Design Notes for Ductwork, 1983, Table 6.3, CIBSE.
4. W. T. Cory, *Energy Savings with Centrifugal Fans and Disc Throttle Variable Volume Controller*, Woods of Colchester Ltd, (inc Keith Blackman).
5. BS 848: 1980, Fans for general purposes, Part 1, Methods of testing performance.
6. BS 6583: 1985, Methods for volumetric testing for rating of fan sections in central station air handling units (including guidance on rating).
7. BS 5141: Specification for air heating and cooling coils: Part 1: 1975 (1983), Methods of testing for rating of cooling coils.

7 Ductwork Design

7.1 DUCTWORK SIZING

Ductwork is normally sized on the basis of the required airflows being delivered through a system of distributing ductwork without excessive fan, duct or terminal pressures and without exceeding air velocities at which airflow generated noise becomes a problem. This is usually achieved by identifying the index terminal which is the terminal with the highest duct resistance from the fan to room served (with all its dampers fully open). This is called the index circuit and is usually the terminal with the longest run (allowing for the equivalent length of fittings). The index run is then sized on a target pressure drop per metre using the CIBSE Chart Figure C3.1[1] but subject to velocity limits, space restraints and the need to rationalize duct sizes for economic reasons. The cost of a reducing fitting is often more than the extra cost of continuing the same size duct past a branch so that single size header ducts are often used to connect a group of terminal branches. By using the table format for manual duct sizing (see example Figure 7.1), and by starting at the index terminal and working back to the fan the accumulative total of resistance at a branch is the pressure available for overcoming the resistance of that branch.

7.1.1 General practical notes

(1) The target constant pressure drop per metre run should be regarded only as a general guide, there are often good reasons for selecting much lower pressure drops as mentioned above. Fan energy is also an important consideration as a major portion of the fan head is duct resistance.

(2) Duct connections to equipment such as fans, attenuators, dampers, grilles, diffusers and other terminals should always be kept at the equipment duct connection size and run straight for at least three duct diameters (or maximum widths if rectangular). Failure to meet this requirement will affect the manufacturers' air performance and noise data.

(3) Problems with poor detail design and poor construction workmanship are more common with ductwork than other service distributions. Ductwork should always receive a high priority for space allocation and specialist site supervision.

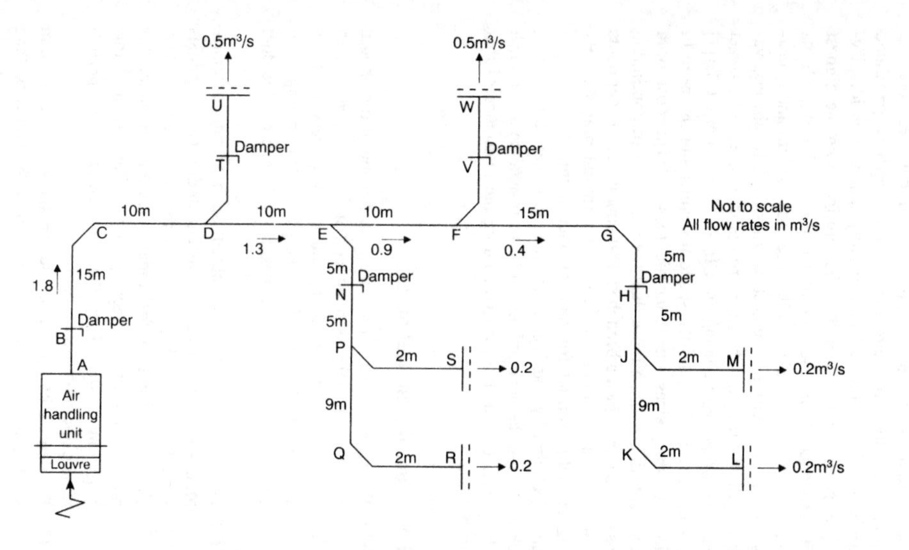

Figure 7.1 System pressure loss example (external to A H U)

7.1.2 Maximum air velocities

These are dependent on the following factors:

(i) The noise sensitivity of the area through which the duct passes.
(ii) The noise transmissibility of the shaft or ceiling separating the duct from the occupied space.
(iii) Whether the system air terminal includes attenuation. (Low velocity air distribution systems usually have no terminal attenuation whereas higher velocity air distributions usually have attenuation after automatic balancing dampers which have high pressure drops and consequently generate noise.)
(iv) The width of the flat surfaces of the duct (flat sheet metal flexes and transmits in duct noise). Conversely circular ducts have stiff surfaces with a low noise transmission.
(v) The degree of air turbulence, high aspect ratio rectangular ducts create more air turbulence than circular ducts especially at fittings and changes in direction.

These factors are taken into consideration in the table of recommended maximum air velocities, Figure 7.2.

7.1.3 Non-circular ducts

The duct sizing chart gives velocities and resistance for circular ducts only. For other shape ducts the size must be derived from the circular equivalents for equal friction rate and flow.

CIBSE Table C4.30[2] gives data for rectangular ducts and Table CV.32[3] gives data for flat oval ducts but air velocities in non-circular ducts must be calculated from the volume flow m^3/s divided by cross-section area m^2.

7.2 STATIC REGAIN SIZING

This procedure is not often used as it involves more design work and results in larger, more expensive ducts. Its merits are that it gives nominally the same static pressure at each terminal branch thereby facilitating air balancing and reducing terminal damper noise. Static regain sizing also results in a lower fan pressure and energy compared with the equivalent constant pressure drop sized system.

The principle of static regain sizing uses the fact that apart from frictional losses the total pressure of moving air remains constant so that if the air velocity is reduced the static pressure is increased by the difference in velocity pressures. In static

(1) Component	(2) Size (m)	(3) Area (m^2)	(4) Air volume (m^3/s)	(5) Velocity (4)/(3) (m/s)	(6) Velocity pressure (Pa)	(7) k Value for fitting	(8) Equivalent diameter (m)	(9) Friction rate Pa per metre (Target 1.0 Pa/m)	(10) Length (m)	(11) Friction Item (6) × (7)or (9) × (10) (Pa)	(12) Totals accum. (Pa)
Grill L	0.6 × 0.15	0.09	0.2	2.2	2.9					30.0	
Enlarge L	0.3 × 0.15	0.045	0.2	4.4	11.6	0.2				2.3	
Bend K	0.3 × 0.15	0.045	0.2	4.4	11.6	0.22				2.6	
Straight JL	0.3 × 0.15	0.045	0.2	4.4	11.6		0.231	1.2	11	13.2	
Reduction J	0.3 × 0.15	0.045	0.2	4.4	11.6	0.07				0.8	
Branch J	0.3 × 0.3	0.09	0.2	2.2	2.9	1.0				2.9	51.8
Damper H	0.3 × 0.3	0.09	0.4	4.4	11.6	0.2				2.3	
Bend G	0.3 × 0.3	0.09	0.4	4.4	11.6	0.14				1.6	
Straight FJ	0.3 × 0.3	0.09	0.4	4.4	11.6		0.33	0.8	25	20.0	
Reduction F	0.3 × 0.3	0.09	0.4	4.4	11.6	0.07				0.8	
Branch F	0.5 × 0.3	0.15	0.4	2.7	4.4	2.3				10.1	86.6

Straight EF	0.5 × 0.3	0.15	0.9	6.0	21.6		0.424	1.0	10	10.0	
Reduction E	0.5 × 0.3	0.15	0.9	6.0	21.6	0.07				1.5	
Branch E	0.5 × 0.4	0.2	0.9	4.5	12.2	0.25				3.0	101.1
Straight DE	0.5 × 0.4	0.2	1.3	6.5	25.4		0.492	1.0	10	10.0	
Reduction D	0.5 × 0.4	0.2	1.3	6.5	25.4	0.07				1.8	
Branch D	0.5 × 0.5	0.25	1.3	5.2	16.2	0.09				1.5	114.4
Bend C	0.5 × 0.5	0.25	1.8	7.2	31.1	0.14				4.4	
Damper B	0.5 × 0.5	0.25	1.8	7.2	31.1	0.2				6.2	
Straight AD	0.5 × 0.5	0.25	1.8	7.2	31.1		0.551	1.0	30	30.0	
Enlarge Fan Conn	0.4 × 0.3	0.12	1.8	15.0	135	0.22				29.7	
Inlet Louvre 0.5 FA	1.0 × 1.0	1.0	1.8	1.8	1.94	4.5				8.7	193.4

Figure 7.2 Worksheet for calculation of fan total pressure

regain sizing (working in the direction of air flow) the air velocity after a branch take-off is calculated to give sufficient increase in static pressure to overcome the resistance of the duct and fittings to the next terminal branch (see Figure 7.5).

The procedure is given in some detail in CIBSE Technical Memoranda TM8 Design Notes for Ductwork.

It should be borne in mind that some element of static regain occurs in normal duct sizing wherever the velocity reduces past a branch.

7.3 TYPES OF DUCTS

Circular ducts should always be selected in preference to rectangular or flat oval for the following reasons: their resistance is lower for a given cross sectional area, perimeter or weight; heat loss or gain is lower than for the equivalent rectangular or flat oval ducts; noise breakout is much less and is usually negligible; leakage is usually much less and is more predictable.

Circular fittings are usually factory made of pressed steel to standard dimensions and are more predictable in performance. Circular ducts are usually lighter and are easier to insulate effectively as the absence of flanges means the full insulation thickness can be maintained throughout.

Construction work should be faster and require less skilled labour particularly if push fit self-sealing joints are used but these require a close manufacturing tolerance on the duct outer diameter. Double skin pre-insulated ducts are factory produced as a standard for different applications such as external use and attenuation.

External connecting flanges and angle stiffeners are not required. In situation where the void depth is insufficient to accommodate a conventional layout of circular ducts consideration should be given to increasing the number of ducts serving an area in order to reduce the maximum duct size.

All ducts and duct fittings should comply with the construction standards set out in HVCA Specification DW142[4] or its equivalent.

7.4 FLEXIBLE DUCT CONNECTIONS

Fabric flexible duct connections are often used to take up misalignment and to allow limited movement such as occurs when a fan is mounted on anti-vibration springs. They are also used in some applications for connecting from an octopus outlet of a

Criterion NC/NR	40			35			30			25		
	Circ.	Rectangular or flat oval		Circ.	Rectangular or flat oval		Circ.	Rect. or flat oval		Circ	Rect. or flat oval	
Aspect ratio W/D		2:1	3:1		2:1	3:1		2:1	3:1		2:1	3:1
a) Systems with no terminal attenuation, i.e. low velocity with fixed damper.												
Riser	12	8	6	10	7.5	6	8	6	5	6	5	5
Main branch and small ducts	10	6	5	7.5	5	4	6	4	3	5	3	3
b) Systems with terminal attenuation e.g. constant vol. boxes or VAV boxes.												
Riser	15	13	8	15	12	8	12	9	7	Not applicable		
Main Branch (in False ceilings)	12	10	8	10	10	7	7	6	6			

The limiting velocities are a guide to normal good practice with moderate sized systems to avoid noise generation in ductwork and high pressure in ductwork.

Figure 7.3 Recommended maximum duct velocities

Criterion NC/NR	40			35			30			25		
	Circ.	Rectangular or flat oval		Circ.	Rectangular or flat oval		Circ.	Rect. or flat oval		Circ	Rect. or flat oval	
Aspect ratio		2:1	3:1		2:1	3:1		2:1	3:1		2:1	3:1
Small ducts (less than 200 l/s												
(i) Limiting factor pressure drop	6	6	5	6	6	6	6	5	4			
(ii) Limiting factor noise breakout	12	7.5	6	10	6	5	8	5	4			

Figure 7.3 continued

Criterion NC/NR	40			35			30			25		
	Circ.	Rectangular or flat oval		Circ.	Rectangular or flat oval		Circ.	Rect. or flat oval		Circ	Rect. or flat oval	
Aspect ratio		2:1	3:1		2:1	3:1		2:1	3:1		2:1	3:1
c) Final connections (all systems)												
Grilles (neck)	3	3	2.5	2.5	2.5	2.0	2	2	1.5	1.5	1.5	1.5
Diffuser (neck)	2.5	2.5	2.5	2	2	2	1.5	1.5	1.5	1	1	1
Stub ducts (above ceilings)	4	4	4	3	3	3	2	2	2	1.5	1.5	1.5

Manufacturers' data for air terminals should be consulted but beware of damper noise. Always use manufacturers' neck sizes.

NOTES:

(1) In practice most noise problems emanate from excessive pressure drops across dampers and other devices.
(2) It is possible to interpolate between the above figures.
(3) Airways in rectangular or flat oval ducting should not excess 3 to 1 aspect ratio. If necessary fit internal splitters in shallow ducts or use multiple circular ducts.
(4) The use of high velocities on index circuits will cause high pressure loss at fittings, e.g. the velocity lead at 15m/s is 141 Pa. This may lead to selecting a fan which will have high pressures giving excessive noise, motor power rotational speeds and commissioning problems. It is, however, possible to size non-index main riser ducts to higher velocities up to the limits given above.
(5) Fixing of ductwork to lightweight structures, e.g. partitions can lead to vibration and noise re-generation and should thus be avoided.

Figure 7.4

box to a number of diffusers. This latter use is really only suitable at very low air pressures and velocities. The installation requires care to ensure that the flexibles are reasonably taut and unable to collapse. Generally speaking, flexibles should not be used to take up axial misalignment. Their best and most economical application is in short lengths not exceeding one diameter for the purpose of creating a slight change in direction (see Figure 7.6).

It must be realized that fabric flexibles are a common source of leaks, they are acoustically transparent, and unless they are kept taut they concertina losing free area which creates resistance and noise.

7.5 FAN APPROACH AND LEAVING CONDITIONS

7.5.1 Centrifugal fan inlets

Most fans are rated with a free inlet drawing from a large plenum so that the air approaches the inlet with negligible velocity. Any obstruction or fitting likely to cause an uneven approach velocity will have an adverse affect on the fan rated performance. A bend or change in duct direction close to a fan inlet is likely to cause a rotation of the air which will derate the fan if air rotation is in the same direction as the impeller rotation (motor amps low) or if in the opposite direction will increase the motor load (motor current high) and increase fan noise (see Figure 7.7).

7.5.2 Centrifugal fan outlets

Because centrifugal fan outlet velocities are high and the peak of the velocity profile is much higher than that of a normal duct profile it is good practice to provide about three widths of straight duct after the connection. This enables a normal duct velocity profile to develop without significant pressure loss. This straight duct should preferably expand gradually to regain some static pressure and reduce the duct velocity. Unfortunately in practice, limited space often prevents the use of this straight duct. The consequences of any form of fitting or disturbance close to the fan outlet are considerable increase in noise generation and the resistance of any fitting or device. This increase in pressure loss can be considerable for example, an additional four velocity heads for a 90° change of direction at the fan outlet turning the air in the opposite direction to that of the impeller rotation. Even a discharge into a plenum with no straight duct can incur an additional two velocity heads loss over and above a straight duct connection (see Figure 7.8).

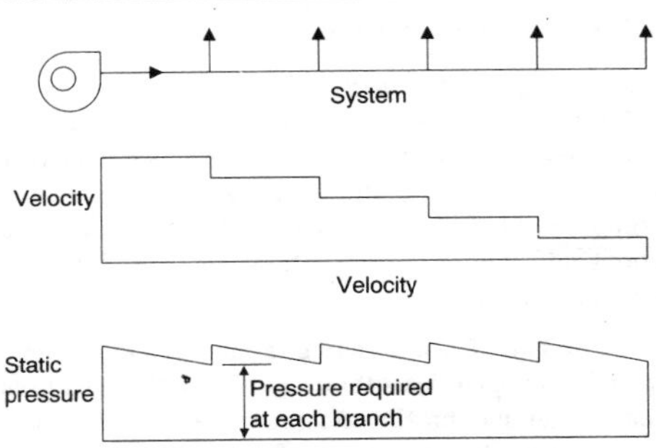

Figure 7.5 Static regain pressure diagram

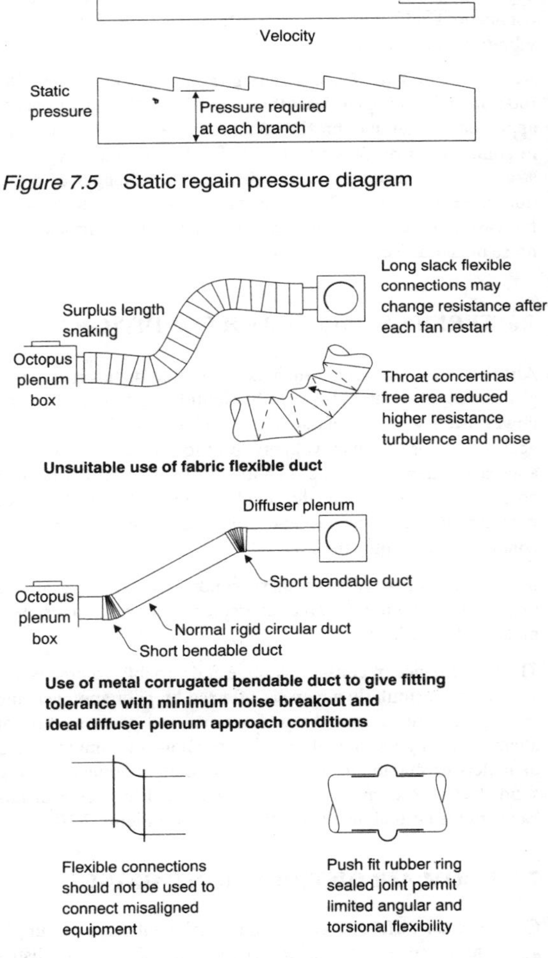

Figure 7.6 Flexible connections

7.5.3 Axial fan connections

These have a significant effect on the fan and system performance. Most fans are rated on straight inlet and outlet duct connections and any departure from this arrangement adversely affects the noise and fan performance. Any disturbance of air approaching the blades of the impeller can create shock entry connections and with large axial flow fans can set up blade vibrations and failure.

Air leaving an axial fan is spinning in the direction of impeller rotation and requires at least three diameters of straight duct to approach normal duct profile conditions. This can be corrected to some extent by fans fitted with fixed outlet guide vanes to straighten the air but it is advisable to avoid changes of direction close to the fan. Flexible connections directly to an axial fan casing are undesirable because of air flow disturbance and noise breakout (see Figure 6.8).

7.6 TERMINAL APPROACH CONDITIONS

Air distribution terminals such as grilles and diffusers are usually tested and rated for air distribution pattern and sound power levels with a manufacturers' neck size straight duct connection, a normal duct velocity profile and a noise-free air approach; alternatively the manufacturer will supply a plenum connection to turn the air through 90° and the terminal is rated with the plenum connected but with the same duct approach conditions to the plenum inlet.

Installation departure from these conditions adversely affects performance; Figure 7.9 gives examples of connection arrangements to be avoided.

The performance of constant volume and variable volume boxes and valves is similarly adversely affected by poor approach and leaving conditions as is the performance and pressure drop of attenuators. A good rule of thumb is to allow a minimum of 2.5 diameters or maximum duct widths of straight connection size solid duct before and after any device which causes a disturbance to the normal duct velocity profile (see Figure 7.10).

7.7 ROOM AIR DISTRIBUTION (MIXING)

Conventional diffusers, grilles and nozzles introducing supply air to the space served need to achieve the following air distribution performance:

(i) Mix the supply air with room air before entering the occupied zone.

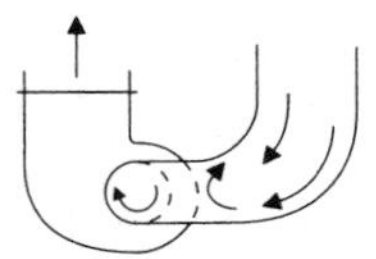

Bend creates air spin in direction of impeller rotation - fan derated power required lower than design

Bad

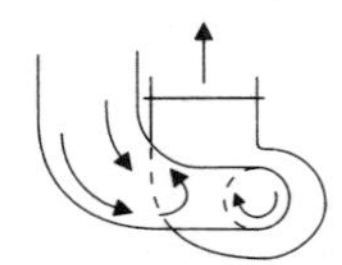

Bend creates air spin in opposite direction of impeller rotation - noise and power required increased

Bad

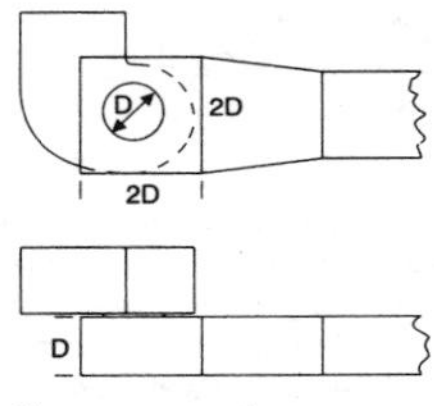

Plenum approach

Acceptable

Turning vane - gradual transition approach

Acceptable

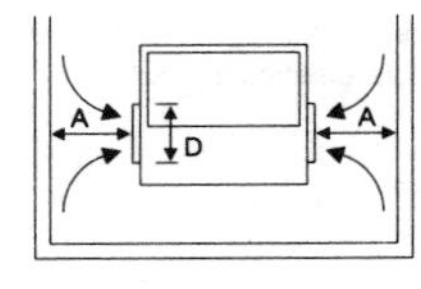

DIDW fan inlets should not be less than 0.75D from wall and not less than 1.5D from another fan inlet. Space between fan and wall should be unobstructed

Figure 7.7 Centrifugal fan inlets

(ii) Impart sufficient momentum to the air stream to ensure that it reaches the boundaries of the space served and gives a uniform space temperature in the occupied zone.
(iii) Ensure that the maximum velocity of the air stream decays to less than 0.25 m/s before entering the occupied zone.

See Figure 7.11

The extract from a room served by a mixing air supply system has negligible affect on the air distribution provided the supply air has mixed with room air before being extracted. In practice this rarely imposes any restraint on the position of the extract points indeed when supply and extract points occur in the same plane they can be adjacent without short circuiting taking place (see Figure 7.12).

With usual office room heights about 2.7 m there is an upper

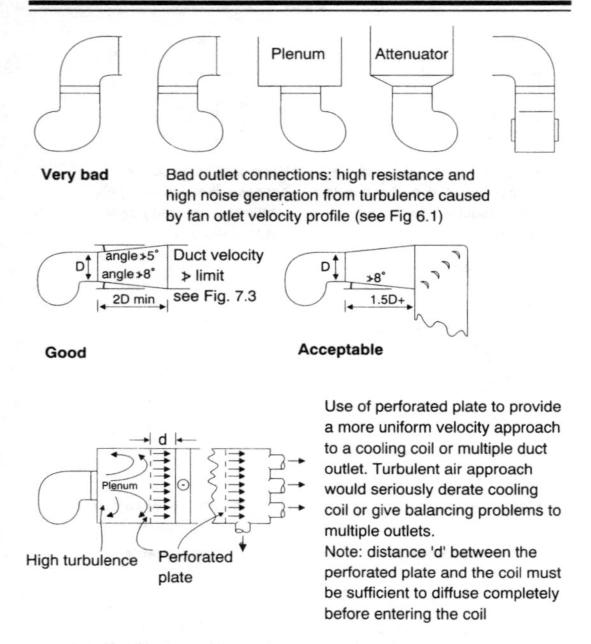

Figure 7.8 Centrifugal fan outlets

limit to the quantity of supply air that can be introduced by different types of supply air terminals without excessive air velocities in the occupied space. This limit is affected by the type and number of terminals but is approximately 10–12 air changes per hour for sidewall grilles and 20–25 air changes per hour for ceiling diffusers. Perforated surfaces such as ceilings can be used to supply air at much higher air change rates without discomfort and are used for applications where the required air change rate is beyond the scope of conventional diffusers also where non-turbulent laminar air flow is required in the room.

The throw of a jet of air before its velocity decays to a defined limit is given by the formula:

$$T = C\sqrt{V \times Q} \tag{7.1}$$

where

T = the throw in open space (m)

C = coefficient dependent upon the shape of the outlet and residual velocity at the end of the throw.

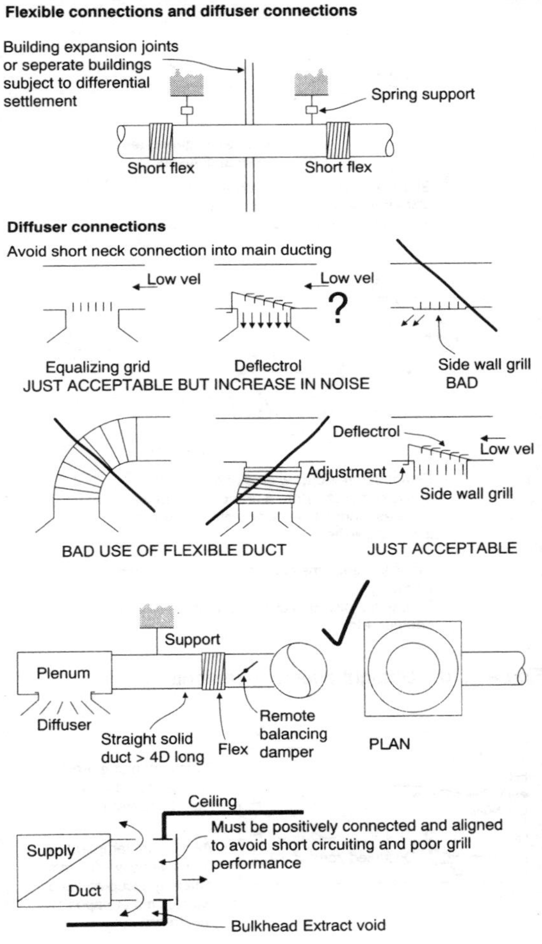

Figure 7.9 Terminal connections

(Typical values for a residual velocity of 0.4 m/s (0.75 m/s) are: Nozzles C = 14.0 (7.0) Bar Grilles C = 12.4 (6.2))

V = the velocity at the outlet (m/s)

Q = the air flow rate at the outlet (m^3/s)

If the air stream is close to a surface such as a ceiling it will attach to the surface and the throw will increase.

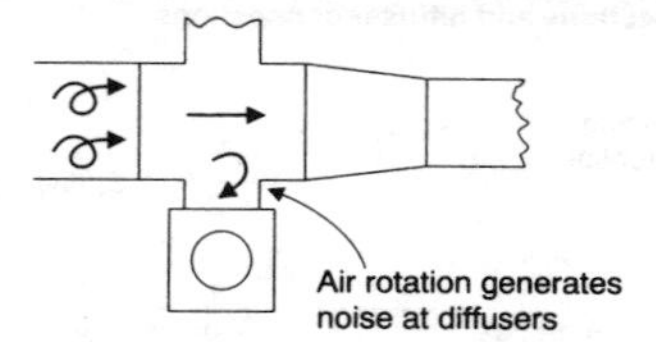

Short branch close to upstream air disturbance will be subject to poor airflow and noise generation - AVOID

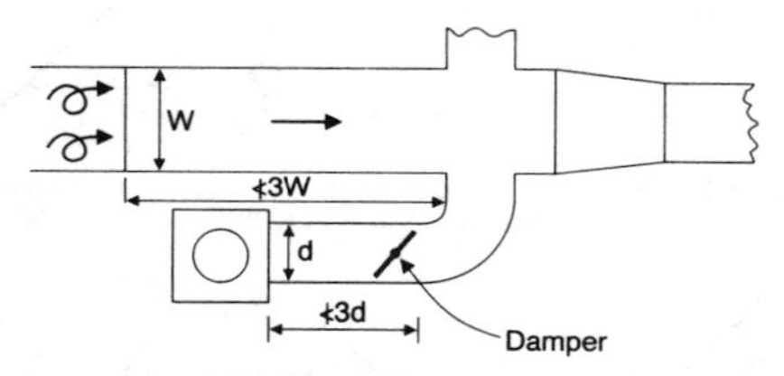

Planning three duct widths or diameters of straight duct before a branch or trerminal ensures a normal velocity profile approach is re-established

Air disturbance may be caused by any of the following:-
Fan, bend, direction change, branch attenuator, damper, etc.

Figure 7.10 Straight duct requirements

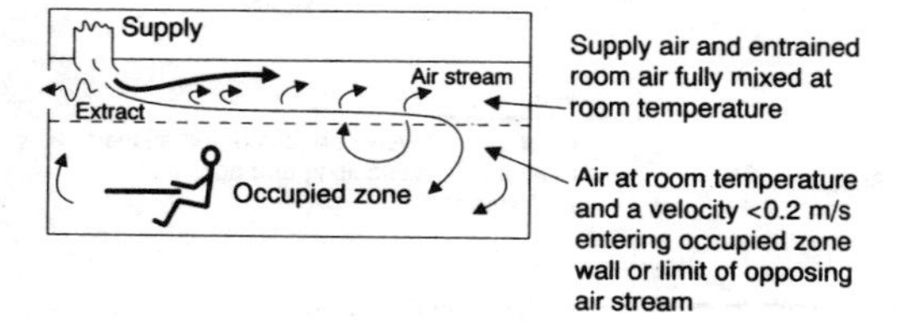

Figure 7.11 Air mixing distribution

Note that the residual velocity of 0.75 m/s is only appropriate for comfort cooling by air movement at high room temperatures.

The above formula applies to supply air at room temperature and is generally not seriously affected by variations in supply air temperature within ± 10°C of room temperature. The density differences between supply air and room air outside this Δt are likely to cause buoyancy or dumping effects before mixing is complete when using conventional grilles or diffusers.

However, special highly inductive terminals are available which are able to effectively mix at higher Δts. Manufacturers' performance data should be used for these terminals which usually require a much higher upstream pressure than conventional diffusers.

7.8 ROOM AIR DISTRIBUTION (DISPLACEMENT)

Developed in Scandinavia, this form of air distribution is designed to promote air stratification in the room by introducing the air at low velocity at occupant level. Some mixing takes place in and close to the terminal but the air distribution motive force is mainly due to convection air currents from internal warm surfaces. Air being extracted at high level.

The advantages claimed are:

(i) Convection currents carry room produced contaminants to high level where they are extracted and not recirculated, the quality of the air at occupant level is higher than that obtained by conventional air mixing distribution.
(ii) The supply air temperature typically 16°C to 18°C is much higher than that of an equivalent mixing air distribution system giving a much longer or period of the year when outside air can be used without mechanical cooling, or alternatively better comfort conditions are obtained without mechanical cooling.
(iii) The higher extract temperature typically 26°C to 28°C enables more effective heat recovery to be obtained from exhaust air to fresh air.

See Figure 7.13

Similar air distribution advantages are obtainable with floor supply systems using inductive swirl diffusers. Using floor voids to convey air to the diffusers reduces the amount of ductwork required.

On the debit side such systems are less effective in low office spaces as it is more difficult to obtain the air stratification advantages. Generally the higher the space the greater the temperature gradient and lower the air flow required to remove a given heat load at occupant level. The displacement air terminals, connecting ducts and extract system also take up significant space. A separate perimeter heating system is usually required to offset the heat losses.

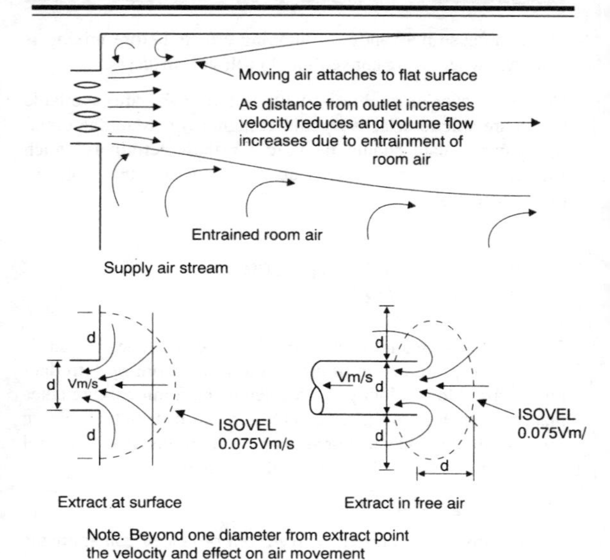

Figure 7.12

There is a small area around the front of a displacement supply air terminal or around a floor supply diffuser where the air temperature and velocity are uncomfortable for long-term occupation. Manufacturers' selection data provides comprehensive information including the extent of these areas, the vertical temperature gradients, the supply air temperatures and the room sensible heat gains that can be handled by these air terminals.

7.9 NOISE CONTROL NOTES

Excessive air flow generated noise probably accounts for more problems with air systems than any other cause. The following notes suggest how many of these problems may be avoided.

(1) Terminal equipment is often rated in room NC sound pressure level; this is often based on equipment laboratory tested sound power levels, less a room effect of 8 dB, whereas most office rooms with hard surfaces are closer to zero room effect when empty. Terminal equipment ratings are based on equipment in perfect condition with a normal duct velocity profile and noise free

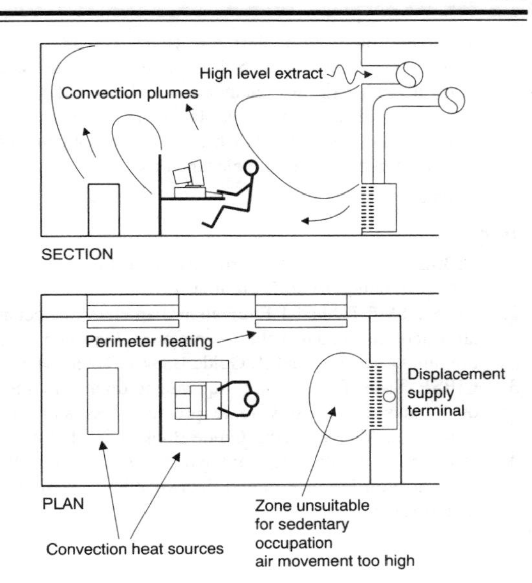

Figure 7.13 Displacement air distribution

air approach. These conditions are rarely achievable under site conditions and suitable allowances must be made.

(2) In low velocity system balancing dampers generate a surprising amount of noise when subjected to any appreciable pressure drop. Reputable manufacturers now give catalogue data which enables this sound power to be evaluated and if necessary attenuated. Generally balancing dampers should be remote from the room terminal.
(3) A straight duct of approximately 3 duct diameters or maximum widths beyond any fitting or item of equipment will usually allow the turbulence to dissipate and a normal stable velocity profile to resume. This straight duct should be allowed before a branch is taken off, or a terminal is connected (see Figure 7.9).
(4) Fabric flexible ducts are notorious for producing air disturbance and noise if incorrectly applied (see Section 7.4).
(5) High aspect ratio ducts should not be used except at very low air velocities (say below 1 m/s). Where unavoidable splitters should be fitted to create airways with aspect ratios not exceeding 3:1.

(6) Circular ducts are preferred to either rectangular or flat oval. They have a much lower noise breakout because of their high rigidity. Standard circular duct fittings are more predictable in air resistance and noise generation than tailor-made fittings which are often not constructed to recognized dimensional standards.

References

1. CIBSE TM8 Design notes for ductwork Figure C3.1 also in CIBSE Guide Book C Figure C4.2.
2. CIBSE TM8 Table 4.1 Equivalent diameters for rectangular ducts for equal volume, pressure loss and surface roughness. Also in CIBSE Guide Book C Table C4.30
3. CIBSE TM8 T Table 4.2 Equivalent diameters for flat oval ducts for equal volume pressure loss and surface roughness. Also in CIBSE Guide Book C Table C4.32.
4. Heating and Ventilating Contractors Association, DW 142 Specification for sheet metal ductwork low, medium, and high pressure.

8 Refrigeration

8.1 VAPOUR COMPRESSION CYCLE [1,2]

Figure 8.1 illustrates a simple vapour compression cycle of refrigeration.

The essential mechanical components of the system are shown in Figure 8.1 (a) and the related thermnodynamic changes in Figure 8.1 (b). The numbered points from 1 to 4 correspond.

For the process to occur a suitable refrigerant must be chosen. This is a volatile fluid that can change readily between the liquid and vapour states as its pressure varies, with related changes in its temperature (see Figure 3.12).

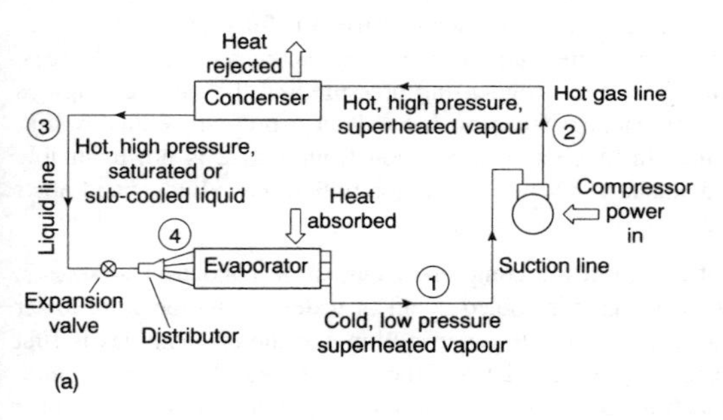

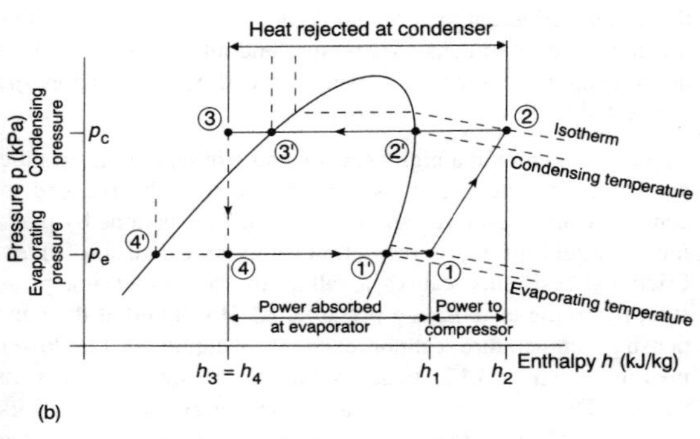

Figure 8.1 Simple vapour compression refrigeration process: (a) mechanical components and piping; (b) pressure – enthalpy diagram

The refrigerant in the evaporator boils at a low pressure and temperature, the value of which must be less than the temperature of the air or water being cooled. The heat absorbed from the air or water causes the liquid refrigerant to evaporate to a saturated vapour. For practical reasons the saturated vapour absorbs a little more heat than is necessary to complete the evaporation and hence the state of the refrigerant leaving the evaporator is superheated by about five to eight degrees. This state is shown by the point 1 in Figure 8.1(b). Low pressure in the evaporator is achieved by connecting it to the suction port of the compressor. Cold, low pressure, superheated vapour flows from the evaporator through the suction line and enters the compressor, which compresses and discharges it at state 2, as a hot, superheated, high pressure gas.

Theoretically, the change of state from 1 to 2 is reversible, adiabatic compression and occurs at constant entropy. Entropy is a concept that need not concern us in this text, except to say that it is useful in establishing the position of state 2. The entropy of state 1 is known because it is for vapour at the evaporating pressure, superheated by a known number of degrees. State 2 is at the condensing pressure and the same entropy as state 1. Hence it can be located on a pressure–enthalpy diagram. (In practice, compression from 1 to 2 is not reversible and adiabatic because of friction within the cylinders and other factors.)

Hot gas from the compressor enters the condenser at state 2. The condenser is cooled by air or water at a temperature lower than that of the refrigerant within and the entering gas is first desuperheated (2 – 2'), and then condensed (2' – 3') to a liquid. If additional heat transfer surface is provided in the condenser the liquid refrigerant may be sub-cooled (3' – 3). The process in the condenser is at constant pressure and all the heat absorbed at the evaporator, plus the power provided by the compressor, is rejected to the condenser coolant.

Liquid refrigerant at a high pressure and temperature leaves the condenser at state 2 and its temperature must be reduced to achieve a subsequent refrigerating effect. This is done by passing the liquid through a restriction (often an expansion valve). Frictional resistance causes a fall from the condensing pressure, p_c, to the evaporating pressure, p_e. Hot liquid at the condensing temperature cannot exist as a liquid at the lower pressure (Figure 3.12) hence some of the liquid flashes to vapour. The parent body of liquid loses energy as some of its molecules change from the lower energy liquid phase to the higher energy gaseous phase and this appears as a temperature drop in the liquid. No heat is supplied to or rejected from the expansion valve and it does no work on the refrigerant, hence the process from 3 to 4 occurs at constant enthalpy and is termed a throttling expansion. Referring to Figure 8.1(b) it is seen that state 4, entering the evaporator is a mixture of saturated liquid at state 4' and dry saturated vapour at state 1'. The process is completed in the evaporator, at constant pressure, as the heat absorbed from the air or water being cooled boils the liquid refrigerant to a dry saturated vapour and provides a few degrees of superheat, as mentioned earlier.

The heat flow rates for the processes described above are as follows, in terms of the notation used in Figure 8.1 and with $\dot{m}$ representing the mass flow rate of refrigerant in kg s^{-1}:

Refrigerating effect (kW):

$$\begin{aligned} Q_r &= (h_1 - h_4)\,\dot{m} \\ &= (h_1 - h_3)\,\dot{m} \end{aligned} \qquad (8.1)$$

Power absorbed by compressor (kW):

$$Q_w = (h_2 - h_1)\,\dot{m} \qquad (8.2)$$

Rate of heat rejection at condenser (kW):

$$\begin{aligned} Q_c &= Q_r + Q_w \\ &= (h_2 - h_3)\,\dot{m} \\ &= (h_2 - h_4)\,\dot{m} \end{aligned} \qquad (8.3)$$

The effectiveness of a refrigeration process is expressed by its coefficient of performance (COP):

$$\text{COP cooling} = \frac{\text{Refrigerating effect}}{\text{Power absorbed by compressor}} \qquad (8.4)$$

$$= \frac{(h_1 - h_4)}{(h_2 - h_1)} = \frac{(h_1 - h_3)}{(h_2 - h_1)} \qquad (8.5)$$

Similarly for heat pumping:

$$\text{COP heat pumping} = \frac{\text{Rate of heat rejection at the condenser}}{\text{Power absorbed at the compressor}}$$

$$= \frac{(h_2 - h_3)}{(h_2 - h_1)} \qquad (8.6)$$

Hence the numerical value of the COP for heat pumping will always be one more than the COP for cooling.

8.2 EXPANSION VALVES AND FLOAT VALVES [3,4]239

8.2.1 Thermostatic expansion valve

The functions of the valve are to reduce the pressure between the condenser and the evaporator and to regulate the flow of refrigerant so as to maintain a few degrees of superheat at the

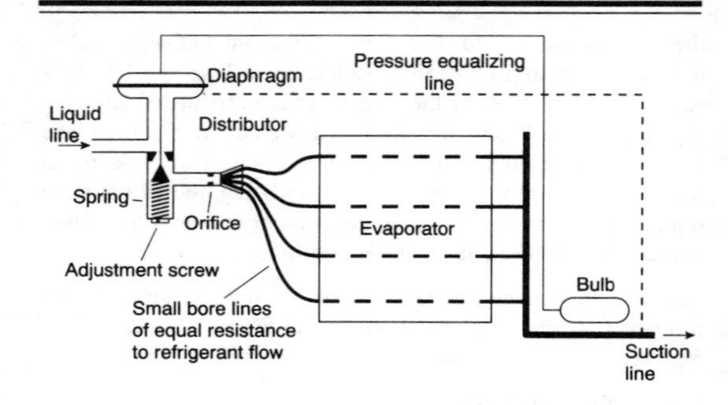

Figure 8.2 Simplified diagram of a thermostatic expansion valve

evaporator outlet. Figure 8.2 shows how a thermostatic expansion valve works. Liquid passes through the space between the valve pin and its seating, suffering a pressure drop and some flashing to vapour with a fall in temperature. The mixture of cold, low pressure, saturated liquid and bubbles then enters the distributor and flows through an orifice. A further pressure drop occurs and the bubbles and liquid are thoroughly mixed. In order to feed the mixture uniformly over the depth and height of the evaporator coil the refrigerant then flows out of the distributor through small bore tubes, of equal resistance, to the evaporator. The cold liquid refrigerant absorbs heat and boils, to provide the refrigerating effect.

Under partial load conditions the position in the evaporator where boiling is completed varies and there is a risk that liquid might enter the compressor and do damage, because it is incompressible. In order to protect the compressor the refrigerant vapour is superheated about 5 to 8 degrees (from 1' to 1 in Figure 8.1). A bulb is strapped to the outside of the suction line, close to the evaporator outlet (Figure 8.2), and connected to the upper side of the diaphragm in the expansion valve. The bulb contains saturated liquid and saturated vapour at the same temperature as the superheated vapour leaving the evaporator. Hence the saturated vapour exerts a pressure above the diaphragm corresponding to a temperature five to eight degrees above that of the saturated pressure within the evaporator. The force exerted on the underside of the diaphragm equals the saturated vapour pressure from the evaporator plus the effect of the spring. As the load on the evaporator diminishes the amount of superheat reduces and the force above the diaphragm also reduces. The valve pin moves closer to the seating and less refrigerant flows. The evaporation ends earlier in the evaporator and the amount of superheat increases until a balance is achieved. The amount of superheat may be adjusted manually by means of a screw on the valve. Most valves have internal pressure equalization to ensure that the evaporating pressure is exerted beneath the diaphragm, but for extensive installations an external pressure equalizing line is provided (Figure 8.2).

8.2.2 Electronic expansion valve

Different forms of this are available but, in principle, it is an expansion valve that has been motorized. It is then possible to introduce refinements to the form of control exercised over the valve. In particular, temperature and pressure sensors can be located in various places in the refrigeration circuit and information from them fed to a microprocessor, yielding closer control over the performance of a system with a reciprocating compressor than is possible with other methods. A thermistor located near to the suction valves in a hermetic compressor can compare the temperature it senses with that of another thermistor in the evaporator, and ensure the minimum amount of superheat at the compressor to give safe performance. There is a considerable temperature rise through the stator of a hermetic compressor (Section 8.4.1) and, taking account of this, superheat occurring in the evaporator can be virtually eliminated, giving either a greater cooling capacity or a smaller evaporator for the cooling capacity required.

8.2.3 Float valves

On larger refrigeration plant, a flooded evaporator (Section 8.3) is used and the expansion valve is replaced by a high-side or a low-side float valve. In a shell-and-tube condenser only the tubes above the liquid level of the refrigerant are effective in desuperheating and condensing the hot gas and the high-side float valve maintains a nominally constant liquid level beneath the tubes. As the load increases more hot gas enters and is condensed. The liquid level rises, the float lifts, opening the valve and letting liquid flow from the condenser into the evaporator, with a drop in pressure and temperature. The behaviour is similar to that of a steam trap. In a flooded shell-and-tube evaporator only the tubes beneath the refrigerant liquid level in the shell are effective in chilling the water passing through the tubes and the low-side float maintains a refrigerant level above the tubes. As the load increases more liquid refrigerant is evaporated and the level sinks. The float falls, opening the valve and admitting more liquid from the condenser, with a drop in pressure and temperature.

8.3 EVAPORATORS [2]

Direct-expansion air cooler coils are similar in appearance to chilled water cooler coils (Section 3.6). The major difference between the two is that the performance of the direct-expansion coil is dependent on the performance of the condensing set to which it is connected.

Figure 8.3 shows the performance characteristic for a direct-expansion air cooler coil. The coil is selected to give its design cooling duty, Q, for a specific entering wet-bulb temperature, t'_1, with an evaporating temperature of t_e. This establishes the design load point. If the coil is entirely wetted with condensate and the evaporating temperature equals the entering wet-bulb, no heat transfer will take place and the load will be zero. This establishes the zero load point. If it is then assumed that the characteristic is a straight line, joining the design and zero load points gives the characteristic. If the entering wet-bulb temperature falls, the zero load point moves to the left and a new characteristic can be drawn from the zero load point by assuming it is parallel to the design characteristic. If the face velocity over the coil reduces, the characteristic tilts, as shown by the broken line in Figure 8.3. The refrigeration duty achieved depends on where the characteristic for the condensing set intersects that for the cooler coil.

When a direct-expansion system is switched off the compressor should be allowed to run on, the low pressure cut-out being bypassed. in order to pump the refrigerant into the condenser or a liquid receiver, whence it will be able to give a subsequent

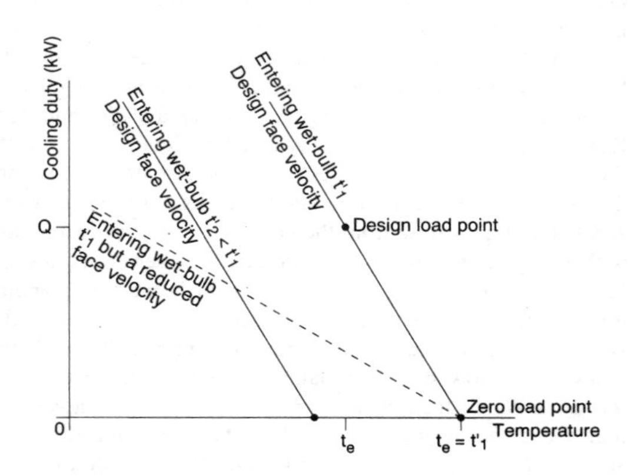

Figure 8.3 Characteristics of a direct-expansion air cooler coil

safe start. This is termed pump down. Pump down must never be used with water chillers because of the risk of freeze-up and expensive damage.

Liquid chillers are usually shell-and-tube heat exchangers with water flow inside the tubes and boiling refrigerant in the shell, and a float valve regulates the flow of refrigerant (Section 8.3). For smaller duties it is also possible to have dry-expansion chillers. These comprise a bundle of tubes, in hairpin shape, fed from an expansion valve and distributor. The boiling refrigerant is within the tubes and water being chilled flows through the shell. With extended surface on the tubes high rates of heat transfer can be obtained.

Plate heat exchangers are also sometimes used for water chilling systems. Heat transfer rates are good and manufacturers claim that a measure of flexing is possible for the plates, with less risk or immediate damage to the heat exchanger if the chilled water is inadvertently frozen. With shell-and-tube and dry-expansion evaporators the tubes will be burst or crushed (respectively), with expensive consequences, if the system is badly controlled or carelessly operated.

For industrial applications, where lower temperature chilled water or brine is required, submerged evaporators are sometimes adopted. These comprise a bundle of tubes submerged in a lagged tank of water or brine. Agitators ensure good circulation of the water or brine over the tubes. If aqueous glycol is used it must be remembered that the flow characteristics and heat transfer coefficients are not the same as those of water or other brines. There may also be corrosion problems and glycol lost through glands. Mechanical seals are needed for pumps.

Characteristics for water chillers can be constructed in the same way as for air cooler coils by assuming that the return chilled water temperature is the indication of the load, instead of the entering wet-bulb temperature.

8.4 COMPRESSORS [5]

8.4.1 Open and hermetic machines

Compressors may be driven by an external motor through a belt drive (occasionally) or a direct coupling. This is termed an open arrangement. The advantages are:

(i) The electric driving motor is cooled like any other motor and runs less risk of being burnt out through maloperation or inept control of the refrigeration plant.

(ii) The motor and compressor are easily accessible for maintenance or replacement.

(iii) The arrangement may be better suited for heat pumping than some closed machines by permitting slightly higher condensing temperatures.

A disadvantage is that leakage of refrigerant out of the compressor, or air into it, can occur at the shaft seal where the crankshaft emerges from the compressor casing.

A closed arrangement is more favoured and this may be hermetic or semi-hermetic. In the hermetic form the compressor and driving motor are close coupled in the same casing and access for maintenance or repair is impossible. Hermetic machines are used for the smaller range of cooling duties. For larger duties semi-hermetic machines are used and these are similar, the major difference being that a bolted flange divides the casing into two parts and allows occasional access for maintenance or repair. The motors of closed machines operate in a refrigerant atmosphere and this permits them to be more highly loaded than the motors of comparable open machines. There is no shaft seal and hence no air or refrigerant leakage.

The motor stators in closed machines are usually cooled by cold suction gas. The cooling effect is directly proportional to the mass flow rate of refrigerant and suction pressure (and hence density) must not be allowed to fall too much under partial load conditions. If the suction pressure does reduce excessively, the stator will overheat and burn out. High limit temperature sensors, embedded in the stator windings, cannot be relied upon for protection: continual overheating bakes the insulation on the windings and this ultimately fails before the high limit set point is reached. Starting the machine too frequently exacerbates matters and recommended maximum starts per hour are 8 to 10 for open reciprocating machines, 4 to 6 for hermetic or semi-hermetic reciprocating machines, and 2 for hermetic centrifugal machines.

8.4.2 Reciprocating compressors

Most compressors are reciprocating and refrigeration capacity depends on the mass flow rate of refrigerant pumped by the machine. Hence, for a given suction pressure (and so density) and a given condensing pressure, the capacity depends on volumetric displacement: cylinder size, number of cylinders and running speed. Not all the cylinder volume can be used because there must be a clearance space at the top of the stroke to avoid damage to the valves, cylinder head or piston. Gas trapped in this clearance space is at the discharge pressure and it must expand to less than the suction pressure, as the piston moves away from top dead centre, before fresh gas can enter through the suction valve. This gives rise to the concept of volumetric efficiency:

$$\text{Volumetric efficiency} = \frac{\text{Actual volume of fresh gas}}{\text{Swept volume}} \times 100$$

Some typical volumetric efficiencies [6] are given in Table 8.1.

Table 8.1 Typical volumetric efficiencies for R12

Compression ratio	2.0	3.0	4.0	5.0	6.0
Volumetric efficiency (%)	78	74	70	66	62

The relationship between capacity and suction pressure is almost a straight line for a given volumetric displacement and constant condensing pressure (Figure 8.4). It is not a practical proposition to try to construct a characteristic for a compressor — it must be obtained from the manufacturer. The behaviour of a compressor depends on condensing pressure and it is customary to give the behaviour of both the compressor and the condenser to which it is piped (termed a condensing set).

Figure 8.4 shows the characteristics for a condensing set operating at a constant speed and condensing temperature: curve A is with four cylinders in use whereas B is for two cylinders. Two cylinders displace half the volumetric flow rate of four cylinders and since the suction density is constant for a given suction pressure, curve B is located beneath curve A but with half the capacity for a given absolute suction pressure. Using four cylinders but reducing the compressor speed by 50 per

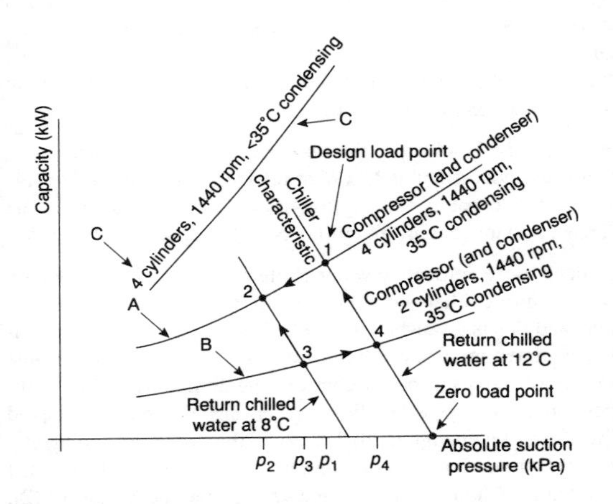

Figure 8.4 The intersection of characteristics for a water chiller and a condensing set with cylinder unloading to control capacity as the cooling load varies

cent has a similar effect. If the condensing pressure falls, the machine finds it easier to pump gas and the characteristic curve for four cylinders and a constant speed would move from A to occupy the position shown by C. Condensing pressure must be controlled at a stable value for reciprocating machines in order to provide enough pressure for the proper operation of the expansion or float valve.

If a water chiller is piped to a condensing set it might chill water from a design return temperature of, say, 12°C to a flow temperature of, say, 6°C. The design performance is shown in Figure 8.4 by the intersection of the condensing set curve with the chiller characteristic at the point 1.

Cylinder unloading is often used to reduce capacity at partial load and is achieved by lifting the suction valve off its seating. Gas then flows in and out of the suction valve without compression taking place. Lubricating oil pressure is used to operate the valve lifting mechanism and the signal for cylinder unloading is generally taken from the suction gas pressure, although it is possible to use return chilled water temperature. At partial load the return chilled water temperature might fall towards 8°C, say. Figure 8.4 shows that the characteristic for the chiller slides down the condensing set curve, from 1 to 2. Suction pressure, p_2, would be the signal to unload a pair of cylinders and the operating point would shift from 2 to 3, with a rise in suction pressure to p_3. If the load then increased, the chiller characteristic would move to the right and the point of operation would move from 3 towards 4, as the return chilled water temperature increased and the suction pressure rose. At a suction pressure of p_4 the other pair of cylinders would be loaded and the point of operation would move from 4 to 1, with a fall in suction pressure to p_1.

At a speed of 1440 rpm the maximum practical capacity of a single cylinder is about 35 kW of refrigeration, using R22. This limits the capacity of multi-cylinder piston machines, the largest possible having 16 cylinders and a capacity of 560 kW.

In the past, reciprocating water chillers have been controlled by using return chilled water temperature to unload pairs of cylinders and this is a satisfactory method. With speed control possible over reciprocating compressors and the use of electronic expansion valves, capacity control from chilled water flow temperature may be possible. Whatever means of control is adopted there must be enough chilled water in the system of piping etc. to prevent starting the compressors too frequently at low load conditions. Chilled water storage vessels may be necessary [2]. It is also vital to ensure constant chilled water flow rate through the evaporator: if the flow rate reduces the evaporating temperature will drop and the plant will freeze, with disastrous results.

Reciprocating compressors are positive displacement machines: they will continue to pump gas as long as sufficient power is fed into the crankshaft, within the limits of the strength of the materials involved.

8.4.3 Centrifugal machines

These are available with refrigeration capacities from 280 kW upwards and are commercially viable from about 500 kW. They are not positive displacement machines but use aerodynamic lift, generated by the aerofoil vanes of a rotating impeller, to pump refrigerant from the suction to the discharge pressure. Condensing pressure must be controlled to prevent it from getting too high but a limited fall can be allowed, when possible, to conserve energy consumption and improve performance at low load. This would be related to the flow temperature of the water from a cooling tower being permitted to drop from 27°C in summer design weather to a limit of about 18°C.

At partial load conditions suction pressure tends to fall and if this is coupled with a high condensing pressure the impeller may find it impossible to pump refrigerant against the pressure difference. The flow then momentarily reverses, the suction and discharge pressures equalize, the impeller again pumps gas and the cycle repeats. This is called surging. A considerable noise results and the bearings and other components of the machine are stressed. A certain amount of surge generally occurs during commissioning but it is not a condition that should be tolerated as a feature of low load operation. The solution is to reduce the condensing pressure and increase the suction pressure. Figure 8.5 compares the performance of centrifugal and reciprocating machines at constant condensing pressure, and shows the surging. Multi-stage compressors may be used to develop the pressures needed but the same can be achieved by running the impellers at higher speeds and this has tended to be preferred. Automatic control is by using variable inlet guide vanes and, with proportional plus integral plus derivative action, close control over the chilled water temperature leaving the evaporator can be obtained.

8.4.4 The screw compressor

Available in twin or single screw form this is a positive displacement machine available in the range 140 to 2000 kW, using refrigerants at pressures above atmospheric. Most machines are water cooled although an air-cooled version is possible.

The twin screw compressor comprises a pair of intermeshing rotors, one male, with four or five lobes and the other female,

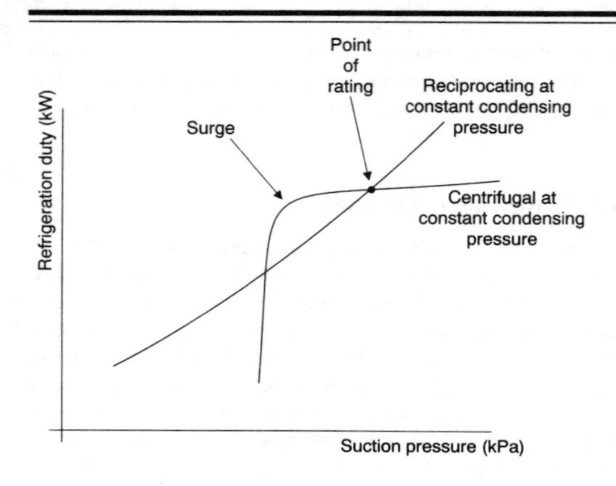

Figure 8.5 Characteristic curves for reciprocating and centrifugal machines compared at constant condensing pressure

with six or seven lobes. The male rotor is usually driven and is separated from the female rotor by a thin film of oil. The screws are contained within a casing that has a suction opening at one end and a discharge opening at the other. As the screws rotate, the space between a pair of male and female lobes is presented to the suction opening and gas enters until the interlobe space has passed beyond the suction opening. The full length of the space is occupied by suction gas, contained by the two lobes, the suction end plate, the discharge end plate and the compressor casing. As rotation proceeds the volume of the interlobe space is reduced as the gas is compressed against the discharge end plate. Eventually, the interlobe space is opposite the discharge opening and the trapped, compressed gas is discharged. Multiple interlobe spaces pass the discharge opening in rapid sequence and the flow is virtually continuous.

Capacity control is by a sliding valve which allows gas to escape to the suction side of the screws, without being compressed. Machines may be open or semi-hermetic and most of the latter are cooled by suction gas, although one manufacturer uses discharge gas, with less effective stator cooling but avoiding the need for an oil separator at discharge.

A single-screw compressor is also available using a pair of planet wheels to separate the high and low pressure sides. Like the twin screw, some liquid feedback is tolerable. Frictional losses are low and operating lives of 200 000 hours are claimed.

8.4.5 The scroll compressor [5,7]

Air-cooled, closed versions are available for 3 to 50 kW of refrigeration with capacity control by speed variation. Compression is achieved by a pair of interleaved scrolls. One scroll is fixed at both parallel edges but the other is fixed only at its peripheral edge. The central edge of this scroll is driven and orbits about the centre-line of the fixed scroll. The scrolls are contained within a cylinder having peripheral, suction openings at one end and a central discharge opening at the other. As orbiting starts, suction gas enters the interscroll space through two peripheral suction slots. As orbiting proceeds the scrolls bear against one another, the two spaces between the scrolls are closed and reduced in volume as they move towards the centre, compressing the trapped gas. Ultimately, the gas is discharged at high pressure from a central port at the opposite end of the casing.

8.5 THERMOSIPHON COOLING [8]

Water chilling may be accomplished without using the compressor under favourable, cool, outside conditions. If the condenser is at a higher level than the evaporator the compressor may be bypassed and the evaporator connected directly to the condenser. Superheated refrigerant vapour leaves the evaporator and migrates upwards to the condenser. If the condenser coolant (air or water) is cold enough, the refrigerant will condense and sub-cool. The expansion valve or the float valve is also bypassed and sub-cooled liquid refrigerant drains by gravity into the evaporator and is available for chilling water. The performance is like a heat pipe and the compressor is not used.

8.6 ABSORPTION CHILLERS [1,2,5]

The vapour absorption refrigeration cycle uses two fluids in solution, one a refrigerant and the other an absorbent. The affinity of the absorbent for the refrigerant on the one hand, and the application of heat on the other, are used to vary the strength of the concentration of refrigerant in the absorbent. An absorber and a generator take the place of the compressor (heat being supplied at the generator instead of power at the compressor), but the evaporator and the condenser remain as the pieces of equipment where heat is removed from the water being chilled and where surplus heat in the process is rejected to the outside.

Water chillers using lithium bromide as the absorbent and water as the refrigerant are commercially available with capacities from 350 kW to 6000 kW of refrigeration, employing steam at absolute pressures from 115 kPa to 1293 kPa as the source of thermal energy for the generator. MTHW and HTHW have been used, instead of steam, but have not given satisfaction because of the thermal expansion associated with the difference between the flow and return temperatures. Because of fundamental thermodynamic limitations, coefficients of performance exceeding unity are not really practical. Realistic coefficients at design conditions are 0.6 to 0.72 and steam consumptions are about 0.9 g/s for each kW of refrigeration. The refrigerant may crystallize because of control malfunction, air in the system, failure of the expansion valve, or interruption to the electrical supply (needed to operate certain pumps in the plant). Crystallization is a nuisance rather than a disaster but water-cooled machines are preferred over air cooled because there is less risk of crystallization.

References

1. W. B. Gosney, *Principles of Refrigeration*, Cambridge University Press, 1982.
2. W. P. Jones, *Air Conditioning Engineering*, 4th edition, Edward Arnold, 1994.
3. ARI Standard 750-87, Thermostatic Refrigerant Expansion Valves, 1994, ANSI/ARI 750-87.
4. W. P. Stoecker, *Refrigeration and Air Conditioning*, McGraw-Hill, 1958.
5. *ASHRAE Handbook*, 1992, Systems and Equipment, SI Edition
6. *Trane refrigeration Manua*l, The Trane Company, La Cross, Wisconsin, USA.
7. E. Purvis, Scroll compressor technology, 1987, Heat Pump Conference, New Orleans, USA.
8. S. F. Pearson, *Thermosyphon Cooling*, The Institute of Refrigeration, 1 March 1990.

9. Heat rejection [1,2]

9.1 WATER-COOLED CONDENSERS

With small plants, the water-cooled condenser is usually coil-in-coil or coil-in-shell but shell and tube arrangements, with water inside the tubes and refrigerant in the shell, are used for better quality plant and larger duties. The shell is commonly used to store refrigerant during maintenance. In the case of direct-expansion air coolers it may also be used to store the refrigerant when the plant is switched off and a pump-down cycle is used, but water chillers must not use a pump down cycle (Section 8.3).

A typical cooling water flow rate in the UK is 0.06 l/s for each kW of refrigeration, with a five degree temperature rise. Water from a cooling tower is dirty and, when ordering a water-cooled condenser, a fouling factor of 0.000 175 6 $m^2 kW^{-1}$ should be used. With an enhanced internal surface area (longitudinal straight or spiral grooves) this must be greatly increased.

Condensing pressure is sometimes controlled directly by throttling the cooling water flow, or bypassing it around the condenser. Alternatively, the flow water temperature is controlled.

In laying out the plant, the full length of a shell-and-tube condenser must be allowed for tube withdrawal.

9.2 AIR-COOLED CONDENSERS

Mechanical airflow over finned tubing condenses the refrigerant within but very little space is available for storing refrigerant during maintenance. With larger plant, a special vessel, termed a liquid receiver, is provided for this purpose. Typical airflow rates are from 140 to 200 l/s for each kW of refrigeration with a temperature difference of about 15° to 20° between the entering air temperature and the condensing temperature. Figure 9.1 shows a preferred arrangement that is not affected by the direction of the wind. Vertical condensers are much influenced by wind direction and are to be avoided.

The outside air temperature used to select an air-cooled condenser should be two degrees higher than the design value chosen for the rest of the air conditioning system. Furthermore, since the refrigeration plant will fail on its high pressure cut-out as soon as the outside air temperature rises above the value used for selection purposes, the capacity of the refrigeration

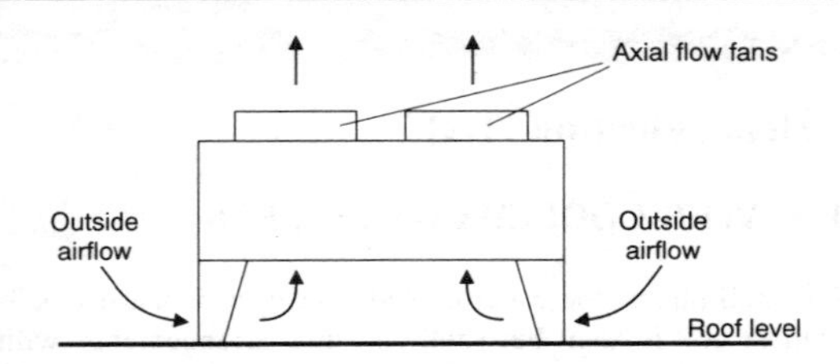

Figure 9.1 Horizontal air-cooled condenser — not affected by wind direction

plant should be reduced before this happens (say by cylinder unloading). The refrigeration plant and air conditioning system will continue to operate beyond the design temperature limit, with some benefit, but at reduced capacity.

Free, unimpeded outside airflow through the condenser is vital. Condensers must not be located in plant rooms, even if these are ventilated, unless there is ducted, fan-assisted airflow into the condenser itself.

Except for very small plants, where propeller fans may be used, most air-cooled condensers have axial flow or centrifugal fans. Noise is present in the airstream of the fans themselves and is radiated from the casings of the fans and the condenser. If silencers are added to fan outlets, the fans must develop enough extra pressure to overcome the additional air resistance of the silencer. It may be necessary to fit acoustic louvres to deal with noise radiated outward. The roof beneath the condensers must not transmit noise or vibration into the building.

Condenser pressure is commonly controlled by varying fan speed but in cold weather this may not be enough to prevent pressure falling to an unacceptably low value. Since it is only the surface of the tubing above the liquid level that is effective for desuperheating and condensing, a solution is to regulate the level of the liquid within the condenser (Figure 9.2). As the pressure sensor C1 detects a fall in condensing pressure it throttles a two-port motorized valve, R1. Less liquid leaves the condenser and the level rises. A smaller area is available for heat transfer and the condensing pressure increases. A pressure equalizing line must be provided between the condenser and the liquid receiver to ensure that condensing pressure acts on the upstream side of the expansion valve. A larger than usual liquid receiver is needed to accommodate the extra quantity of liquid refrigerant needed for this method of control.

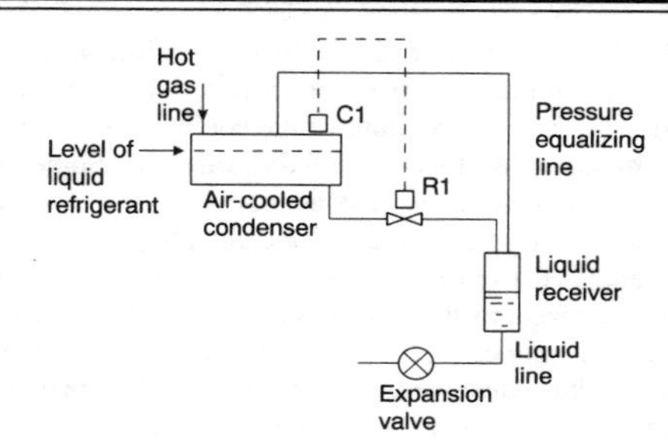

Figure 9.2 Condenser pressure control by varying the level of the liquid inside the consenser; C1 is a condensing pressure sensor; R1 is a motorized throttling valve

9.3 COOLING TOWERS [1–3]

It is very uneconomic to use mains water, discharged to waste, hence cooling water is generally provided by a cooling tower. Relatively warm water is fed from the condenser to the top of the tower and distributed over an inert packing material within the body of the tower. A large surface area of wetted packing is presented to the airflow, and heat transfer, with evaporative cooling, is counterflow between the descending water and the rising air (see Figure 9.3).

C1 is a pressure sensor that controls condensing pressure

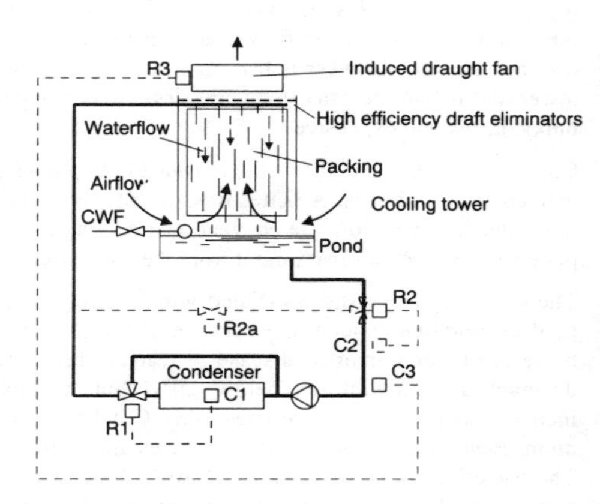

Figure 9.3 An induced draught cooling tower

directly by diverting part of the cooling waterflow around the condenser, through three-port mixing valve, R1. Alternatively C2 controls the flow temperature of the cooling water onto the condenser by means of three-port mixing valve R2. Sometimes, with centrifugal machines, instead of using the three-port valve, R2, a butterfly valve, R2a, located in a bypass across the tower, is regulated from the temperature sensor, C2. The butterfly valve is sized so that the pressure drop across it, when fully open, corresponds to the static lift of the water to the top of the tower. Thus when the butterfly valve is fully open all the water bypasses the tower. Another temperature sensor, C3, cycles the fan in colder weather.

Forced draught cooling towers that blow air through the packing are also used and it is possible to obtain cross-draught towers that have a lower silhouette. These towers contain the same amount of wetted packing as conventional towers. Axial flow or centrifugal fans are possible, the latter tending to be the quieter.

Instead of emptying them down in winter frost protection should be provided for cooling towers. This involves: controlling the fans from a thermostat in the pond (Figure 9.3), immersion heaters in the pond near the cold water feed and, on piping that can contain static water in winter, the provision of lagging, with thermostatically controlled electric tracer heating cable between the pipe and the insulation.

Closed cooling towers have sometimes been used. A serpentine arrangement of piping replaces the packing of an ordinary cooling tower and an aqueous solution of glycol is circulated. Spray water is fed from a collection pond to the top of the piping and upward airflow is induced or blown over the wetted pipe surface to give counterflow heat exchange and evaporative cooling. The closed tower is less efficient than a conventional tower and offers the same hygienic risks. It is comparatively bulky, heavy and expensive.

Corrosion, scaling and dirt on the outside of the piping presents maintenance problems. A better way of providing clean cooling water would be to use a conventional cooling tower with a plate heat exchanger separating it from the condenser.

There are hygiene risks associated with cooling towers [4,5] (and evaporative condensers) because of the presence of the bacteria of Legionnaires' disease in water. The bacteria are dormant at temperatures below 20°C but multiplication increases as the temperature rises to 37°C, when it is a maximum. Beyond this, multiplication reduces and ceases at 46°C. The bacterium is killed instantly at 70°C. Since most cooling towers used for air conditioning plant in the UK operate in the range 27°C to 32°C, steps must be taken to prevent contaminated spray from becoming a health risk. High efficiency drift eliminators must be fitted to limit the carryover of liquid droplets outside the cooling tower and the best recommended maintenance procedures must be followed [4,5].

9.4 EVAPORATIVE CONDENSERS

Evaporative condensers are the most efficient way of rejecting condenser heat. Hot refrigerant gas is fed through a serpentine arrangement of piping, cooling water flows downwards over the piping to a collection pond and is recirculated to the top of the piping. A fan induces airflow upwards to give counterflow heat exchange and evaporative cooling. Control over condensing pressure is by varying the airflow, using either motorized modulating dampers or fan speed control. The pipes are closely nested and problems with dirt, scale and corrosion have greatly discouraged the use of such condensers. They are not a practical solution in the UK for heat rejection.

9.5 DRY COOLERS

These are very similar to air-cooled condensers but have an aqueous solution of glycol from the condenser circulated through the tubes instead of refrigerant. No evaporative cooling is involved and condensing temperatures are a few degrees above those obtaining with cooling towers. Hence compression ratios are higher and compressor motor powers correspondingly greater. Dry coolers are a convenient method of heat rejection for small refrigeration loads but do not compete with other methods for larger duties: net plant areas, working weights and absorbed fan powers are approximately two or three times greater than with conventional cooling towers [6].

References

1. W. P. Jones, *Air Conditioning Engineering*, 4th edition, Edward Arnold, 1994.
2. *ASHRAE Handbook* 1992, Systems and Equipment, SI Edition.
3. J. Jackson, *Cooling Towers*, Butterworth Scientific, 1951
4. Report of the Expert Advisory Committee on Biocides, Department of Health, HMSO 1989.
5. Minimising the Risk of Legionnaires' Disease, Technical Memorandum TM13: 1991, CIBSE
6. Legionnaires' Disease Study, Contract 4642, Report No 1, BSRIA 1987.

10 Pipework design for closed water circuits

This section deals with the theoretical and practical aspects of heating and chilled water pipework systems.

10.1 PRIMARY CIRCUITS

The primary circuit contains the main thermal plant such as boilers, chillers, etc. and its main purpose is to enable the thermal producing plant to operate at their design flow rates and most efficient and safe operating temperatures and pressures.

The primary circuit may also be the distribution circuit. Different primary circuit arrangements are required for boilers to avoid standing losses, whereas for water chillers there are no standing losses associated with units which are circulating but not operating. There is, however, a need to avoid degrading the temperature of the common chilled water delivered to the primary circuit by mixing flow water from operating units with water at return temperature from non-operating units.

Figure 10.1 shows a typical boiler primary circuit arrangement, and Figure 10.2 shows the boiler connection arrangement for large boilers or oil-fired boilers where there is a requirement to bring boilers to normal operating temperature as quickly as possible to avoid fireside corrosion.

Figure 10. 3 shows the alternative pump position for water tube boilers which usually have a high water resistance, low water capacity and can tolerate high working pressures.

10.2 SECONDARY CIRCUITS

The secondary circuit connects the thermal output terminals such as radiators, air handling units, fan coils, etc. Its main purpose is to deliver water at the temperature and flow rate required by the terminals (which should be determined by the terminal manufacturer not by assuming a temperature difference between flow and return water).

In cases where the terminal flow rate and temperature requirements can be matched to those of the thermal producers (boilers, chillers, etc.) a common primary and secondary circuit may be used. Usually the secondary circuit is also the distribution circuit, but on extensive multi-building systems it may be more

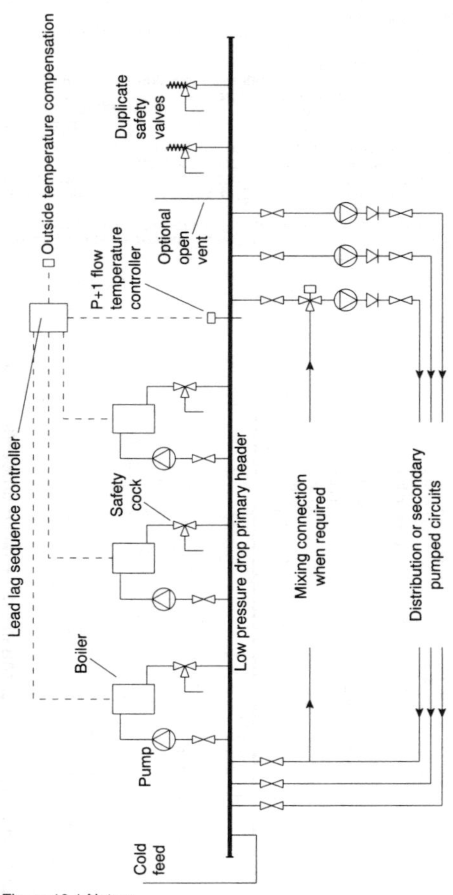

Figure 10.1 Heating shunt-connected primary circuit arrangement

Figure 10.1 Notes:

1. Anti-flash margin — This governs the minimum cold feed pressure. The pressure at the highest point in the boiler flow must be above that at which water boils when at a temperature equal to system design flow temperature plus temperature rise through the boiler plus a margin of 15°C to allow for temperature swings.
2. Boiler high temperature cut-out setting — This must be at least 8°C below the flash temperature and 7°C above the maximum boiler flow temperature.
3. This primary system arrangement is not suitable for water chillers.
4. Many manufacturers provide a package including header, controls, boiler pumps and connections.
5. Boiler units and distribution pumps should be interlocked so that the boiler units run only when at least one distribution pump is running.
6. Example determination of pressure and temperatures:
   (For a design, full load, distribution circuit flow temperature = 85°C)

(a) Pressure:	Maximum boiler return temperature	85°C
	Plus boiler temperature rise	10°C
	Plus margin	15°C
		110°C

Figure 10.1 Notes continued

110°C requires 0.433 bar g. minimum pressure at highest point of the boiler flow main.

(b) High limit thermostat setting:

Maximum boiler return temperature	85°C
Plus boiler temperature rise	10°C
Plus margin	7°C
	102°C

(c) Individual boiler control thermostats control band:
85°C (at max. output) to 95°C. (at min. output)

7. The boiler pumps may be on the boiler inlet for high resistance, water tube boilers or on the outlet for shell boilers selected close to their design pressure.
8. If a mix of condensing and non-condensing boilers are used then the condensing boilers should be upstream of the non-condensing boilers to receive the lowest water temperature.

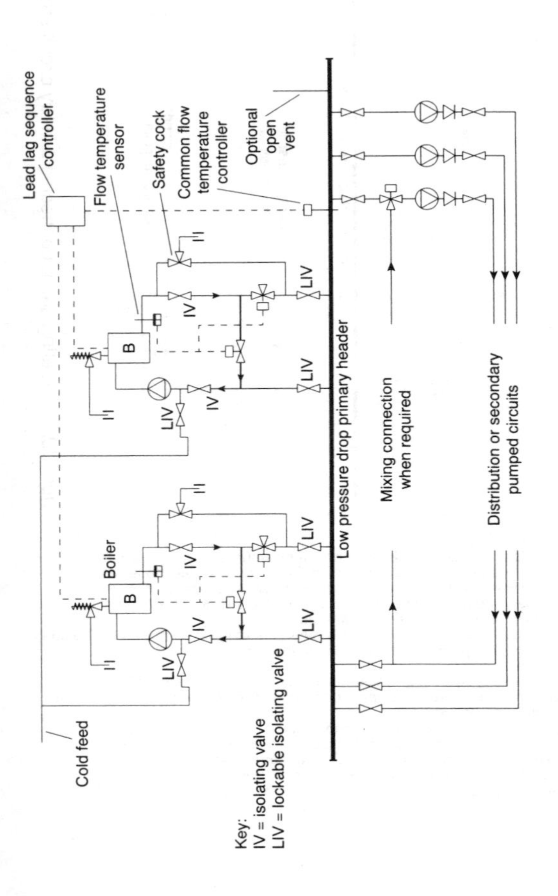

Figure 10.2 Heating shunt-connected primary circuit arrangement — oil fired boilers

Figure 10.2 Notes

1. Additional flow and return isolating valves are required outside the main isolating valves on boilers operating at or above 100°C, and on boilers having manhole access.
   These additional isolating valves, and the lockable isolating valves on the cold feed connections, must be locked closed during boiler maintenance or inspection and locked open at all other times.
2. Boiler units above 3 MW size should have small bypass valves round the main boiler return isolating valve.
3. Boiler units above 3 MW size should have independent cold feed connections with locked open isolating valves.
4. Each boiler must either have an open vent on the boiler side of the main flow isolating valve or an open vent sized three-port safety cock bypassing both the main flow isolating valve and any control valve, with the discharge piped to a safe position.
5. Primary circuit water capacity and the possible need for added storage should be investigated.
6. Each boiler should have its own, self contained, temperature and safety control system, complying with all relevant regulations and British Standards.
7. Each boiler pump should have timer controlled run-on after boiler shut down.
8. Boilers should have outlet temperature controlled recirculation until design outlet temperature is approached.
9. The common flow temperature controller should start the next boiler unit only when the operating units have failed to maintain the common flow temperature for a preset time (15 to 30 min).
10. Boiler units should be interlocked to run only when at least one distribution pump is running.
11. Refer also to the notes for Figure 10.1.
12. The boiler pumps are shown in the position for water tube boilers, but, for shell boilers it is usually preferable for the pumps to be on the boiler outlet to avoid imposing the pump pressure on the boiler shell.

economical to use a separate pumped distribution circuit operating at the largest temperature differential possible greater than either the primary circuit or the final secondary circuits, this reduces pump energy and mains losses (see Figure 10.3).

10.3 MIXING CIRCUITS

These are secondary pumped circuits which require to operate at a lower (heating) or higher (cooling) flow temperature than the primary sources temperature. In some applications this temperature may be varied by scheduling the controller set-point against some other variable, e.g. outside air temperature. Mixing circuits have the flow temperature regulated by mixing a proportion of the return water into the flow (see Figure 1.3). Where the maximum (heating) or minimum (cooling) secondary flow temperature is lower (heating) or higher (cooling) than the primary sources temperature a fixed bypass may be required in addition to the automatically controlled bypass. The primary sources may be the return of another circuit, a primary ring main or a vessel. The primary source pressure drop between the secondary flow and return connections should be quite low, say not more than 5 per cent of the primary or secondary pump head, whichever is the lower.

Large systems and distribution circuits
Separate pumped secondary circuits (injection or heat exchanger type), remote from the primary circuit, may be justified where:

1. Terminals and secondary circuits need to operate at different times
2. Terminals and secondary circuits need to be separated from high differential pressures in large distribution circuits
3. Terminals and secondary circuits need to be separated, with a heat exchanger, from high static pressures in large distribution circuits
4. Some or all secondary circuits are constant flow type, whereas the distribution circuit is variable flow to reduce pumping costs.
5. Terminals and secondary circuits are under separate management or ownership from the distribution circuit

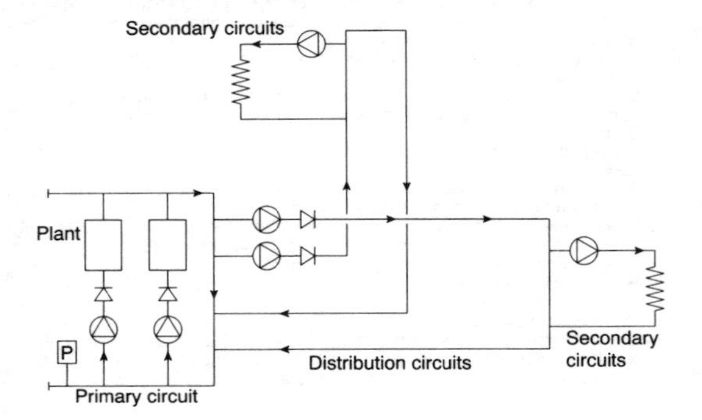

Notes:

1. The primary circuit flow rate must be equal to or greater than the sum of the distribution circuit flow rates to avoid degrading the flow temperature by return mixing.
2. The primary circuit may be commissioned and set running before completion of distribution circuits provided there is sufficient thermal load to fully load one plant unit.
3. To avoid reverse flow in non-operating distribution circuits, line sized non-return valves should be fitted on distribution pump outlets.
4. Distribution circuits can be designed for maximum temperature differences to reduce pump and pipe sizes.
5. Distribution circuits can be variable flow.
6. The same principles can be applied to large cooling or heating systems.

Figure 10.3 Large system arrangements

10.4 INJECTION CIRCUITS

These are used where a constant or variable flow pumped secondary circuit is fed from the flow and return of a constant or variable flow primary or distribution circuit. The main heating application is to provide a lower or variable temperature for the secondary pumped circuit fed from a constant or variable flow primary heating circuit. The secondary circuit must be suitable for operating at the same pressure as the primary circuit. See Figures 10.4 and 10.5.

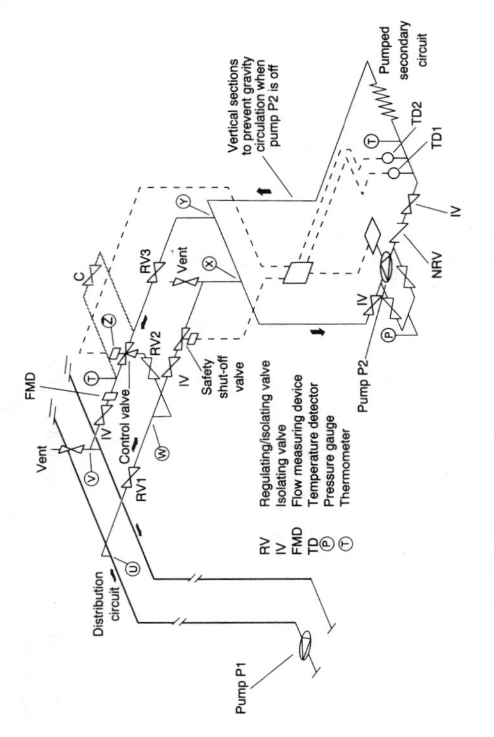

Figure 10.4 Injection secondary circuit connected to a constant flow distribution circuit

Controls operation:

1. TD1 modulates the control valve (3-port mixing, linear, symmetrical characteristic) when pump P2 is running.
2. When pump P2 is switched off and/or electric failure occurs, the control valve goes to full bypass.
3. Where the distribution flow temperature could cause damage to the secondary circuit (e.g. embedded panels) or cause a safety hazard, TD2 should be fitted, to shut a manual-reset safety shut-off valve in the injection flow.

Setting up procedure:

1. System full and cold, RV1 closed, P1 and P2 running.
2. Set the temperature control to give no bypass through the control valve.
3. Regular RV1 to give design flow rate as measured at the FMD.
4. Set the temperature control to give full bypass through the control valve.
5. Regulate RV2 to give design flow as measured at the FMD.
6. Put the control valve to automatic operation.

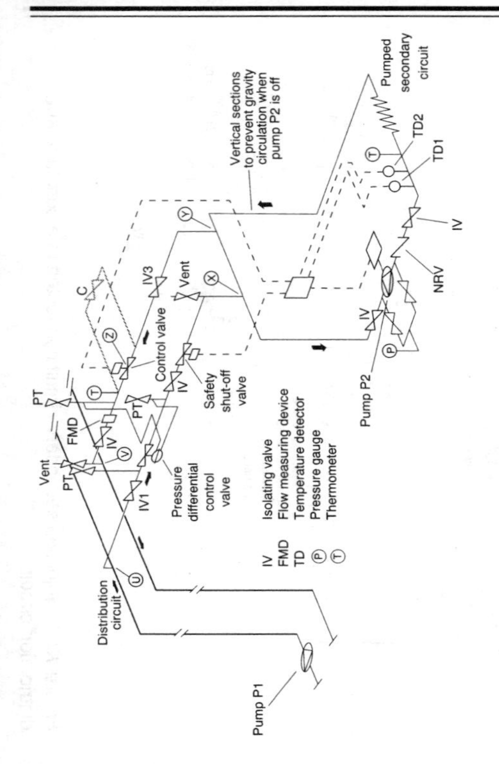

Figure 10.5 Injection secondary circuit connected to a variable flow distribution circuit

Controls operation:
1. TD1 modulates the control valve (2-port linear characteristic) when pump P2 is running.
2. When pump P2 is switched off and/or electric failure occurs, the control valve goes to fully closed.
3. Where the distribution flow temperature could cause damage to the secondary circuit (e.g. embedded panels) or cause a safety hazard, TD2 should be fitted, to shut a manual-reset safety shut-off valve in the injection flow.

Setting up procedure:
1. System full and cold, RV1 closed, P1 and P2 running.
2. Set the temperature control so that the control valve goes to full open.
3. Adjust the pressure difference control valve to give design flow rate as measured at the FMD.
4. Check that the pressure difference control valve holds the pressure difference across the injection circuit approximately constant for any control valve opening.
5. Put the control valve to automatic operation.

Component functions (refer to Figures 10.4 and 10.5)

RV1 (constant flow) or PDR (variable flow) absorbs the surplus pressure differential of the distribution circuit.

The control valve regulates the injection flow rate to give the required secondary circuit flow temperature as detected by TD1.

Injection and mixing take place at X.

Optional bypass C (not normally fitted, for safety reasons), enables hand control to be effected if the control valve has to be removed and its connections have been blanked.

Optional high limit thermostat TD2 and the safety shut-off valve are only necessary where the distribution flow temperature can cause a serious hazard in the secondary circuit.

The NRV on the secondary circuit pump is to prevent the possibility of reverse gravity circulation in the secondary circuit when pump P2 is off.

The anti gravity circulation loop prevents circulation in the secondary circuit when pump P2 is off in the case of heating systems.

The isolating valves on the connections to the distribution circuit must be closed before draining any part of the secondary circuit.

The re-filling of any major portion of the secondary circuit should be from an external water source. Equalize the pressures after refilling and/or venting the secondary circuit, by very slowly cracking open the isolating valve on the injection return connection. Pressure must be equalized and the secondary circuit pump operating before opening the injection flow isolating valve.

Sizing

T_{DF} = distribution circuit flow temperature (°C)
T_{SF} = secondary circuit flow temperature at full design load (°C)
T_{SR} = secondary circuit return temperature at full design load (°C)
W_S = Secondary circuit flow rate at full design load (kg/s)
W_D = Injection flow rate required from distribution circuit at full load (kg/s)

Then: $W_D = W_S \times (T_{SF} - T_{SR}) / (T_{DF} - T_{SR})$

The pressure difference available in the distribution circuit, i.e. between U and V, is absorbed in sections UX and VY.

The small pressure difference between X and Y can be assumed to have no significant effect on the secondary circuit.

The control valve at Z should have a linear characteristic and should be sized to have a high authority, i.e. not less than 50 per cent.

or

The control valve should have a resistance equal to or more than the resistance of:
(a) pipes WX plus YZ in the constant flow distribution case
(b) pipes UX plus YV in the variable flow distribution case

Pipe XY should be the same size as the adjacent secondary circuit pipes.

In the case of constant flow distribution, the resistance of bypass ZW including the control valve should be equal to or slightly greater than the resistance of WX plus YZ, including the control valve. In cases where the control valve authority is very high (>0.75), it is permissible to omit RV2, and to substitute an isolating valve for RV3, provided that the control valve has symmetrical characteristics.

10.5 DISTRIBUTION CIRCUITS TO TERMINALS

10.5.1 Single-pipe connections (see Figure 10.6)

These are sometimes used to connect low pressure drop terminals such as radiators and continuous convectors, they must not be used to connect terminals with a high pressure drop such as fan coils.

The pressure difference causing circulation is the resistance of the bridge pipe between the connections.

The mean water temperature of the terminal is lower than that of an equivalent terminal connected two-pipe, hence the terminals need to be larger. The flow temperature to each terminal is degraded by terminals upstream, limiting the circuit temperature drop to a maximum of about 12°C. The pipe connection arrangement is much neater than that of an equivalent two-pipe system and is quite suitable for exposing to view.

10.5.2 Two-pipe connections (see Figure 10.7)

This is the most common method of connecting terminals, equipment and single-pipe circuits. The terminal and its connections are part of the circulating pump head and can have a high water resistance. If the terminals on a circuit all have the same resistance and that resistance is high compared with the resistance of the mains between the first and last terminals on that circuit the terminals may have satisfactory water balance without regulation (see Figure 10.7).

The flow temperature to each terminal is degraded by the heat losses or gains of the flow main. With most single building systems the increased water flow required to offset the effect of the lower terminal flow temperature can be ignored provided the flow main is well insulated. With large systems serving several buildings the water flow increase on remote terminals required to achieve the same mean terminal water temperature becomes significant and must be calculated. Alternatively, the terminal manufacturer is given the revised flow temperature and is asked to re-select the water flow rate required.

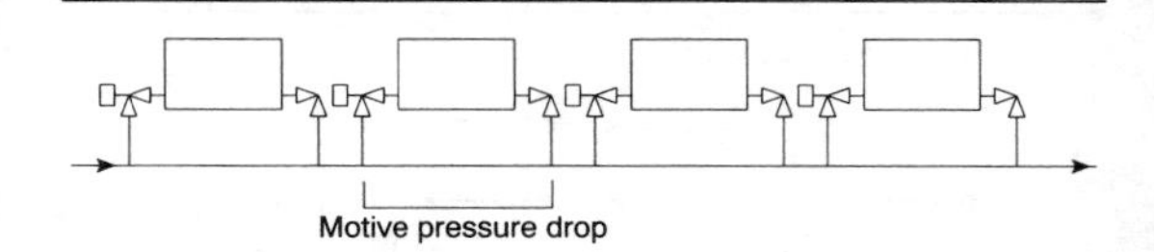

Figure 10.6 Single-pipe connections (only for low pressure) (drop terminals)

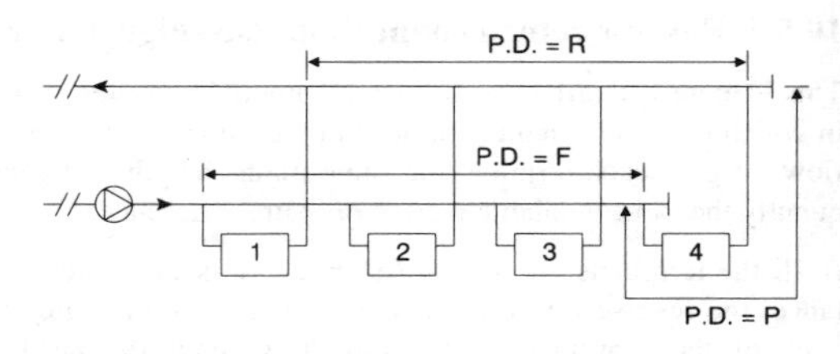

Header and Plant Connection Sizes for Parallel Connection

Plant may be connected in normal 2 pipe arrangement as Figure 10.7 or in reverse return as Figure 10.8.

The resistance of each plant (Boiler, Chiller etc), plus the resistance of its connections to the common header must be equal and must be high compared with the pressure drop of the header between the first and last plant item.

For example, if plant items 1 to 4 in Figure 10.8 are identical and have a resistance at design flow of P, then, in order for the plant flow rates (without regulation) to be within + 5%–0% of the design flow, the resistance F of the flow header, plus the resistance R of the return header, must not exceed 10% of P.

In practice, this means that with most modern high resistance boilers and chillers there is no practical advantage in using the reverse return arrangement. If the plant units on a primary circuit header are identical, it is unlikely that regulation of water flow through the units will be necessary with normal sizing methods.

Figure 10.7 Normal two-pipe plant arrangement

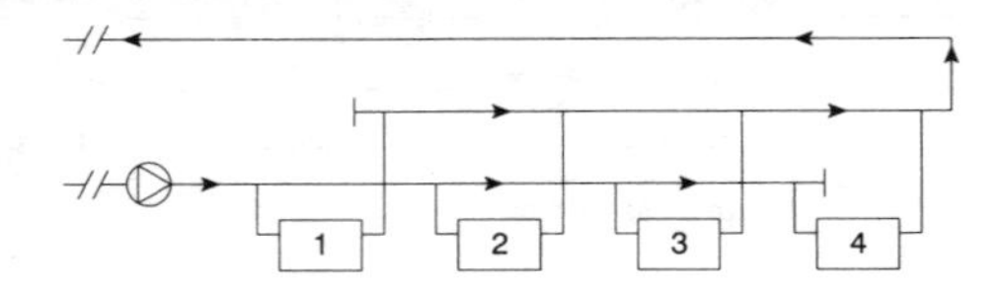

Figure 10.8 Reverse return plant arrangement

10.5.3 Reverse return connections (see Figure 10.8)

This is merely a variation of the conventional two-pipe system, in which the return main continues in the same direction as the flow, to give each terminal the same mains length and consequently the same available motive pressure difference.

If all the terminals are selected to have the same water resistance, the reverse return arrangement will ensure a water balance of the terminals on the circuit, without the need for regulating valves. However, if the terminals and their connections have different resistances to flow, the reverse return is of no benefit. If terminal connections are fitted with dynamic automatic constant flow regulators (CFR), the reverse return arrangement offers no benefit over the simpler conventional two-pipe arrangement.

10.5.4 Series connections with bypass (see Figure 10.9)

This is a radiator connection arrangement, developed mainly for domestic heating and is limited to a maximum of about eight radiators per circuit, determined by the maximum flow rate of the 15 mm pipe and valves. It is usually associated with single-entry radiator valves, which make a very neat connection arrangement and are available with thermostatic heads.

10.6 PIPE SIZING

Determining the sizes of the pipes in a water distribution system is governed by the following factors:

(1) Ensuring that the resistance of the circuit at the design flow rates matches the pressure difference available across the circuit.
(2) Ensuring that water velocity in each pipe does not exceed the erosion limits set out in CIBSE Table B1.13[1].
(3) Rationalizing the pipe sizes (subject to the above

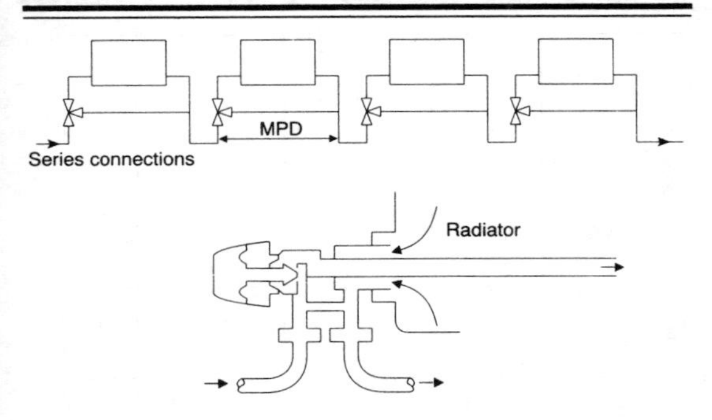

Figure 10.9 Single-entry thermostatic radiator valve for series connection

restraints) in order to maximize standardization and prefabrication, e.g. it is usual to keep risers a constant size top to bottom and to avoid horizontal mains less than 25 mm size (terminal connections excepted).

(4) Pipe sizes arranged to facilitate removal of debris by water flushing at high water velocities with terminals and plant isolated and bypassed.

Pipe sizing is usually carried out by establishing a target mean pressure drop per metre run for each circuit based on pressure drop available across the circuit and selecting the nearest commercially available pipe size which will carry the required mass flow rate without exceeding the water velocity limits and rationalizing the sizes to aid prefabrication and flushing.

10.6.1 Manual pipe sizing procedure (see example Figures 10.10 and 10.11)

(1) Prepare simple circuit diagrams starting with the circuit which has the highest resistance from the thermal source to the terminal, this is called the index circuit. With large systems it is usually necessary to construct separate diagrams for the branch circuits.
(2) Starting at the thermal source or branch number reference each pipe.
(3) Starting with the index terminal, fill in pipe sizing form working back to the thermal source, assume the index regulating valve is fully open and record its resistance separately. Use manufacturers data for the resistance of plant and pipe line components. Use CIBSE Pipe Sizing Tables C4.11-25[2] for the appropriate pipe specifica-

Circuit	Pipe Ref	Flow (Kg/s)	PD Available (Pa)	Target (Pa/m)	Size (mm)	V (m/s)	Fittings, valves, etc. ζ Factors	Factor Σζ	EL le	Σζ x EL	L (m)	L+ΣζxEL Total equiv. length	Actual (Pa/m)	PD (Pa) (Section)	PD (Pa) (Cumulative)	PD (Pa) Remaining	Regulating valve	
																	Size	Setting

Figure 10.10 Pipe sizing chart

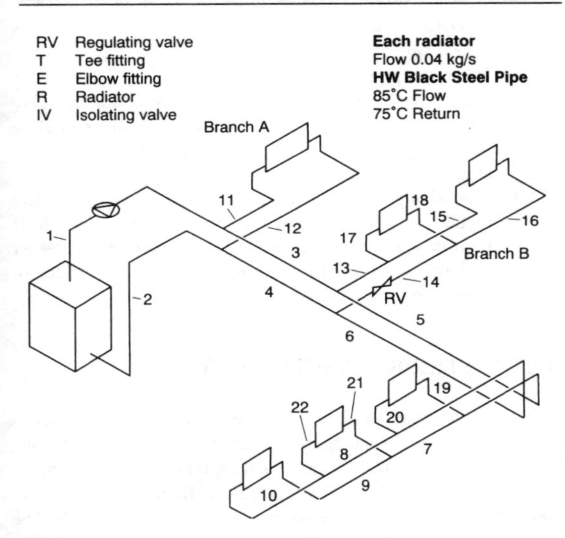

Figure 10.11 Pipe sizing example

tion and water temperature. The water temperature only has a minor effect on resistance and it is usual to choose the 75°C tables for all heating applications and 10°C for all chilled water applications. Pipe sizing tables give equivalent length figures (l_e) which when multiplied by the ζ (zeta) values for fittings given CIBSE Table C4.36 give the length of straight pipe with the same resistance as the fittings.

(4) The pipe sizing table needs to be filled in for the index circuit, only up to and including the PD cumulative column. The last figure in the cumulative column is the resistance of the index circuit which is also the theoretical head required from the circulating pump.

(5) Branch circuits are dealt with in order working from the main index terminal. The PD cumulative in the index circuit at a branch is the PD available for the branch and also provides the basis for calculating a target Pa/m for the branch: Target Pa/m = PD available less the resistance of branch index terminal and connections divided by the branch length of flow and return pipework including fittings.

Note at this stage it is necessary to make an arbitrary allowance for fittings by examining a part of the index circuit with similar flow rates and establishing the percentage addition for equivalent length of fittings over length of straight pipe.

(6) When the pipe sizing form is complete for a branch up to the final cumulative PD. The branch regulating valve size and setting may be selected from a valve manufacturers' chart: required valve pressure loss = PD available – PD cumulative for the branch.

(7) Pipe sizing, pump head and regulating valve requirements are often carried out by computers but these require similar input data and it is sometimes necessary to have several runs in order to assess the affect of rationalizing pipe sizes.

10.7 WATER BALANCING METHODS

In all constant water flow circuits the correct water flow needs to be allocated to each branch to ensure that every terminal receives its design flow within an acceptable tolerance. This is normally achieved by regulating valves fitted on each return and adjusted to add the required resistance to the circuit. The resulting flow is measured by a flow measurement device (FMD). This equipment is used to manually balance in accordance with procedures detailed in BSRIA application Guide 2/89[3] and CIBSE Code W [4].

10.7.1 Manual proportional balancing (site)

Briefly the procedure is:

(1) Scan the flow rates achieved with all regulating valves and control valves full open.

(2) Express all flow rates as a percentage of the design flow (% DFR).

(3) If the % DFR of any intermediate terminal is lower than the last on the leg then the last terminal or branch regulating valve is throttled down until is % DFR is the lowest on the leg being balanced.

(4) Start with the leg which has the terminal with the highest % DFR on the circuit.

(5) Working back from the last terminal or branch each regulating valve in turn is adjusted until the % DFR achieved is the same as the CFR measured at the last terminal which is used as the reference for balancing the others.

Note: It is necessary to continually monitor the % DFR of the reference as this changes as other valves on the system are adjusted.

(6) When all legs are in proportional balance (i.e. all their

terminals have the same % DFR), the same procedure is used to balance each leg against the last leg again using the same reference terminals.

10.7.2 Manual compensated balancing (site alternative method)

See Figure 10.12

(1) Any leg is selected for balancing by ensuring that all its regulating valves and control valves are fully open and valving off all the other legs.
(2) In the unlikely event of the last terminal on the leg having a % DFR less than 1.0, the first few terminal regulating valves are throttled down until % DFR of the last terminal (reference) exceeds 1.0.
(3) Operator A at the leg main regulating valve observes the flow rate of the last terminal via remote reading instrument (or telephone message from Operator C) and adjusts the main regulating valve to keep the reference % DFR constant.
(4) Operator B starts at the terminal next to the reference and works back adjusting each terminal regulating valve in turn to achieve the same % DFR as the reference (which is being kept constant by Operator A).
(5) When the leg is in balance the next leg is opened up for the same procedure before shutting the circulation of the balanced leg.
(6) When all legs are in balance the same procedure is used to balance each leg against the last leg.

Note: This method avoids the need for a complete preliminary scan of the flow rates and gives higher flow rates and flow measuring signals.

10.7.3 Balance by calculation

If the plant, terminal equipment, control valves and regulating valves, etc. actually installed have the same flow and resistance

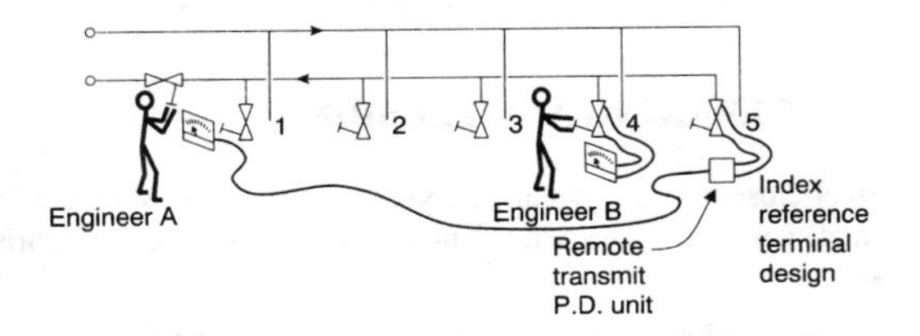

Figure 10.12 Compensated balancing method

characteristics as the corresponding equipment selected as the basis of the pipe sizing calculations (see Section 10.6) it is usually sufficient to adjust the regulating valves to the settings predicted in those calculations. Unfortunately this is rarely the case and changes in equipment suppliers usually means that the changed hydraulic characteristics of the alternative equipment render the original regulating valve predicted settings invalid requiring a site balance.

Note: Minor changes in the number of fittings and pipe lengths usually have negligible affect on the water balance of systems with high resistance terminals.

10.7.4 Constant flow regulators (dynamic)

These automatically ensure an accurate water balance without manual regulation. It is important to confirm that the units are correctly fitted, the water system is clean, isolating valves are open and flushing bypasses closed. The pump and circuit flow rate should be within ± 5 per cent of the sum of the CFR unit nominal flow ratings. The pressure differential across each CFR should be within the specified limits (typically between 14 kPa and 220 kPa) although it is usually only necessary to check this in the event of obvious problems with lack of terminal thermal performance.

10.7.5 Variable flow systems

These systems are also dynamically self-balancing provided they are not overloaded, and the controls are working correctly. It is important to confirm that each pressure differential regulator controls the pressure difference across its circuit at a level which will just provide the circuit design flow under all operating upstream differential pressures. The control of pump speed should give adequate, but not excessive differential pressures across the index circuit and nearest circuit at both maximum and minimum flow. These flows may be simulated by arranging the appropriate percentage of terminal control two-port valves to be fully open uniformly over the system and confirming the flow diversity by measuring the pump duty.

10.8 CLEANING AND FLUSHING

Steel closed water systems are very vulnerable to blockage of small valve apertures with debris. The main sources of debris are:

(a) Surplus jointing material and building debris
(b) Millscale and corrosion products formed and released

when sections of a large system are pressure tested then drained and left empty and wet for long periods until the system is complete.

(c) Blockage problems are exacerbated by the very small valve orifices necessary with the high pressure drops and low terminal flow rates associated with modern large systems.

To eliminate these problems rigorous cleaning and flushing routines are required followed by careful water treatment.

10.8.1 Chemical cleaning

This falls broadly into two types:

(i) The acid clean which removes iron oxides leaving the steel tube internally clean but vulnerable to flash corrosion in the event of any lapse in subsequent water treatment. Some types of seals are vulnerable to acid attack.

(ii) The dispersent clean which ensures that sludge and loose corrosion products are released from the internal pipe surface and kept in suspension to enable them to be flushed out. The dispersent, however, leaves a layer of corrosion on the pipe wall to assist in inhibiting further corrosion. This corrosion layer may be subject to further erosion. With all cleaning procedures it is necessary to completely flush out all solids in suspension and immediately inhibit the steel surfaces against corrosion.

10.8.2 Flushing

Flushing steel piping systems at high water velocities without putting dirty water through terminals or plant items requires a number of special design features. These are specified in detail in BSRIA application Guide 8/91[5] Pre-Commission Cleaning of Water Systems. Briefly the features required are:

(a) Isolating valves and valved bypasses for all terminals and plant items with steel pipe connections. Strainers preceding main plant.

(b) Mains size dirt pockets with removable caps, drain cocks and isolating valves at the base of risers.

(c) Valved connections on primary and secondary circuits to connect temporary flushing tanks and pumps.

(d) Valved bypasses from flow to return at the end of each steel circuit and at main size reductions.

(e) Drain cocks on terminal flow connections to facilitate back flushing and proving terminal and connections free of blockages.

(f) Valved cross-over connections at main circulating

pumps to enable the flow direction to be reversed in the pumped circuits for the flushing routine.

These requirements are shown diagrammatically in Figure 10.13.

Note: There is a strong case for using clean non-ferrous pipework for connecting terminals thus providing greater mechanical flexibility to deal with positional adjustment and expansion movements but also eliminating the need for terminal flushing bypasses.

10.9 AIR REMOVAL

Experience has shown that air removal from large pressurized water systems is difficult and requires special techniques. Failure to remove free air results in corrosion, noise, pump derating, inaccurate flow measurements and sometimes reduced performance of heat exchangers.

Conventional air bottles at high points and circuit extremities have a useful function in enabling a system to be filled effectively, but are useless at collecting air when the system is circulating. Some air is always trapped in parts of the system during the initial filling and thus air is broken down into very small bubbles under pumped circulation. These bubbles are swept past air bottles and can only be removed by reducing the water velocity to below approximately 0.2 m/s and by reducing the pressure below the lowest normal system pressure to ensure that the maximum amount of air comes out of solution.

The most effective way of achieving this with a system having an open feed tank is by arranging a temporary flow of system water to be diverted via a submerged hose into the feed tank until air ceases to be discharged (see Figure 10.14).

With a pressurized system the following procedure can be used:

(a) Fill the system slowly from the bottoms up releasing air at high points and extremities.
(b) Valve off and switch off the pressurizing set.
(c) Fit a temporary small open tank about the highest point in the circulation and connect by temporary hose pipes to convenient valved connections on the system with sufficient pressure differential to circulate approximately 10 per cent of the system flow through the temporary tank (see Figure 10.15).
(d) if necessary throttle valve A to avoid negative system pressures.

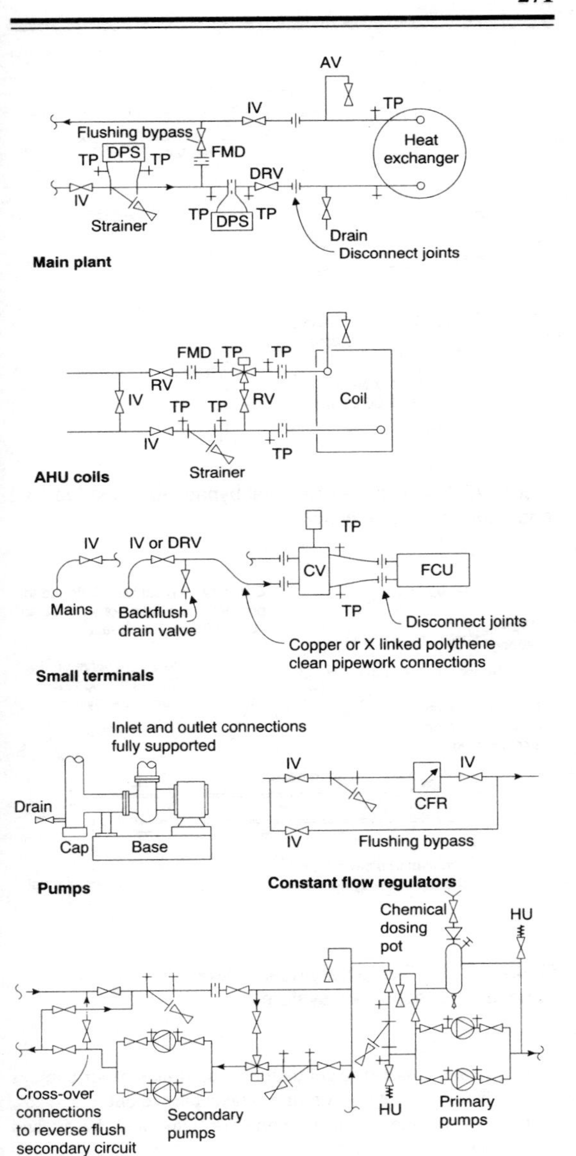

Figure 10.13 System cleaning and commissioning facilities

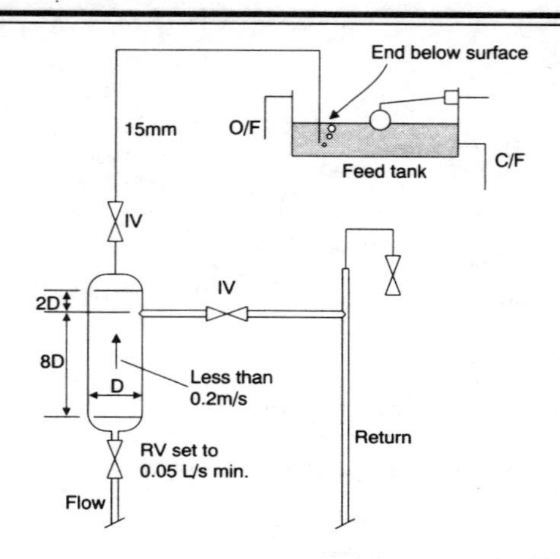

Figure 10.14 Initial air removal bypass for open feed and expansion tank systems

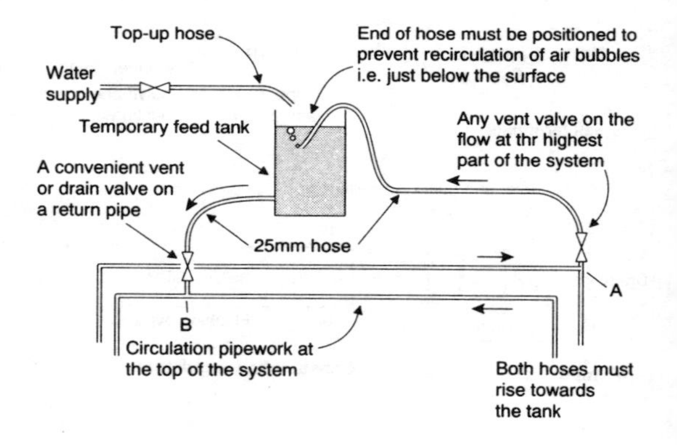

Figure 10.15 Temporary header tank for initial air removal on pressurized systems

(e) Leave circulating continuously until lack of air bubbles indicates that all free air has been purged out.
(f) Isolate and remove temporary tank and hoses and restore pressurizing set.

10.10 FEED ARRANGEMENTS AND SYSTEM PRESSURES

All closed water systems need to be kept full of water and at sufficient pressure to provide anti-flash margins and to avoid sub-atmospheric pressures throughout the system under all operating conditions.

Water in systems expands with temperature increase and contracts when cooling. This change in volume must be accommodated in the feed tank or pressurization set without loss of water or unacceptable change in system static pressure.

Temperature change (°C)							
5–20	10–60	10–75	10–100	10–125	10–150	10–175	10–200
Water volume change (%)							
0.17	1.67	2.6	4.4	6.5	9.1	12.1	15.7

Note: Chilled water volume changes are very small compared with heating water.

Open feed and expansion tanks (see Figures 10.16 and 10.17) provide the simplest effective means of feeding a system causing negligible change in system static pressure and enabling open vents to be used where appropriate, but care is needed to ensure that the make-up level is correctly set and that the feed and vent arrangements prevent any continuous interchange of water between the tank and system as this will introduce oxygen into the system causing venting and corrosion problems. The tank must be high enough to give the system adequate pressure. A pressure switch should be fitted in the cold feed near the system connection arranged to shut the plant down and register alarm in the event of water loss causing a drop in pressure.

Pressurizing sets achieve the same function via a system of gas-filled diaphragm vessels, pressure-switched pumps and spill valves (see Figure 10.18). Their main advantage is that they do not have to be mounted above the highest point of the system.

It is advisable to fit low flow water meters to the make up connection to open feed tanks or pressurizing sets. These record the water make-up to a system over a period of time and give warning of undetected leaks or other faults.

Cold feed connections into heating systems should be into the bottom of the pipe to minimize convective circulation within the cold feed pipe. Similarly cold feed connections into chilled water systems should be into the top of the main.

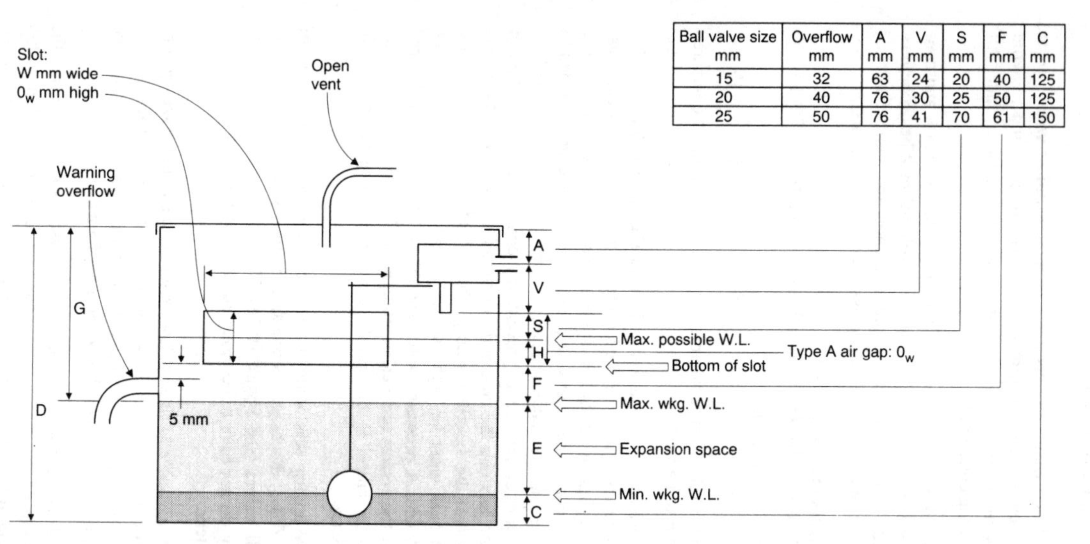

Ball valve size mm	Overflow mm	A mm	V mm	S mm	F mm	C mm
15	32	63	24	20	40	125
20	40	76	30	25	50	125
25	50	76	41	70	61	150

Figure 10.16 Tank with a type A air gap and weir overflow

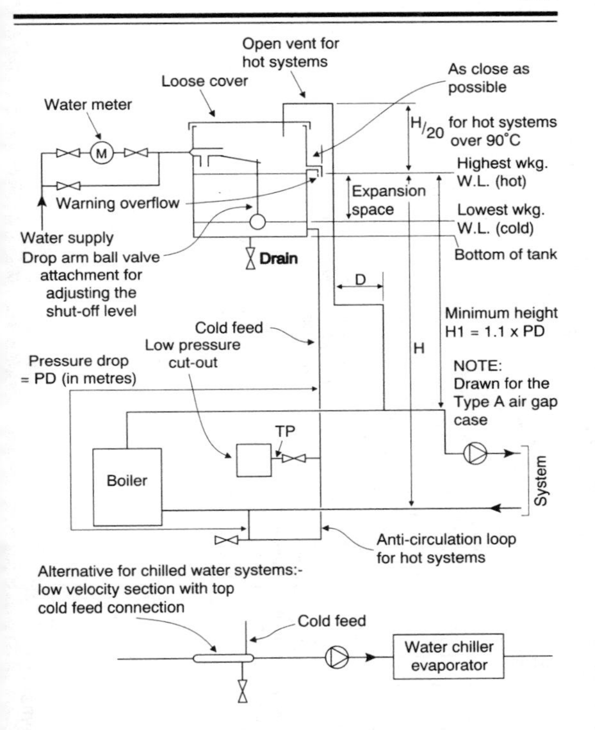

Figure 10.17 Feed and expansion tank connections

10.10.1 Pressure at a point in a system

It is often necessary to examine the pressure at critical points in a system to ensure that they are above atmospheric or do not exceed the design working pressures of equipment or cause open vent discharge.

The pressure at any point in a system can be calculated by taking the pressure of the cold feed at its connection into the system and adding or subtracting the following components as appropriate. It is convenient to use metres wg (1 mwg = 9.8 kPa at 15°C) (1 mwg = 9.64 kPa at 60°C).

(1) Level difference from point to cold feed connection add if point below subtract if above cold feed connection.
(2) Pump head; add only if pump occurs in direction of flow from cold feed connection to point, subtract only if pump occurs in direction of flow from point to cold feed connection.

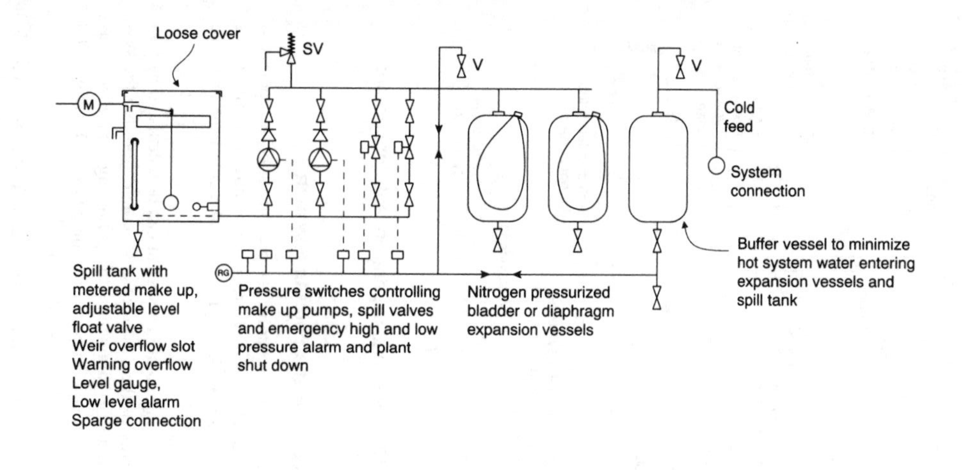

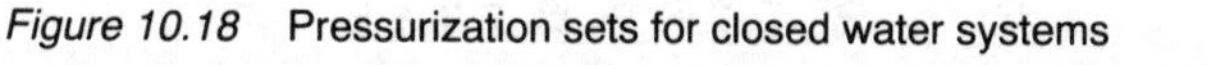

Figure 10.18 Pressurization sets for closed water systems

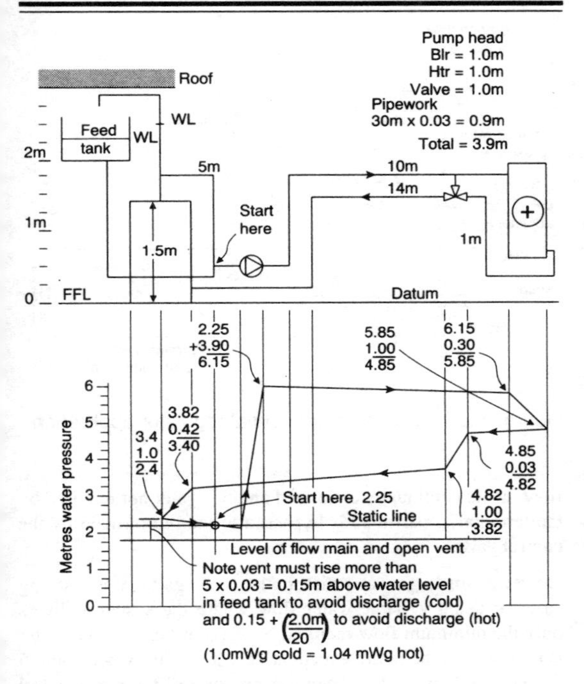

Figure 10.19 Pressure diagram

(3) Pipe and equipment resistance to flow between cold feed connection and point. Add if flow is from cold feed connection to point, subtract if flow is from point to cold feed connection.

It is sometimes useful to show the pressure distribution graphically be plotting only the dynamic changes of pressure starting at the cold feed connection pressure and working round the system in the direction of flow and plotting the pressures relative to a datum level usually the plant room floor. The actual pressure is the vertical distance in metres between the level of the point and the pressure line (see example Figure 10.19).

10.11 CONSTANT FLOW AND VARIABLE FLOW SYSTEMS

Most heating and chilled water distribution systems for single buildings are designed for constant flow where the maximum

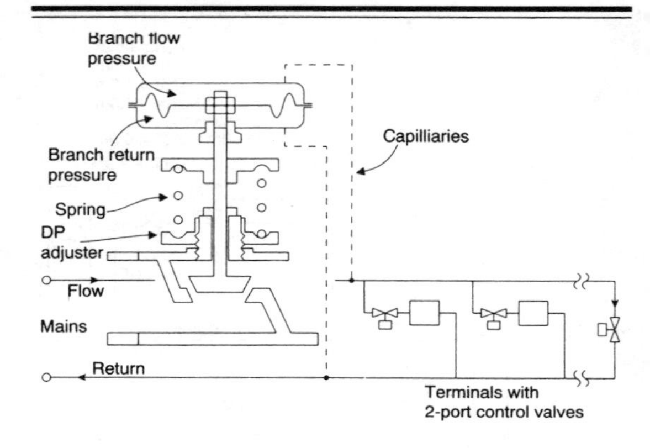

Figure 10.20 Pressure differential regulator application

flow rate is continually circulated and the water not required by the terminal at part load is bypassed to the return main by the control valve.

However, for large water distribution systems there is a strong case for using variable water flow where the system delivers only the minimum flow required by the terminals and no water is bypassed to the return except at very low loads when about 5 per cent is bypassed to keep the pressure differential control valves floating, and enough water flowing through the circulating pumps to prevent them overheating (see Figure 10.20).

10.11.1 Constant and variable flow comparison

	Constant flow	Variable flow
Design work	Normal procedures	Load diversities need to be established. Part load flow rates need to be obtained from terminal characteristics before normal procedures applied
Pump size	Flow rate is sum of terminal peak flows plus margins for balancing inaccuracies	Flow rate is sum of terminal actual flow at design load (typically 50 to 60 per cent of equivalent constant flow)

Pump energy	Full pump power X operating time	Pump power reduces with thermal load (pump energy typically less than 10 per cent of equivalent constant flow system)
Hydraulic control	Fixed water balance by manual or automatic flow regulators	Pressure differential regulators and pump capacity control required
Economics	Lower first cost except with very large systems	Extra cost of design work and hydraulic controls is offset to some extent by smaller pumps, pipes and lower operating costs
Commissioning	Water balancing required unless automatic constant flow regulators are used for terminals	No water balancing required but hydraulic controls need to be set and demonstrated. Full load performance is difficult to demonstrate

10.11.2 Variable flow design notes

In the past variable flow systems have often given problems of noise and poor terminal control. These problems are usually the result of oversized pumps and lack of hydraulic controls to protect the terminal control valves from excessive pressure drops. The assessment of the diversity factors to be applied to the flow requirements of the system is complex and the following procedure is suggested:

(1) The maximum flow requirements of each terminal must be obtained from the terminal manufacturer's selection to deal with the maximum module heat gain or heat loss.
(2) The whole building peak heat gains and heat losses must be calculated allowing for diversity of internal heat gains and the changed solar gains, etc. The whole building heat loss must only allow infiltration heat loss on the facades facing the wind (the leeside facades have no infiltration).
(3) For terminals with two-position water control valves the pump and mains flow rate can simply be calculated: Diversified flow rate

= sum of terminal maximum flow rates × whole building thermal load divided by sum of module peak thermal loads

(4) For terminals with modulating water control valves this calculation will give too high a flow rate as terminals at part load have a much lower flow requirement per watt thermal output than at full load. (See Figure 5.1).
Terminal duties at the time of building peak load should be calculated from the module heat gains and losses but with adjustment to the internal gains to allow for diversity, and heat losses with and without air infiltration losses.
(5) The terminal manufacturer should be asked to select water flow rates required to give these part load terminal duties. Note: this information is often not readily available from catalogue data. Figure 5.1 shows the approximate relationship between water flow and output for a typical low temperature hot water heater.
(6) The pump and mains flow rate is the sum of the terminal flow rates.
(7) The overall flow diversity (OFD) = the pump duty calculated divided by the sum of the terminal flows required for peak terminal duty.
(8) Pipe sizing is carried out as described in Section 10.6 but some judgement is required in deciding the maximum flow rates to be handled by each pipe in the distribution.
The flow rate for pipes connecting each terminal or small group of terminals should be based on the flow rates required for peak terminal duty. The flow rate for pipes feeding one aspect of the building are also likely to be loaded to deal with the sum of the peak terminal duties. For uniform internal loads the associated flow rate could be estimated from a flow diversity for the circuit involved calculated from the following formula:

$$CFD = OFD + (1 - OFD)\,(X - x)/(X - 1) \qquad (10.1)$$

Where CFD = circuit flow diversity factor
OFD = overall flow diversity factor
X = total number of terminals in system
x = number of terminals in circuit

This grades the CFD from OFD at the pump to unity at the terminal. The flow rate for each pipe = number of terminals served × peak flow per terminal × CFD. Unlike constant flow systems the flows are not additive at branches.

Important
This formula is only applicable to the flow rate associated with uniformly distributed internal loads such as heat gains from

people, lights, and small power. The balance of the flow rate i.e. that associated with external environment such as building envelope heat gains and losses should be calculated as appropriate.

The theoretical procedure described is involved and time consuming. For practical design work the following arbitrary method may be considered, it is simple, does not involve additional modular heat gain calculations but does result in oversizing the pump and mains to some extent.

Simplified procedure

1. Calculate module peak and whole building peak thermal loads.
2. Select terminals to meet the module peak loads and obtain the terminal flow/output curves.
3. Assume 50 per cent of terminals are fully loaded. Calculate their output and flow requirement.
4. Assume that the other 50 per cent share the remaining whole building peak thermal output. Select their flow requirement.
5. The pump and main header flow rate is the sum of the terminal flow rates (no additional margin is required).
6. All branch pipework from main headers being sized on the basis that the terminals served are at full load (flow diversity factor 1).

   Note Where different size terminals are involved. The 50 per cent assessed at full load should be a proportionate mix of the sizes.

10.11.3 Comparison of constant flow and variable flow pump duties

If a normally designed constant flow system pump duty is taken as 100 per cent the maximum pump duty for an equivalent variable flow system is reduced by the following factors:

(1)	No margin is required for water balancing inaccuracies typically	–10%
(2)	Diversity factor thermal load whole building load divided by sum of terminal peaks (about 70 per cent) Part load terminal flow per output characteristics (typical 40 per cent flow for 70 per cent output)	–60%
(3)	Terminal manufacturers' flow selection margins (not known) say	–10%
(4)	Design data and performance specifications margins (not known) say	–10%

These margins are indicative and need to be assessed for each application. However, it can be seen that the accumulative effect may easily give a variable flow maximum pump capacity less than half of an equivalent constant flow pump.

10.12 CONTROL VALVES

Control valves for closed water systems are used for several functions:

(i) Terminal thermal output control where the valve regulates the water flow through the terminal in response to a temperature controller (see Figure 5.2). These valves usually have an equal percentage characteristic which gives the controlled terminal an approximately linear output response against valve lift.
(ii) Mixing control valves regulate two flows of water at different temperatures to give a mixed outlet flow at the required temperature (see Figure 1.3) mixing applications require valves with a linear flow to lift characteristic.
(iii) Pressure control valves throttle the water flow to give the required pressure or differential pressure downstream. For this application valves are often direct acting using the water pressure difference across a diaphragm against adjustable spring pressure to move the valve which usually has a near linear characteristic.

10.12.1 Valve authority

A valve can only change the flow rate in a circuit by changing the total resistance of that circuit. It therefore follows that the valve must have a significant resistance in its open position if it is to regulate the flow with its initial movement. It is usual to aim at having the valve about the same resistance as the remainder of the circuit in which it regulates the flow (valve authority = 0.5).

Valve authority = the full open valve resistance/the resistance of the whole circuit (including the valve) in which the valve regulates the flow.

Valves with insufficient resistance will move without significantly changing the flow, and will tend to hunt and give unstable control. Rangibility is a term used to define the proportion of total valve movement over which the valve gives stable regulation of flow. When a valve approaches the fully closed position its flow control becomes unstable.

Cavitation is the usual cause of valve noise (if the system is free of air). Valves with a large pressure drop in near closed position are more prone to cavitation. The valve porting geometry is also a significant factor.

10.13 VALVE CHARACTERISTICS AND SELECTION

10.13.1 Isolation valves

Isolation valves should be able to give a tight shut off after several years of service. They need to be selected for the maximum pressure and temperature of the system with some margin to allow for higher test pressures and operating temperature swings. Up to and including 50 mm size valves are usually of gunmetal or similar alloy with screwed or copper compression ends. 50 mm and larger sizes are more usually cast iron or steel with flanged or mechanical grooved connections. Gate valves have been generally used for water but there is now a growing trend to use ball valves in smaller sizes and butterfly valves in the larger sizes, these quarter-turn valves with resilient seals tending to give more reliable tight shut off with the dirty water conditions commonly found in closed water circuits.

There should be facilities for locking isolating valves in closed or open position. Larger quarter-turn valves should preferably be gear operated to avoid the hydraulic shock possible when lever valves are opened or shut suddenly.

10.13.2 Regulating valves

Regulating valves are used to add resistance to a circuit to achieve water balance. They are usually characterized so that when the memory stop is preset the resistance to flow can be predicted for the required flow rate. Most regulating valves also serve as isolating valves, when re-opened to the memory stop the flow rate obtained should be repeatable, this is unlikely to be achieved if the regulating valve is too large which results in the memory stop being near the closed position. It is important to size the valves to be as near the full open position as selection will permit and they may need to be smaller than line size.

Regulating valves are usually of the screw down Y type with characterized plugs, but in larger sizes butterfly valves are often used.

10.13.3 Instrument or control isolation valves

Instrument or control isolation valves need to give a tight shut off and may also be required to damp out pressure fluctuations. Normal taper plug gauge cocks are not suitable for water systems, they tend to seize and leak. Screw-down needle valves are a good choice.

10.13.4 Constant flow regulators

Constant flow regulators are an alternative to manual regulating valves for constant flow systems giving tight flow control over a wide range of differential pressure. They give a dynamic water balance under varying conditions. Unlike manual regulating valves they should be fitted only to the terminals. They are unstable if fitted in series.

10.13.5 Flow measurement devices (FMDs)

These are usually necessary for confirming or adjusting flow rates through all heat exchangers and pumps. They are often combined with regulating valves for the purpose of manual balancing. (CIBSE Code W gives procedures and advice on commissioning closed water systems.)

They usually take the form of orifice plates, but other types such as averaging pitots, venturis and water turbines are available.

Care must be taken to ensure that the specified straight pipe lengths are fitted upstream and downstream of the device. Low pressure drop orifice plates with screwed connections are particularly vulnerable to serious measuring errors if subject to slight upstream disturbances such as local excess jointing or pipe burr intruding into the pipe bore.

10.14 PUMP CHARACTERISTICS AND SELECTION

Centrifugal pumps are usual for most closed water circuits, Figure 10.21 shows typical characteristics.

10.14.1 Selection guide

Flow margins are necessary to allow for inaccuracies in system water balance and the possibility of system extension or change. Typical margins to be added to calculated flows are 10 per cent for manually balanced water systems and 5 per cent for systems balanced by constant flow regulators. Head margins are necessary to allow for balancing inaccuracies causing flows and hence resistance above design figures. A further margin is necessary to allow for practical contingencies such as additional pipe fittings unforeseen at design stage. Typical margins to be added to calculated resistance are 15 per cent for manually balanced systems and 10 per cent for systems balanced dynamically by constant flow regulators.

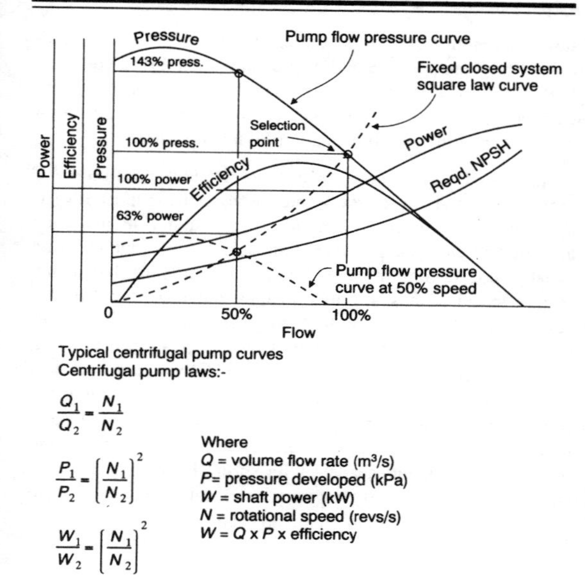

Figure 10.21 Typical centrifugal pump curves

Because variable flow systems are subject to load diversity and rarely operate at full design load it is usual to add little or no margin to the flow and only about 5 per cent to the head.

Pumps should preferably be selected so that the operating point is close to the point of maximum efficiency. For variable flow systems the operating point should be selected to the right of the point of maximum efficiency.

10.14.2 Net positive suction head (NPSH)

This is the pressure drop from the suction connection to the lowest pressure in the impeller. It is also the pressure increase at the pump suction above that at which the water handled by the pump boils.

To avoid cavitation occurring in the pump the minimum suction pressure should be the pressure at which the water would boil (obtained from steam tables) plus the pump manufacturers NPSH read from the pump curves plus a margin of 10 kPa.

The pump should be suitable for the maximum working pressure, maximum temperature and cleanliness of the water handled. At high temperature and/or pressures monoblock construction pumps may not be suitable and a bedplate mounted pump with coupling to motor may be required to relieve the motor bearings from the end thrust and/or high temperature.

Motor speed and balance need consideration. Four-pole AC motors run at about 24 rev/s and are normal for most HVAC applications. Two-pole AC motors run at about 48 rev/s and are smaller and cheaper than four-pole motors but they are usually noisier and require higher standards of balancing. Higher speed pumps are also smaller and cheaper. Standard commercial electric motors have rotors which are statically balanced only, but for any noise sensitive application it is worth the extra cost of having the rotor dynamically balanced as this reduces noise and vibration considerably, often removing the need for mounting the pump set on springs and fitting flexible pipe connections. It is always advisable to support the pump pipe connections and headers with spring hangers adjusted to take the full weight of the pipework and water content to avoid transmitting any weight to the pump casing.

10.15 INSULATION AND VAPOUR SEALING

The British Standard specifies the thickness and K value for insulating chilled water pipes and hot water pipes not used as controlled heating or cooling surface. All chilled water mains and long or exposed runs of heating mains need insulated support clips to avoid the high conduction and convection losses of uninsulated supports. In the case of chilled water under certain conditions condensation will occur on uninsulated supports. Chilled water and cold water mains carrying water at a temperature below the dew-point of the surrounding air must have a continuous vapour barrier insulation finish to prevent the air reaching the cold surface and the water vapour condensing. When condensation does occur within the insulation because of flaws in the vapour seal it will corrode steel pipe and if the insulation is not of closed cell material it will eventually become saturated and ineffective, when this occurs it is impossible to dry it out and the insulation must be replaced.

10.16 PIPEWORK EXPANSION

Data for linear expansion of metal pipes is given in CIBSE Guide Table B16.4[6]. Care must be taken with expansion of plastic pipes which are likely to expand six to twenty times as much steel and which are also seriously derated in working

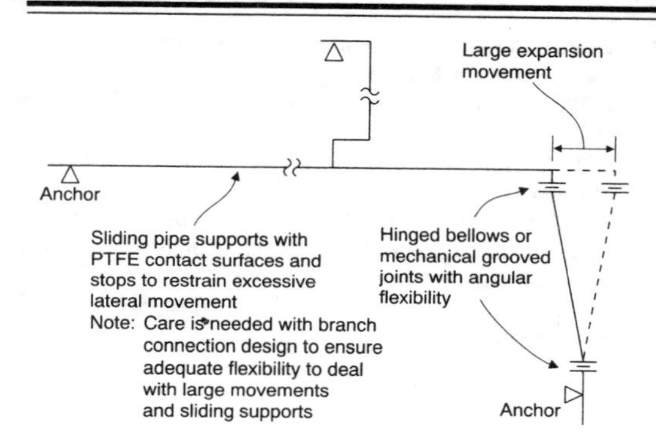

Figure 10.22 Dealing with large expansion movements

stress by modest increases in temperature, see CIBSE Guide Table B16.2 [7].

Expansion movement must be absorbed without causing excessive stress in the pipe material, excessive loads on pipe anchors, excessive strain on equipment connections (heating and cooling coil connections are particularly vulnerable) or causing buckling of long runs of pipework.

Wherever possible expansion movement should be absorbed in the flexibility of pipe offsets and changes in direction CIBSE Guide Section B16 [8] gives procedures and tables for determining the geometry of offsets, calculation of anchor loads, and the position of guides for expansion loops.

In cases where it is not possible to design sufficient flexibility into the pipework other forms of expansion joint are available. Axial bellows joints may be used but these impose high anchor loads and require careful guiding, manufacturers give advice on these requirements.

Large expansion movements from long straight runs of pipework are best absorbed by articulated joints or hinged bellows or mechanical grooved joints with limited angular flexibility (see Figure 10.22). Long runs need guides to prevent buckling. The friction of the pipe supports give significant anchor loads.

References

1. CIBSE Guide Book B Table B1.13 Limiting water velocities in pipework.
2. CIBSE Guide Book C Tables CR.11 to C4.25 Flor of water in pipes.

3. BSRIA Applications Guide 2/89 the commissioning of water systems in buildings.
4. CIBSE Code W Commissioning water distribution systems.
5. BSRIA Applications Guide 8/91 Precommission cleaning of water systems.
6. CIBSE Guide Book B Table B16.4 Expansion of pipes.
7. CIBSE Guide Book B Table B16.2 Comparative properties of piping materials.
8. CIBSE Guide Book B Section B16 Pages B16-3 to B16-11 Pipework.

11 Fire protection

11.1 SPRINKLERS

11.1.1 Introduction

Sprinkler systems are a very reliable form of fire protection. The systems are designed to detect automatically and control or extinguish a fire in its early stages. The history of sprinklers has shown that they are extremely effective in performing this duty. The first automatic sprinkler was invented over 100 years ago. Since that time the design and performance has been constantly improved with many different types, sizes, fusing temperatures and finishes available on the market today.

Many countries have their own standards for the design and installation of sprinkler systems. It is essential, at the outset of a project, to determine the standards with which the sprinkler installations must comply to obtain the necessary Authority approval.

The principal standards applicable in the United Kingdom are issued by the British Standards Institution1 [1] (BSI) and Loss Prevention Council [2] (LPC). Any references to the rules will mean these standards unless mentioned otherwise. American standards are issued by Factory Mutual Insurance [3] (FM) and the National Fire Protection Association [4] (NFPA). They are also used in this country, but to a lesser degree.

Prior to 1969, sprinkler protection in the UK was designed to suit the construction of the building only. The exception being in tanneries and corn mills where the standard design was altered to decrease the spacing of the sprinklers, which had the effect of increasing the density of discharge over the protected area.

From 1969 onwards the approach to the design of systems in the United Kingdom changed. It was considered that the occupancy of the building should play the major part in the design criteria for the design but this should still take into account the construction of the building. These rules were known as the 29th edition Fire Offices Committee (FOC) Rules for Automatic Sprinkler Installations [5].

In 1985, the Loss Prevention Council came into being incorporating the Fire Offices' Committee (FOC), Fire Insurers' Research and Testing Organisation (FIRTO) and Fire Protection Association (FPA).

In 1990 the British Standards Institution issued the revised edition of BS 5306:Part 2 which incorporated fully the content of the 29th edition FOC rules. At the same time the Loss Prevention Council issued their rules which consisted of two parts: the British Standard 5306:Part 2:1990 and a series of Technical Bulletins.

The Technical Bulletins clarify and in some cases slightly modify certain sections of the British Standard to meet the LPC requirements. One of the Technical Bulletins lists equipment that has been tested and approved by the LPC for use in installations designed to meet their rules.

The BSI/LPC rules[1,2] give a full list of definitions in section 2 but the following are a few that will be referred to later in the text.

Assumed maximum area of operation (AMAO) – the design caters for the maximum floor area, in which, it is assumed that possibly all sprinklers could operate in a fire.

Design density – the rate of water application over the protected area.

The units used in the BSI/LPC rules[1,2] are in some cases not standard SI units, but are common practice in the industry.

Pressure – bar (b) or millibar (mb)
Water flow rate – litres/min (1/min) or decimetres^3/min.
Design density – millimetres/min (mm/min) or litres/min m^2.

11.1.2 Design criteria

The nature of the business carried out in a building determines the fire risk and therefore the design criteria for the sprinkler installation. At the outset of a project, it is necessary to discuss and agree the proposed design criteria with the Fire Authority and the client's Fire Insurer. Each will have his own views on the design and may have requirements, additional to the sprinkler rules, in order to meet their approval.

The Water Authorities should be consulted at an early stage in the design to obtain information on the suitability of the towns mains for supplying water to the sprinkler installation. In the majority of cases, the towns mains cannot feed the sprinkler installations directly and are therefore used to supply water to a storage tank and pumping system.

The sprinkler rules classify occupancies according to the fire hazard rating, namely, Light, Ordinary or High.

Light hazard occupancies are mainly non-industrial and of low risk, such as hospitals, museums, offices, etc. There are limitations to the use of this hazard: first, areas of rooms should be no greater than 126 m^2 with half-hour fire resisting walls and, second, there should be no more than six sprinklers in a room. If these limitations are exceeded then the design should be to Ordinary hazard group I criteria.

Ordinary hazard is subdivided into four groups depending on the expected severity of the fire. The storage of goods, materials or even a particular process can increase the fire hazard rating, but only in the area involving the actual storage or process. Storage is listed in categories numbered I to IV, the higher the number the higher the risk. For example, if the business was the manufacture of paper goods, the overall hazard would be considered to fall into the Ordinary hazard group III classification. This would give the following design criteria for general coverage:

Minimum design density 5 mm/min (5 l/min m^2)
Maximum area per sprinkler 12 m^2
Maximum distance between sprinklers 4 m (in one direction only)
Assumed maximum area of operation 216 m^2.

The design density is the quantity of water discharged from a sprinkler in litres/min per m^2 of area covered. The units of design density are in mm/min.

The designs are based on a certain number of sprinklers operating in a fire and is related to the maximum floor area assumed to be involved in a fire and is different for each hazard group.

There will probably be an area where the raw materials or finished products are stored ready for use or despatch. Whether the fire risk will be increased in this area will depend on how the goods are stored, for example, the paper products can fall into three categories of storage:

Category I – Sheets of paper stored horizontally.
Category II – Archival and suspended storage, baled waste paper, rolled pulp and paper stored horizontally.
Category III – Bitumen coated or wax coated paper, rolled pulp or paper stored vertically.

The method of storage, namely, whether it is free standing, stored in blocks, on pallets, or on solid or slatted shelves, will have a bearing on the effectiveness of the sprinklers in wetting the stored materials.

In the example of paper goods manufacture, the Ordinary hazard protection at ceiling level would permit block or free-standing storage to be to the following heights without any alteration to the design.

Category I – 4.0 m

Category II – 3.0 m
Category III – 2.1 m

Above these limits the risk is considered High hazard which means reducing the maximum floor area per sprinkler to 9 m², increasing the density of water application, maximum area of operation and water supplies.

There is a maximum storage height for High Hazard with roof only sprinkler protection:

Category I	Height	7.6 m
	Design density	12.5 mm/min
	Area of operation	260 m²
Category II	Height	7.5 m
	Design density	17.5 mm/min
	Area of operation	260 m²
Category III	Height	7.2 m
	Design density	27.5 mm/min
	Area of operation	300 m²

Consideration should also be given to using a combination of sprinklers in the storage racks with the roof sprinklers. This will lower the design density required from the roof sprinklers which when added to the in-rack sprinklers may still make savings in water storage, pumping and pipework.

It is necessary to obtain the following information, before the sprinkler design can be commenced.

(a) The nature of the company business.
(b) Hazard category (to be determined from the sprinkler rules and agreed with the authority having jurisdiction).
(c) Type and location of any process risks.
(d) Type, height and location of storage racking.
(e) Details of the size, pressure and flow characteristics of the Water Authorities towns mains.

11.1.3 Sprinklers

The sprinkler is a heat actuated device for automatically sensing the heat from a fire and spraying water over a given area. The heat-operated element of the sprinkler is either a glass bulb or a fusible link.

The fusible link is constructed of two metal plates with the surfaces joined by solder. At a predetermined temperature the solder will melt and the plates separate, releasing the valve to discharge the water. A fusible link type sprinkler is shown in Figure 11.1.

The sealed glass bulb element contains a liquid which expands with increase in temperature causing the pressure within the

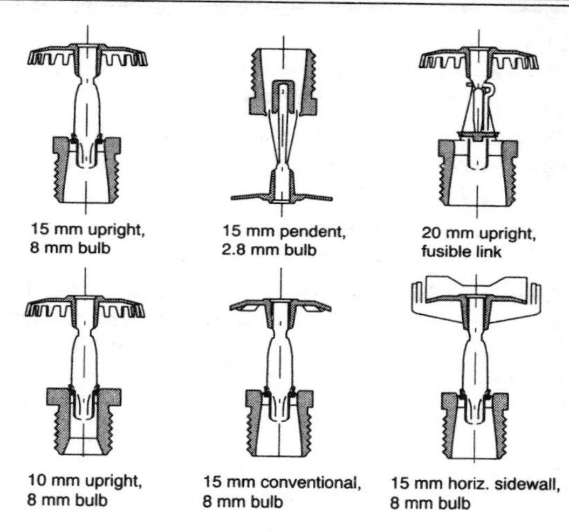

Figure 11.1 Typical sprinklers (courtesy of the Viking Corporation

bulb to rise. At the set operating temperature the bulb will shatter releasing the valve and water.

There is a range of bulbs and fusible links manufactured to cater for the maximum ambient temperatures expected within the protected area. Sprinklers are colour-coded to indicate their operating temperatures, but to prevent the inadvertent operation of the sprinkler, the minimum operating temperature of the selected sprinkler must not be less than 30°C above the maximum expected ambient room temperature.

Over the years research has been done on the time that a sprinkler takes to operate once it reaches its rated temperature and, based on this research, a new range of fast response sprinklers has been developed. Sprinklers can be obtained in both glass bulb and fusible link versions. The fast response bulb can be easily identified because it is slimmer, as can be seen in Figure 11.1, with a 2.8 mm diameter bulb instead of the standard 8 mm diameter.

Once the sprinkler has detected the fire, the water discharged must be at the correct flow rate and the spray pattern over the proper area of protection.

The amount of water discharged from a sprinkler can be calculated using the formula $Q = K\sqrt{P}$ (11.1)

where Q = water discharged in litres/minute
K = a constant for the sprinkler
P = pressure at the sprinkler in bar

The K factors relate to the bore size of the sprinkler and, for the three sprinklers mainly used:

$K = 57 \pm 3$ for 10 mm sprinklers
$K = 80 \pm 4$ for 15 mm sprinklers
$K = 115 \pm 6$ for 20 mm sprinklers

The shape of the sprinkler deflector determines the area of coverage, spray pattern and water droplet size. There are many types of sprinklers designed for different purposes and it is essential that the correct one is installed.

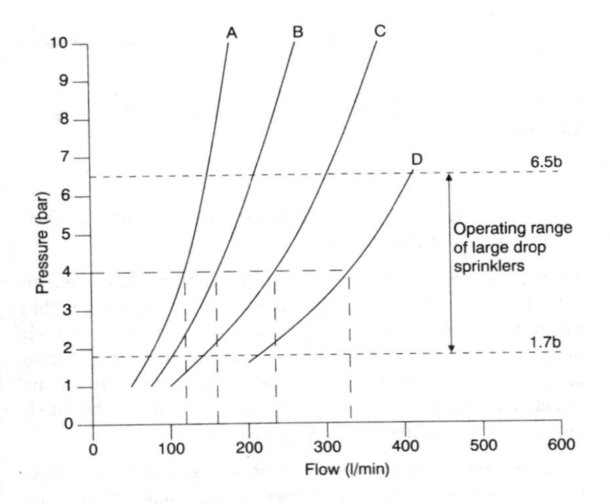

Sprinkler	A	B	C	D
K factor Thread mm	57 10	80 15	115 20	160 20

Sprinkler	A	B	C	D
Pressure (bar)	Flow (l/min)			
1	57.0	80.0	115.0	-
2	80.6	113.1	162.6	226.3
4	114.0	160.0	230.0	320.0
6	139.6	195.9	281.7	391.9
8	161.2	226.3	325.3	-
10	180.2	253.0	363.7	-

Figure 11.2 Comparison of sprinkler discharge rates

Spray sprinkler (Figure 11.1)
The spray pattern from this sprinkler is hemispherical below the plane of the deflector with little water above. These sprinklers can be obtained in pendent and upright versions.

Conventional sprinkler (Figure 11.1)
Water is sprayed above and below the deflector and therefore ceiling wetting is also achieved with this sprinkler. It is used in areas with exposed ceiling steelwork or combustible ceilings. The majority of conventional sprinklers can be used either in the upright or the pendent position.

Sidewall sprinkler (Figures 11.1 & 11.3)
As the name implies, these sprinklers are located along the wall of the protected area. The design of the deflector throws the

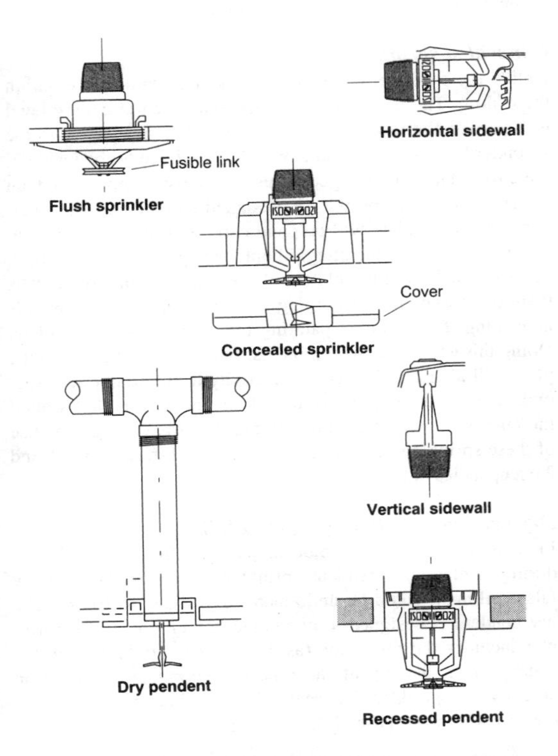

Figure 11.3 Alternative sprinklers (courtesy of the Viking Corporation

majority of the water forward into the room, with a percentage of it backwards onto the wall. These types of sprinkler are only used as a substitute for standard sprinklers in specific cases. For example, in corridors, offices, hoods, or in areas of low headroom where the use of standard sprinklers would be vulnerable. The reason for the restriction in the use of sidewall sprinklers is one of detection of the heat from a fire, rather than the discharge density over the protected area.

Sidewall sprinklers can be obtained for vertical or horizontal mounting.

Flush sprinkler (Figure 11.3)

The flush sprinkler is a spray sprinkler where the body of the sprinkler is above the ceiling with the heat actuating mechanism (fusible link or glass bulb) below the ceiling. The operation of the link or bulb will open the sprinkler, at the same time dropping the deflector below the ceiling line.

Concealed sprinkler (Figure 11.3)

Unlike the flush sprinkler, this sprinkler is completely within the ceiling. The only part of the sprinkler seen at ceiling level is a circular flat cover plate. The cover plate is attached to the sprinkler by a support bracket which is soldered to the inside of the cover plate and arranged to fuse at a lower temperature than the sprinkler. The sprinkler is designed so that, in a fire incident, the cover plate will respond early and fall away prior to the spray sprinkler operating. Should the sprinkler operate first, the water will cool the plate and prevent it falling away. It is therefore essential that nothing must be done to prevent this happening, for example, painting the ceiling and cover plate. Doing this can cause two problems: first, a layer of paint on the plate will alter the thermal conductivity to the solder and, second, there is the possibility that the paint will harden around the circumference of the plate fixing it to the ceiling. The use of these sprinklers are limited to Ordinary Hazard Group 1 and 2 occupancies only.

Dry pendent sprinkler (Figure 11.3)

In areas which are likely to be subjected to freezing conditions during winter, any pendent sprinklers (such as those below false ceilings) would be undrainable and likely to freeze. The dry pendent is designed to overcome this problem. The sprinkler includes the drop pipe (as part of the sprinkler) with the water seal at the top of the pipe preventing the water from entering the sprinkler. The seal is kept in place by a linkage to the sprinkler at the base of the drop pipe.

The K factor of the sprinkler plus the drop pipe (providing it is not too long), is compatible with a standard 15 mm sprinkler and therefore can be used in place of a standard sprinkler.

These sprinklers can be used in either wet, dry, or alternate wet and dry systems.

Large drop sprinklers

The use of this type of sprinkler has been increasing over the years to protect storage from roof only. As the name implies the largest percentage of its spray pattern consists of large water drops. These large drops are more able to penetrate the updraft from this type of fire and, can be very effective. The shape of the spray pattern is similar to the spray sprinkler. It is a 20 mm sprinkler with a large K factor, in the order of 160 (see graph D Figure 11.2). This is a special application sprinkler designed for a specific purpose which, because of its large K factor should not be used just as a large orifice sprinkler to supply an increased amount of water to a risk. The siting and pressure at the large drop sprinkler is more important than the design density.

The rules covering the design and installation of these sprinklers are issued by the National Fire Protection Association [6] and Factory Mutual Insurance [7].

The minimum and maximum pressures at the sprinkler are 1.7 bar and 6.5 bar, respectively, but the above rules should be consulted for the requirements for a particular risk.

As the discharge from one of these sprinklers is twice that of a standard 15 mm sprinkler, careful consideration must obviously be given to the layout and hydraulic calculation of the pipework.

Multiple jet controls

The multiple jet control is a valve, operated by a fusible element of the sprinkler type. The fusible element is usually a glass bulb which, through a linkage, holds the valve shut. The outlets are therefore dry until the valve operates and sends water to open sprinklers, high velocity nozzles or medium velocity sprayers.

Multiple jet controls are used in the following applications:

(i) In electrical switch room, lift motor rooms, etc., where water would be dangerous if sprayed on equipment, but where an alarm is required from the sprinkler system and where water can be discharged externally to protect the enclosure. In these instances, the controls within the room are spaced in the same way as sprinklers, for heat detection purposes.

The outlets from each control are connected together to

discharge through open sprinklers in order to create a water curtain around the outside of the room, if the construction is non-fireproof. If the construction is fireproof, it is only necessary to protect doors, windows, and any other openings.

(ii) To control small groups of open nozzles or sprayers. The controls are spaced relative to the risk, with the nozzles positioned in order to cover completely the equipment being protected.

Examples of this type of protection are: oil fired boiler fronts, oil pumps, process risks, cooling of small chemical tanks, protection of diesel engines, etc.

Controls are classified by the inlet sizes and can be obtained in 20, 40, 50 and 80 mm diameters. The maximum water velocity through a control is in the order of 4.6 m/s.

A later development of the multiple jet control is the addition of an electronic actuator enabling the control to be triggered also by an electrical detection system.

11.1.4 Types of systems

There are several different types of sprinkler system and each type is designed for a specific application. Those most widely used are either the wet type or the alternate wet and dry type.

Wet system (Figure 11.4)

This type is the simplest of all the systems. The pipework is permanently charged with water and therefore instantly available when a sprinkler fuses in a fire situation.

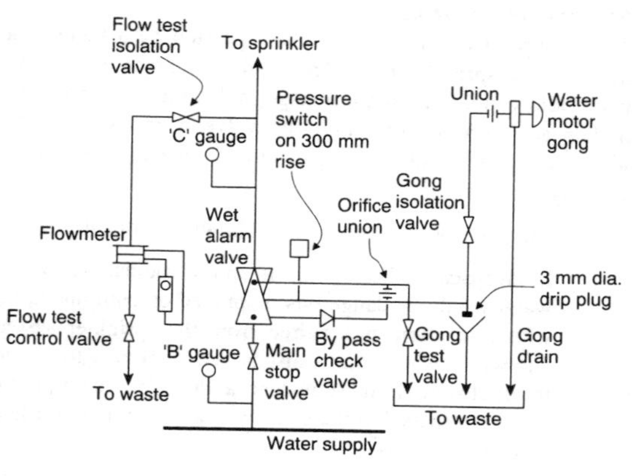

Figure 11.4 Wet system control valves

A wet system is used only in areas where there is no danger of pipework freezing during the winter months. When the majority of a building is heated and only a small proportion subject to freezing, for example, a loading bay, then only this area would be protected by a tail end air valve. The remainder of the building will be on the wet system.

Figure 11.4 is a schematic of a standard set of wet installation control valves. They consist of a main stop valve with a wet alarm valve immediately above. The alarm valve is a specially designed check valve with an additional outlet port piped to an external water motor alarm gong. When the water flows into the installation, the valve clapper lifts and at the same time allows a small percentage of water to be diverted to drive the water motor alarm gong. To give an electrical indication of system operation a pressure switch is connected to the pipe feeding the water motor gong.

Sometimes, the water supplies to the sprinkler system are subject to pressure fluctuations. When this happens false alarms can occur due to the valve clapper lifting momentarily and allowing water to the gong. To prevent this happening on small pressure surges, the alarm valve has either a built-in compensator or a small by-pass pipe as shown in Figure 11.4. However, sometimes the pressure surge is greater than the compensator can cope with and the false alarm still persists. This problem can be reduced by the introduction of a retard chamber in the alarm pipe. This item of equipment increases the capacity of the alarm gong feed pipe and therefore more water has to pass through the alarm valve port before reaching the gong. If the duration of the surge is not long the clapper will reseat, stopping further water entering the alarm pipe. The alarm pipe will automatically drain through a 3 mm drip plug into a tundish.

If the pressure fluctuations of the water supply are so severe that they cannot be overcome using a retard chamber, a small anti-false alarm pump is sometimes fitted to increase the system pressure above the expected surge pressure, but doing this will also delay the operation of the alarm under normal conditions.

It is important to remember that any delay built into the system to overcome false alarms will also delay the alarm in a fire situation. It is therefore essential to discuss and agree any proposal to overcome false alarms with the relevant Authorities/ Insurance Company.

The limitation on the maximum number of sprinklers on one set of wet installation control valves is 1000.

Alternate wet and dry system

Where a wet system cannot be used because there could be a danger of freezing during the winter months, an alternate wet and dry system is used. During the summer months, the pipework is filled with water and behaves in the same way as a wet system, but during the winter the pipework is drained and filled with compressed air.

When the sprinkler operates the compressed air is released: the controlling air valve opens, allowing water flow into the system and out through the open sprinkler head. It can be seen that a delay will occur between the sprinkler operating and the water attacking the fire.

To reduce this delay, the capacity of an alternate system is smaller than a wet system. The maximum number of sprinklers permitted on this type of system is 250, but the number can be doubled to 500, by fitting an accelerator to the air valve. This piece of equipment senses the pressure drop when the sprinkler operates and trips the air valve, allowing the system to charge with water more quickly.

Figure 11.5 is a schematic showing a typical arrangement of an alternate wet and dry set of control valves. Some manufacturers have a combined wet/dry valve, but in the schematic it is shown as two separate valves for clarity. If Figure 11.5 is compared to Figure 11.4, it will be seen that the alternate control valves are a set of wet control valves with an air valve bolted to

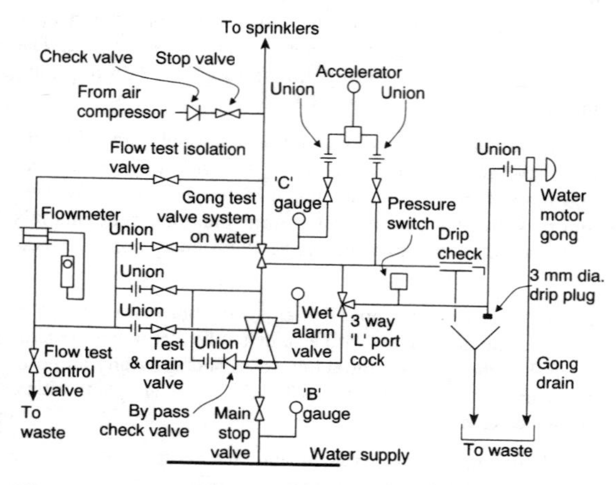

Figure 11.5 Alternate wet and dry system control valves

the outlet of a wet alarm valve. Details of the small bore trim pipework to make the control valve set work correctly are supplied by the valve manufacturer.

The alternate system is set for winter operation as follows. The complete pipework is drained of water and the air valve latched shut. The sprinkler pipework is charged with compressed air. The air pressure depends on the water to air ratio of the valve which is usually about 5 or 6 to 1 and is the tripping point for the valve. The actual air pressure in the system is set at a value high enough to ensure that the valve does not trip in error. When the air pressure is at the correct level, the main stop valve is slowly opened, water pressure applied to the inlet of the air valve and the system is set. Any small losses in air pressure are topped up by a small compressor, controlled automatically by a pressure switch.

Dry systems

The dry system is used in areas where temperatures are permanently below freezing or above 70°C. Unlike the alternate system, the system is permanently charged with air and not changed from air to water (and vice versa) every six months. It is therefore necessary to exercise the dry valve every six months to ensure its correct operation when required.

The maximum number of sprinklers permitted for a dry system is the same as that for an alternate wet and dry system.

Tail-end air systems

As previously mentioned, in buildings where wet systems are installed, there are certain areas that are likely to freeze during the winter. These areas can be catered for by using a tail-end air system. It operates exactly the same as the alternate system, that is, on air in the winter and on water in the summer (see Figure 11.6).

The difference between the alternate and tail-end systems is that the alternate is a complete stand-alone system, whereas the tail-end is an addition to a wet system.

As it is part of a wet system, the number of sprinklers controlled by a tail-end air valve must be included as part of the maximum of 1000 referred to earlier. The maximum number of sprinklers for a tail-end air valve must not exceed 100, with not more than 250 in total on tail-end air valves in a wet system.

Pre-action systems

The pre-action system is a sprinkler system working in conjunction with an electrical detection system. The electrical detection system is designed in accordance with BS 5839 Part 1 [8].

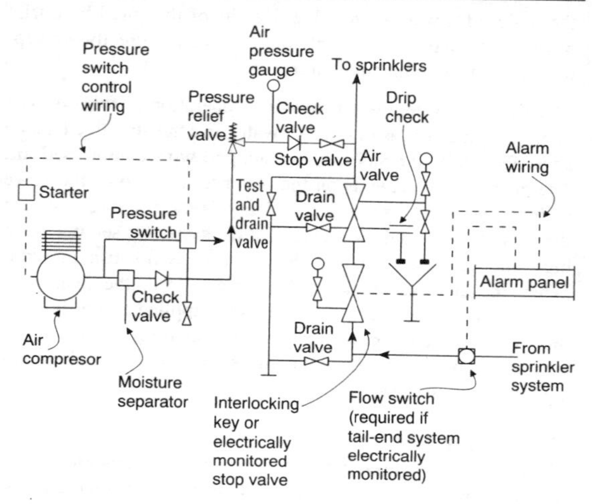

Figure 11.6 Tail-end air valve

The operation of an electrical detector will signal the control panel to open the system control valves and charge the pipework with water. This type of system will discharge water from an opened sprinkler quicker than from a dry or alternate wet and dry system.

There are two types of pre-action system listed in the LPC rules [1,2].

(i) Type 1. This system is filled with air to monitor the completeness of the system and indicate any damage to the sprinklers or pipework. The pipework will only be charged with water following the operation of the electrical detection system.

(ii) Type 2. This type uses an electrical detection system in conjunction with an alternate wet and dry or dry system. The purpose is to operate the control valves earlier than the standard systems described earlier. However, they will still behave as a standard system if a sprinkler operates prior to the detector.

The maximum number of sprinklers permitted on one installation is 500 for Light Hazard and 1000 for Ordinary and High Hazard systems.

Recycling pre-action systems (Figure 11.7)

Where a sprinkler system is the best form of protection for a risk but it is desirable to limit the water discharge as the fire is

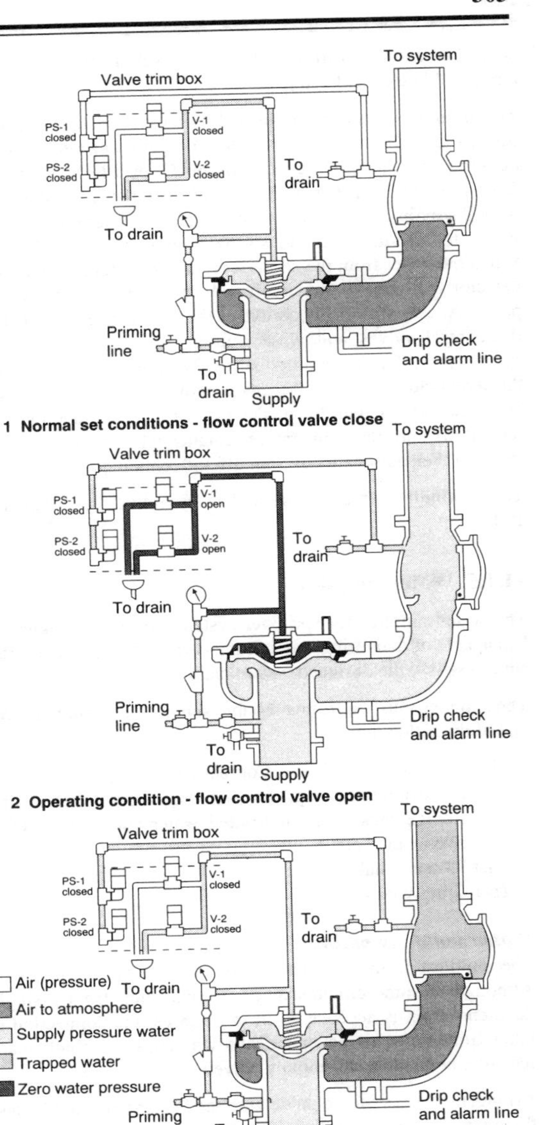

Figure 11.7 Recydling pre-action system (courtesy of the Viking Corporation

controlled/extinguished, the use of a recycling pre-action system should be considered [9].

The sprinkler system is controlled by an electrical heat detection circuit. The detectors are rated to operate at a lower temperature than the sprinkler, usually 60°C for a 68°C sprinkler. Therefore, initially the detector will (via the control panel) open the main control valves and charge the system with water. As the temperature rises, the sprinkler above the fire will open. When the water from the sprinkler reduces the temperature, the detector will remake the circuit and start a timer in the control panel. At the end of this period, if the temperature is still low the control valves will close shutting off the supply to the sprinklers. If during the timer period (between 1 and 5 minutes) the temperature should rise again above 60°C the water will continue and not shut down. The system will continue to cycle on and off until the temperature remains below 60°C. The system will then automatically shut off and remain off.

The maximum number of sprinklers permitted on a system is 1000.

11.1.5 Water supplies

The water supplies for sprinkler systems must be designed to be capable of delivering water at the correct flow rate and pressure to satisfy the designed hazard classification.

The supplies must be from a reliable source and should be from one of the following:

(i) Direct from the Water Company main.
(ii) Duplicate fire pumps drawing water from either a water storage tank or an unlimited water source (e.g. lake, river, canal, etc.).
(iii) Pressure tank.
(iv) Gravity tank.

Water company mains

The simplest form of supply is direct from the town mains. Although in some circumstances, a connection from a single dead-end main is acceptable, in most cases, the connection is taken from either two independent town mains or a single main fed from both ends and suitably valved.

Arrangements have to be made with the Water Company to test the mains pressure and flow characteristics. This is done using two fire hydrants on the main being considered, if possible, one on each side of the proposed connection to the sprinkler installation.

The tests are carried out using a standpipe screwed to each

hydrant. One standpipe, is connected to a flow meter and the other to a pressure gauge. The first hydrant is opened to give a series of flowrates to suit the requirements of the hazard. At each flowrate the corresponding pressure is recorded at the second hydrant. On completion of the set of tests, the position of the flow meter and pressure gauge is reversed and the tests repeated with the water flowing from the second hydrant. Finally, the static pressure in the main is recorded with no flow. Tests should be carried out at the maximum draw-off time of day.

If the pressures obtained during tests are just above the hazard requirements, there is a strong possibility of the pressure being too low in a fire incident (especially if the Fire Brigade have to use road hydrants) and, consequently detrimental to efficient protection. In this case, consideration should be given to the installation of a tank and duplicate fire pumps.

Tank and pumps (Figure 11.8)

If the town's mains are not adequate to supply the sprinkler systems directly. They can be used to feed a water storage tank which in turn will supply duplicate fire pumps (one main pump and the other the standby). Either pump by itself must be able to give the performance characteristics required to suit the system hydraulic calculations.

(i) Main sprinkler pump. The main fire pump, in the majority of cases, is electrically driven from a dedicated electrical supply. This means that the electrical supply must always be available to the pump motor, although power may be switched off to the remainder of premises, in a fire situation.

(ii) Standby sprinkler pump. If a second completely independent electrical supply is available, the standby pump can also be electric driven. In this case, both power supplies must be available to each pump through an automatic changeover facility.

If a second electrical supply is not available then the standby sprinkler pump will be diesel-driven.

If there is no dedicated electrical supply available for the main pump then both pumps would be diesel-driven.

(iii) Pump control. The control of sprinkler pumps is arranged so that they should start automatically when the pressure in the system drops. The pumps should run continuously until stopped manually at the starter/controller.

It must be possible for each pump to operate independently of the other. Should the main pump fail for any reason then the standby pump must be ready. For this reason, the control wiring for each pump is kept completely independent.

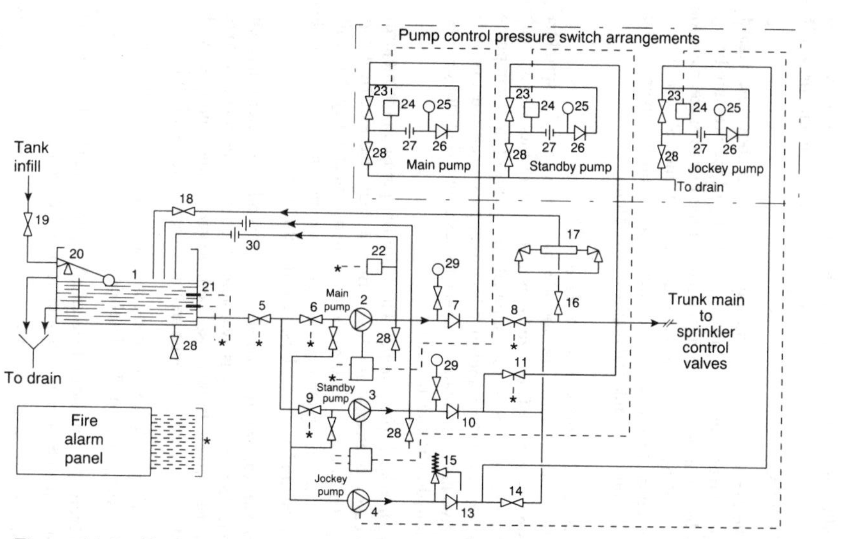

Figure 11.8 Typical pump house (positive head)

Each pump pressure switch arrangement consists of the following components (see Figure 11.8):

23 – pressure switch assembly control valve
24 – pump control pressure switch
25 – pressure gauge
26 – non-return (check) valve
27 – orifice union with a 3 mm diameter hole (non-ferrous plate)
28 – drain valve.

To test the pump manually, close valve (23), gradually open valve (28) to drop the pressure on the switch (24). The pressure at which the pump starts is recorded using gauge (25). When the pump starts, close valve (28) and reopen valve (23). To overcome the danger of the pump not starting automatically, if valve (23) was left closed in error, a non-return valve (26) is installed which bypasses valve (23) and allows the pressure below the switch to drop as the system pressure falls. If valve (23) is not reopened then the pump will restart as soon as the stop control is released, because the pressure is not restored to the switch.

If the pump runs for a long time against closed valve conditions it will overheat. To prevent this happening, the pumping system is designed to discharge a small proportion of water to waste. This can be achieved either by returning a small pipe (complete with orifice union) (30) from the pump delivery back to the storage tank, or by a diaphragm valve fitted between the pump suction and delivery which operates when the pump starts and discharges water to waste. A diesel-driven pump uses this water in the diesel engine cooling system before being discharged to waste.

(iv) Jockey pump. A small jockey pump is fitted to overcome the sprinkler pumps starting unnecessarily because of small losses in the system. Unlike the sprinkler pumps, the jockey pump can cut in and cut out automatically to maintain the system pressure above the cut-in pressure of the main pump.

The closed valve pressure of the jockey pump should be noted, if this pressure is above the maximum allowed in the system then a pressure relief valve should be fitted before the delivery check valve. The reason for this is that if the pressure switch control valve (23) is left closed the pump will continue pumping and not shut off automatically.

(v) Pump testing. The majority of the building services pumping plant is in daily use and therefore any fault will immediately show itself. Obviously this is not the case with sprinkler pumps and it is therefore essential to test the pumps every week to ensure that in the event of a fire they will oper-

ate correctly. This is done using the manual procedure detailed in (iii) above.

The duties of each pump are checked periodically using flowmeter (17) and control valve (18). The control valve (18) is opened until the required flowrate is indicated on the flowmeter and the corresponding pressure on gauge (29) is recorded.

11.1.6 Sprinkler locations

Each sprinkler is a self contained unit and does not rely on the others for its operation.

Sprinklers have to perform two functions, first, they must detect the heat from the fire and, second, they must be able to distribute water over the protected area. Therefore, it can be seen that the correct spacing and positioning of the sprinklers is very important.

All the following will have an effect on the correct spacing of the sprinklers.

(a) length and breadth of the protected area;
(b) the maximum area per sprinkler permitted for the hazard;
(c) shape of the ceiling, namely, flat, curved, sloping, etc.;
(d) roof/ceiling beams, girders, trusses, etc.;
(e) floor to ceiling partitions;
(f) ceiling tile module (where false ceilings are installed);
(g) location of other services, e.g, lighting, ductwork, banks of pipework, etc.

The distance of the sprinkler below the ceiling/roof will have an effect on the time for the sprinkler to detect the heat from the fire. The most sensitive distance is in the zone where the sprinkler deflectors are between 75 and 150 mm below the ceiling/roof. Desirably, the sprinklers should be located within this zone. When this is not possible, the LPC Rules permit the distance to be increased to a maximum of 300 mm below a combustible ceiling or 450 mm below a non-combustible ceiling. The exception is with open joisted ceilings or combustible roofs with common rafters exposed, where the distance must not exceed 150 mm below the underside of the joists or rafters.

For each hazard, the LPC Rules also stipulate the maximum area per sprinkler and the maximum distance between sprinklers, for example:

Hazard	Area/sprinkler (m²)	Max. distance between sprinklers (m)
Light	21	4.6
Ordinary	12	4.0 (standard spacing) 4.6 (staggered spacing)
High	9	3.7

The above maximum distances can only be used in one direction. For example, if the maximum distance for an Ordinary Hazard is 4.0 m, then the maximum distance in the other direction is 3.0 m (12 m² divided by 4.0 m). Some examples of spacing distances are shown in Figure 11.9. If these maximum distances are exceeded the sprinklers will be overspaced. The maximum distance from any wall is half the design spacing (see Figure 11.10).

The minimum distance between any two sprinklers should not be less than 2.0 m, otherwise the spray from the first sprinkler to operate will wash the adjacent sprinklers and, by cooling, prevent their operation. Should it be necessary, for any reason, to install sprinklers closer than 2.0 m apart, then a suitable baf-

ORDINARY HAZARD
Area per sprinkler 12 m²

Distance between sprinklers and ranges

*4.60 × 2.609	*4.20 × 2.857	3.80 × 3.158
*4.55 × 2.637	*4.15 × 2.892	3.75 × 3.200
*4.50 × 2.667	*4.10 × 2.927	3.70 × 3.243
*4.45 × 2.697	*4.05 × 2.963	3.65 × 3.288
*4.40 × 2.727	4.00 × 3.000	3.60 × 3.333
*4.35 × 2.759	3.95 × 3.038	3.55 × 3.380
*4.30 × 2.791	3.90 × 3.080	3.50 × 3.429
*4.25 × 2.824	3.85 × 3.117	3.46 × 3.468

*Extended distances for staggered spacing

HIGH HAZARD
Area per sprinkler 9 m²

Distance between sprinklers and ranges

3.70 × 2.432	3.45 × 2.609	3.20 × 2.813
3.65 × 2.466	3.40 × 2.647	3.15 × 2.857
3.60 × 2.500	3.35 × 2.687	3.10 × 2.903
3.55 × 2.535	3.30 × 2.727	3.05 × 2.951
3.50 × 2.571	3.25 × 2.769	3.00 × 3.000

Figure 11.9 Example of design spacing distances

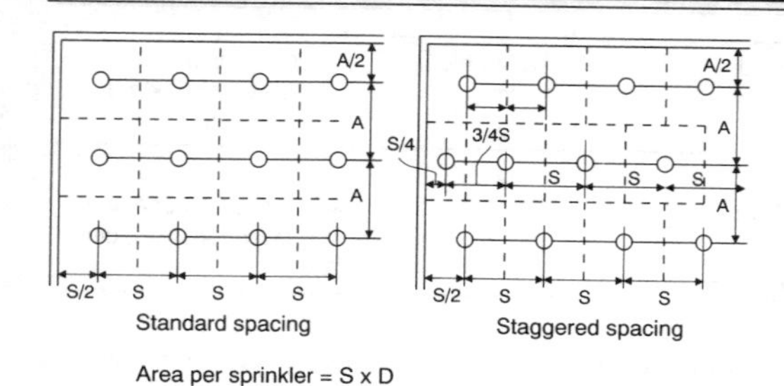

Note: Staggered spacing can only be used in Ordinary hazard
The extended distance to 4.6 m can only be along the range 'S'

Figure 11.10 Sprinkler spacing

fle plate must be installed midway between the sprinklers. The minimum size of baffle is given in the LPC Rules as 200 mm wide by 150 mm high. There can, of course, be other forms of baffle, for example, surface mounted lighting, dropped ceiling bulkheads, etc.

The maximum dimensions between sprinklers and ranges are not always possible due to structural and/or building services obstructions at ceiling level. For instance, the position of structural beams often determine the spacing between sprinkler ranges. The minimum distance from the side of a beam will depend on the type of sprinkler being used and height of the sprinkler deflector above the beam soffit (see Figure 11.11). If the beam is so deep that these dimensions cannot be achieved then it is treated as a wall for the purposes of spacing the sprinklers.

It will be seen later in 11.1.7(b) that the area per sprinkler will determine the water supplies required. The sprinkler rules[1,2] stipulate the minimum pressure permitted at any sprinkler in the area of operation. This means that there is a minimum flow rate from each sprinkler, regardless of the area it covers. The graph in Figure 11.12 shows that it is impossible to achieve a 5 mm/min (5 l/min m^2) density from a 15 mm (K factor 80) sprinkler covering an area less than 9.47 m^2. The smaller the area, the greater the density. Also, to obtain 5mm/min density from a sprinkler covering the maximum 12 m^2, the minimum pressure has to be raised to 0.563 bar.

11.1.7 Pipe layouts

The sprinklers are connected into pipes called ranges. The pipes feeding the ranges are called distribution mains.

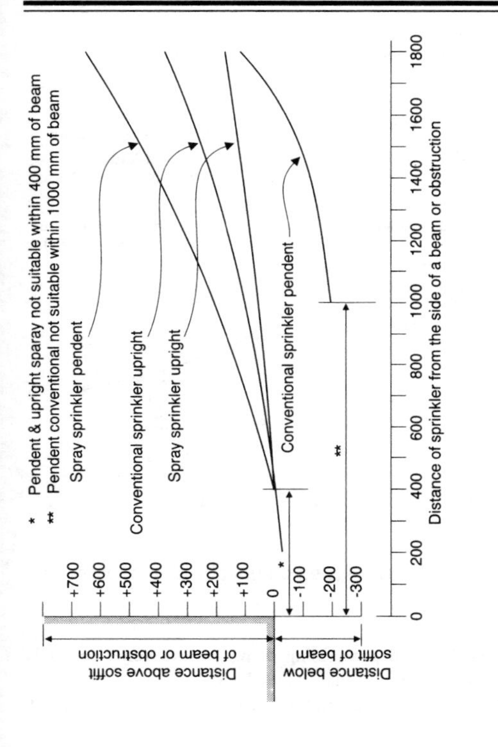

Figure 11.11 Sprinkler location relative to beams

There are three types of pipe layouts described in the sprinkler rules, namely, terminal mains, looped mains, and gridded pipework. Typical examples of pipe layouts are shown in Figure 11.13.

(a) Terminal mains. With this pipe arrangement, there is only one route that water can take from the supply, through the mains and ranges to the open sprinklers.

(b) Looped mains. Terminal ranges are connected into mains, which are looped, therefore, the water can take two paths to the operating ranges.

(c) Gridded mains. Every range is connected at both ends to the mains forming a grid of pipework. With this arrangement the water will flow through all the pipework to the operating sprinklers.

Sizing Methods

There are two methods of sizing sprinkler pipework to comply with the sprinkler rules [1,2].

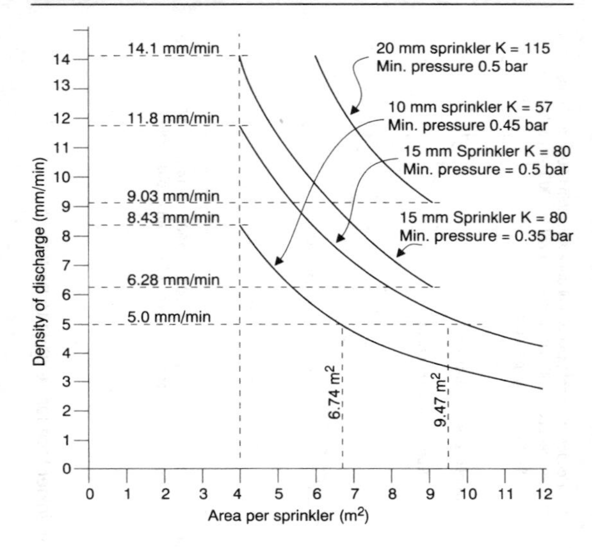

Figure 11.12 Effect of area on density at min. pressure

(a) Precalculated pipe sizing using tables, which can only be used with terminal mains designs.

(b) Full hydraulic calculation, which can be used with all types of design shown in Figure 11.13.

(a) Precalculated pipe sizing. To use this method, an area enclosing a designated number of sprinklers (the number depends on the hazard) is located at the end of very main. The number of sprinklers in these terminal areas are 16 or 18 for Ordinary Hazard and 48 for High Hazard. The point on the main where it enters this area is called the 'design point'.

The mains pipework between the design point and the end of the main, plus all the range pipes throughout the building will be sized to the tables in the sprinkler rule book [1,2].

The mains from each design point back to the control valves are sized by hydraulic calculation, using a specified flow rate (depending on the design density of the risk).

An example of an Ordinary Hazard (OH) project covering two floors is shown in Figure 11.14. There are two design points on each floor. Each design point feeds 18 sprinklers. The maximum permitted pressure loss for this hazard, between any design point and the control valves is 500 mb at a fixed flow rate of 1000 l/min (1000 dm^3/min). The loss per metre in

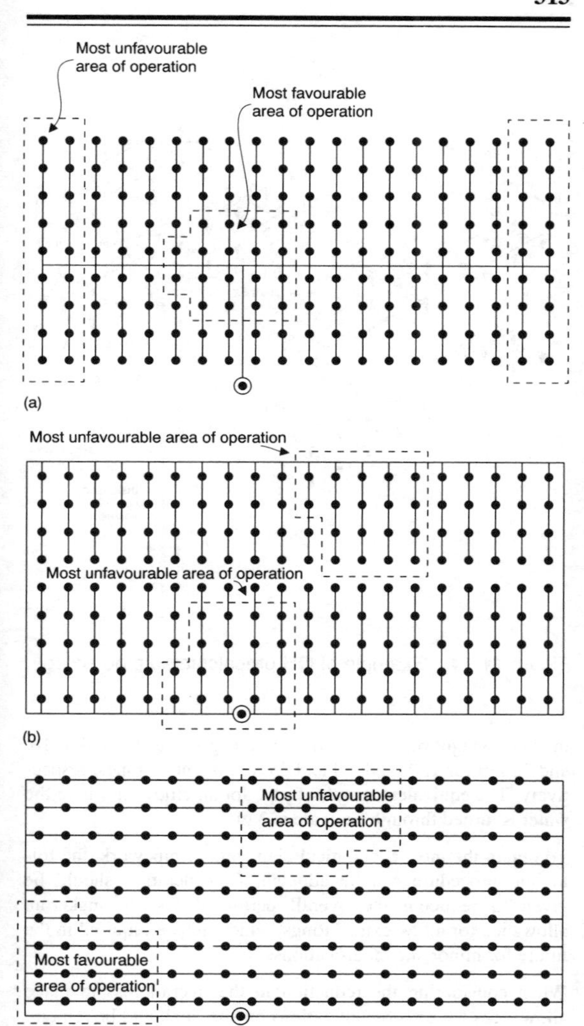

Figure 11.13 Pipework layout. (a) terminal main; (b) looped main; (c) gridded layout

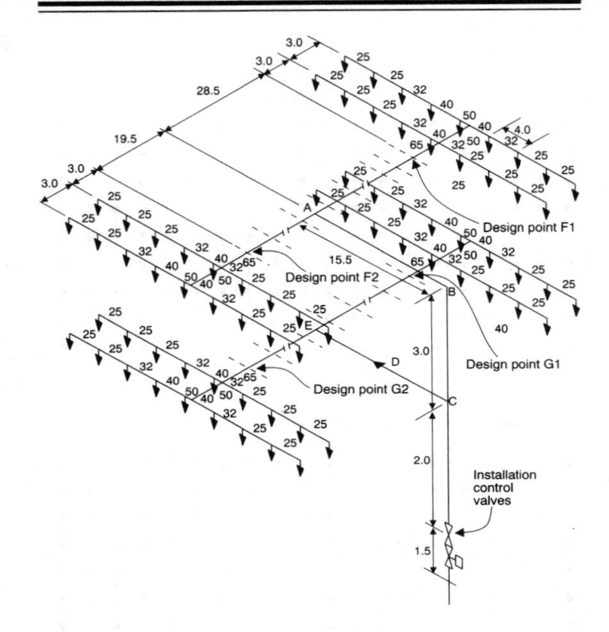

Figure 11.14 Example of OH precalculated pipe sizing

medium weight pipe at this flow rate is given as 0.65, 4.4, 16, and 35 mb for 150, 100, 80 and 65 mm diameter pipes, respectively. The equivalent length of pipe for an elbow or where the water is turned through an angle is 3 m.

To obtain the most economical diameters of pipework, the following procedure can be adopted. Consideration should be given to reducing the overall permitted loss to make an allowance for a few extra fittings, which maybe required in the future for minor site modifications.

When considering the reduction in the overall loss to make allowances for possible alterations on site, it should be remembered that the loss in one bend (i.e. 3 m of pipe) would be:

65 mm	bend	=	105 mb
80 mm	bend	=	48 mb
100 mm	bend	=	13.2 mb
150 mm	bend	=	1.95 mb

Start with the longest route from the furthest design point to the installation control valves. This main should be run in two diameters only. If a third diameter is introduced, then to compensate for every extra metre of the smaller diameter of pipe used, a considerably greater length of the largest pipe is required. This can be seen from the following example.

The total length of pipe from design point F1 to the control valves is 28.5 m + 15.5 m + 3.0 m + 2.0 m + (2 bends = 6 m) = 55 m. If a loss of 450 mb is considered adequate to allow for any future modifications, first try using 80 mm diameter pipe only. This has a pressure loss of:

$$55 \text{ m} \times 16 \text{ mb/m} = 880 \text{ mb, which is too high}$$

Next try using all the pipe in 100 m diameter. This has a loss of 55 m × 4.4 mb/m = 242 mb, which is very low. Hence the most economical sizing will consist of a combination of 100 mm and 80 mm pipe. The actual lengths can be calculated by using a simplified simultaneous equation.

$$y = \frac{(B \times L) - (T + s)}{B - A} \tag{11.2}$$

where y = length of larger pipe (m)
L = total length of pipe route including the equivalent length of bends (m)
A = pressure loss in larger pipe (mb/m)
B = pressure loss in smaller pipe (mb/m)
s = static difference (mb)
T = total design loss (mb).

Therefore in this example:

$L = 55$ m $A = 4.4$ mb/m $B = 16$ mb/m $T = 450$ mb $s = 0$

$$y = \frac{(16 \times 55) - (450 + 0)}{16 - 4.4}$$
$$= 37.07 \text{ m of 100 m pipe}$$

Amount of 80 mm pipe = 55 m - 37.07 m = 17.93 m. The ranges are 3.0 m apart therefore 6 ranges would be:

$$\begin{aligned} 18 \text{ m (of 80 mm pipe)} \times 16 \text{ mb/m} &= 288 \text{ mb} \\ +37 \text{ m (of 100 m pipe)} \times 4.4 \text{ mb/m} &= \underline{162.8 \text{ mb}} \\ \text{total loss} &= 450.8 \text{ mb} \end{aligned}$$

If the distance to the next range (a further 3.0 m is made 80 mm instead of 100 mm), then the total length of 80mm would be 21 m. The loss in the 80 mm is 21 m × 16 mb/m = 336 mb. The loss in the remainder of the pipework (55 m – 21m = 34 m) would be 450 – 336 = 114 mb.

$L = 34$ m $A = 0.65$ mb/m $B = 4.4$ mb/m $T = 114$ mb $s = 0$

By equation (11.2)
$$y = \frac{(4.4 \times 34) - (114 + 0)}{4.4 - 0.65}$$
$$= 9.493 \text{ m of 150 mm pipe}$$

Over three times the length of the extra 80 mm pipe is required in 150 mm to bring the overall loss back to the original figure. The original calculation included no. 150 mm pipe.

Check			
	21.0 m (of 80 mm) × 16 mb/m	=	336.0 mb
	24.5 m (of 100 mm) × 4.4 mb/m	=	107.8 mb
	9.5 m (of 150 mm) × 0.65 mb/m	=	6.2 mb
	total loss	=	450.0 mb

From design point F2 to junction (A) the pipe diameter is unknown but from (A) to the control valves is 100 mm, determined from the above calculation. Therefore loss in the 100 mm pipe from (A) to the control valves is:

$$(15.5 + 3.0 + 2.0 + 3 \text{ (for 1 bend)}) \times 4.4 = 103.4 \text{ mb}$$

Allowable loss between F2 and junction (A) is:

$$450 - 103.4 = 346.6 \text{ mb}$$

Solve for y, using equation (11.2) as before, but this time L = 22.5 m and T = 346.6 mb

Results 1.155 m of 100 mm pipe (say the bend at (A))
21.345 m of 80 mm pipe (say 19.5 m of pipe)

Check			
	23.5 m (of 100 mm) × 4.4 mb/m	=	103.4mb
	3.0 m (of 100 mm) × 4.4 mb/m	=	13.2 mb
	19.5 m (of 80 mm) × 16 mb/m	=	312.0 mb
	total loss	=	428.6 mb

When the design points on the floors below are calculated it is permissible to add the static gain from the highest sprinkler on the top floor to the highest sprinkler on the floor under consideration.

The static gain to design points G1 and G2 is 3.0 m from (B) to (C), which converts to 0.3 bar (300 mb).

The 2.0 m rise from the control valves to junction (C) is in 100 mm diameter pipe, from the previous calculation.

Therefore T = 450 – (2 x 4.4 mb/m) = 441.2 mb
and C to G1 = 15.5 + 28.5 + 6 m (for bends) = 50 m.
Solve for y as before, by equation (11.2),

$$y = \frac{(16 \times 50) - (441.2 + 300)}{16 - 4.4}$$

= 5.069 m of 100 mm diameter
and 44.931 m of 80 mm diameter

Check			
	2.0 m (of 100 mm) × 4.4 mb/m	=	8.8 mb
	5.1 m (of 100 mm) × 4.4 mb/m	=	22.44 mb
	44.9 (of 80 mm) × 16 mb/m	=	718.4 mb
	total loss	=	749.64 mb

static gain	= –300.0 mb
total	= 449.64 mb

Design point G2 is the last point on this small system. Most of the pipe has been sized in previous calculations.

The loss in 7.1 m (including one bend) of 100 mm pipe between the control valves and junction D is 31.2 mb. The loss in 80 mm pipe between D and E is 13.4 m × 16 mb = 214.4 mb. The loss between G2 and E is (450 – 31.2 – 214.4) = 204.4 mb

$$L = 19.5 + 3 \text{ (for 1 bend)} = 22.5 \text{ m}, \quad S = 300 \text{ mb.}$$

$$y = \frac{(35 \times 22.5) - (204.4 + 300)}{35 - 16} = 14.963 \text{ m}$$

Results 14.9 m of 80 mm pipe (say 12 m + 1 bend at E)
7.6 m of 65 m pipe (say 7.5 m)

Check			
	2.0 m x 4.4 mb/m	=	8.8 mb
	5.1 m x 4.4 mb/m	=	22.44 mb
	13.4 m x 16 mb/m	=	214.4 mb
	15.0 m x 16 mb/m	=	240.0 mb
	7.5 m x 35 mb/m	=	262.5 mb
	static	=	–300.0 mb
	total	=	448.14 mb

This completes the sizing of the mains. The installation control valves are sized as 100 mm, the same as the main rise pipe.

The pipe sizes as calculated will be exactly the same for all Ordinary Hazard projects (ie. Groups I to III special) sized to the tables in the rule book.

The water supply is increased to cover the higher groups in the Ordinary Hazard range without increasing the pipe sizing.

(b) Full hydraulically calculated pipe sizing. The second method of pipe sizing is by full hydraulic calculation. As the name implies, it is a completely calculated system, where the total number of sprinklers within an assumed area of operation are considered to be discharging simultaneously.

Every pipe in the area of operation together with the mains route back to the water supplies, must be calculated. As with the precalculated systems, the areas of operation are located at the most remote ends of each main (unfavourable area). The results from the calculation of these areas produce the highest pressure duties on the water supplies.

An area of operation nearest the water supply, (the most favourable area), which will give the highest flow rate required from the supply, is also calculated.

This is unlike precalculated systems, which are limited to terminal mains layouts only. However, designs that are fully hydraulically calculated can be either terminal main, looped, or gridded layouts. The shapes of areas of operation for each of these layouts are given in section 5 sub-section 24 of the rules[1,2].

So that a comparison can be made, the previous terminal main example will be used. The following calculations are manual, but more accurate calculations can be performed using a suitable computer program.

The design criteria for the example are as follows:

Hazard	Ordinary Group III
Design density	5mm/min (5 l/min m^2)
Area of operation	216 m^2
Area per sprinkler	12 m^2
Distance between sprinklers	4 m
Distance between ranges	3 m
Size of sprinkler	15 mm
K factor of sprinkler	80

Step 1
To determine the number of sprinklers to be calculated, divide the ceiling area of operation (216 m^2) by the area per sprinkler (12 m^2) = 18. To this number add any other sprinklers that are necessary within the area, e.g. under ducting.

Step 2
One side of the area of operation is formed by the end range. The number of ranges involved is therefore the number of sprinklers operating divided by the number of sprinklers on a range. The remainder of the sprinklers are grouped close to the main.

Step 3
Draw a schematic (see Figure 11.15) showing a reference number at each pipe junction.

Step 4
The LPC rules call for the design density over any four sprinklers on the corners of a square or rectangle to be never below the minimum required (see Figure 11.16).

Start the calculation to determine the flow from the most remote sprinkler by multiplying the design density by the area covered by the sprinkler: e.g.

$$\text{flow through sprinkler } Q_{(1)} = 5 \text{ mm/min} \times 12 \text{ m}^2 = 60 \text{ l/min}$$

Step 5
To find the pressure at sprinkler (1) use a transposition of equation (11.1) and K factor from page 294.

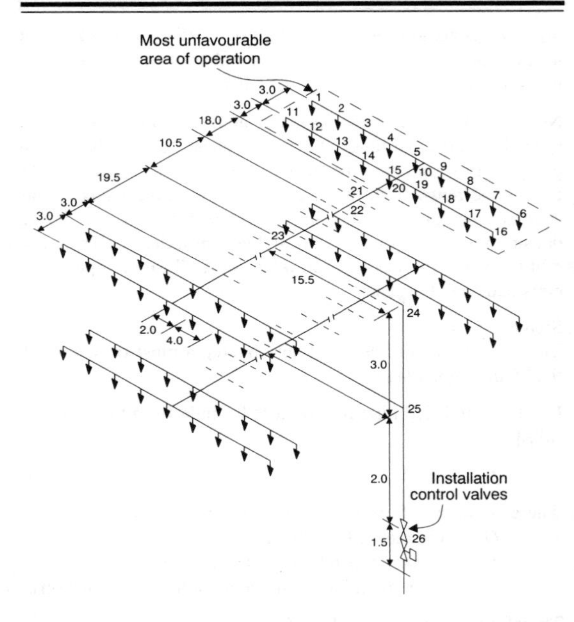

Figure 11.15 Example of OH hydraulically calculated sizing

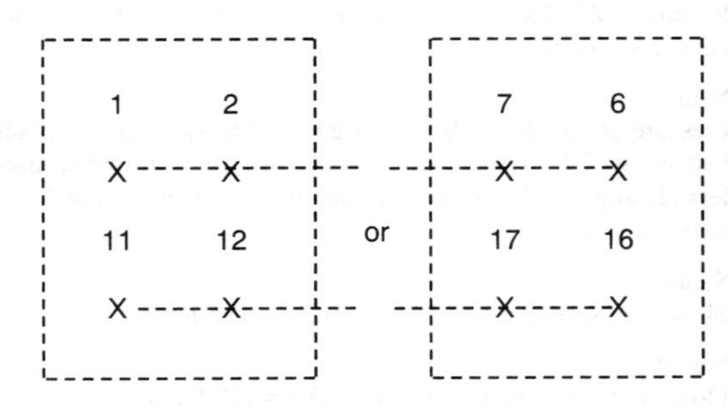

Figure 11.16

$$P_{(1)} = \left[\frac{Q_{(1)}}{K}\right]^2 = \left[\frac{60}{80}\right]^2 = 0.563 \text{ bar}$$

In an Ordinary Hazard system, no sprinkler pressure should be less than 0.35 bar, when all the sprinklers in the area of operation are discharging water (see Figure 11.12). It is sufficient to show flows to the nearest whole number of litres/minute and pressures to two decimal places of a bar.

Note the elevation of the sprinkler above a datum. If the last sprinkler on the last range is not the highest, the one that is will give less flow and therefore less density over the same area. As the second range from the end of the main in our example has smaller pipe diameters than the last range, the minimum design density will be at sprinkler (11) and not sprinkler (1). The minimum would have been at sprinkler (1) if all range pipe diameters had been the same.

Step 6
The flow from sprinkler (1) found in step 4 must flow through the 25 mm pipe (2–1).

Use the simplified pipe friction loss formula from the sprinkler rules[1,2].

$$p = K Q^{1.85} L \tag{11.3}$$

where p = loss of pressure in pipe in bars/metre
Q = flow rate in litres/minute
K = constant depending on size and type of pipe
L = length including equivalent length for fittings (m).

Therefore $p = 8.72 \times 10^{-6} \times 60^{1.85} \times 4$
$= 0.068$ bar

Note: the equivalent lengths for fittings are given in the sprinkler rules[1,2]. The 3 m per fitting is only to be used in precalculated systems.

Step 7
Pressure at sprinkler (2) is pressure at sprinkler (1) + friction loss in pipe (2–1) + or – any static difference between sprinklers (1) and (2). Therefore pressure at (2) = 0.563 + 0.068 + 0 = 0.631 bar.

Step 8
Flow from sprinkler (2) = $80\sqrt{0.631} = 63.5$ l/min.

Step 9
Flow in 25 mm pipe (3–2) = 60 + 63.5 = 123.5 l/min.
Friction loss (same method as step 6) = 0.258 bar.

Step 10
Pressure at sprinkler (3) (same method as step 7)
= 0.63 + 0.258 + 0 = 0.888 bar

Step 11
Flow from sprinkler (3) = $80\sqrt{0.888} = 75.39$ l/min.

Step 12
Flow in 32mm pipe (4–3) = 123.5 + 75.39 = 198.89 l/min.
Friction loss (same method as step 6) = 0.163 bar.

Step 13
Pressure at sprinkler (4) (same method as step 7)

$$= 0.888 + 0.163 + 0 = 1.05 \text{ bar}$$

Step 14
Flow from sprinkler (4) = $80 \sqrt{1.05} = 82.0$ l/min.

Step 15
Flow in 40mm pipe (5–4) = 198.89 + 82.0 = 280.89 l/min.
Friction loss (same method as step 6) = 0.148 bar.

Step 16
Pressure at sprinkler (5) (same method as step 7)

$$= 1.05 + 0.148 + 0 = 1.198 \text{ bar}$$

Step 17
Flow from sprinkler (5) = $80 \sqrt{1.198} = 87.56$ l/min.

Step 18
Flow in 50 mm pipe (10–5) = 280.89 + 87.56 = 368.45 l/min. The length of pipe in this case to be 2 m plus 2.91 m (for the right angle flow through the tee) on inlet from the main, giving a total equivalent length of 4.91 m.
Friction loss (same method as step 6) = 0.095 bar.

Step 19
Pressure at inlet to pipe (10–5)

$$= 1.198 + 0.095 + 0 = 1.293 \text{ bar}$$

Step 20
The pipework feeding sprinklers (9) to (6) is identical to pipework feeding sprinklers (1) to (4) therefore the pressure at (9) = 1.05 bar (as step 13).

Step 21
Flow in 40 mm pipe (10–9) same as (5–4) = 280.89 l/min (as step 15).
Friction loss (same method as step 6) with the length as 2 m plus 2.44 m (for the flow through the 40 mm outlet of the tee), giving a total equivalent length of 4.44 m, for which the pressure loss is 0.164 bar.

Step 22
Pressure at inlet to pipe (10–9)

$$= 1.05 + 0.164 + 0 = 1.214 \text{ bar}$$

Step 23
The pressure at each outlet of tee 10 is the same so it is necessary to balance the lower pressure range (10–6) with the higher pressure range (10–1).

This can be done by using the sprinkler formula equation (11.1) to calculate a K factor for the whole sprinkler range, treating it as one open outlet.

$$K \text{ (for the range (10–6))} = \frac{280.89}{\sqrt{1.214}} = 254.934$$

Step 24
The pressure at pipe (10–5) [step 19] = 1.293 bar, therefore Q (10–6) = 254.934 $\sqrt{1.293}$ = 289.88 l/min.

Step 25
Flow in 50 mm pipe (20–10) = 368.45 + 289.88
= 658.33 l/min
Friction loss [as step 6] but length (20–10) = 3 m
= 0.17 bar

Step 26
Pressure at entry to ranges (20–11) and (20–16)
= 1.293 + 0.17 + 0
= 1.463 bar

Step 27
As already mentioned in step 5 the ranges to be calculated now are smaller than the end ranges and therefore will need to be calculated using the same method as the previous two ranges.

Step 28

Flow from sprinkler	(11)	= 60.0 l/min
Pressure at sprinkler	(11)	= 0.563 bar
Flow in 25 mm pipe	(12–11)	= 60.0 l/min
Friction loss in pipe	(12–11)	= 0.068 bar
Pressure at sprinkler	(12)	= 0.63 bar
Flow from sprinkler	(12)	= 63.5 l/min
Flow in 25 mm pipe	(13–12)	= 60 + 63.5 + 0
		= 123.5 l/min
Friction loss in pipe	(13–12)	= 0.258 bar
Pressure at sprinkler	(13)	= 0.888 bar
Flow from sprinkler	(13)	= 75.4 bar
Flow in 25 mm pipe	(14–13)	= 198.9 bar
Friction loss in pipe	(14–13)	= 0.624 bar
Pressure at sprinkler	(14)	= 1.512 bar
Flow from sprinkler	(14)	= 98.4 l/min
Flow in 32 mm pipe	(15–14)	= 297.3 l/min
Friction loss in pipe	(15–14)	= 0.343 bar
Pressure at sprinkler	(15)	= 1.855 bar
Flow from sprinkler	(15)	= 108.9 l/min
Flow in 40 mm pipe	(20–15)	= 406.2 l/min
Friction loss in pipe (20–15)		
i.e. 2 + 2.44 (equivalent of tee)		= 0.324 bar
Pressure at cross	(20)	= 2.179 bar

Step 29
To calculate range (20 to 16) follow the same procedure described in steps 20 to 22.

Pipework feeding (19) to (16) is the same as (14) to (11), therefore the pressure at sprinkler (19) = 1.512 bar.

Step 30
Flow in 32 mm pipe (20–19) is the same as (15–14) = 297.3 l/min. Friction loss [same method as step 6] with the equivalent length as 2 + 2.13 (for right angle flow through tee) giving 4.13 m = 0.354 bar.

Step 31
Pressure at inlet to pipe (20–19) = 1.512 + 0.354 + 0
= 1.866 bar

Step 32
All range pipes to be balanced to the highest of the range inlet pressures, which is the pressure of 2.179 bar at cross (20) in step 28.

Using the K factor method in step 23. The calculated flow in pipe (20–10) [in step 25]

= 658.33 l/min

The calculated pressure at the inlet to pipe (20–10) [step 26]
= 1.463 bar

K factor range (20–10) $= \dfrac{658.33}{\sqrt{1.463}} = \dfrac{Q}{\sqrt{2.179}}$

Therefore increased flow Q at 2.179 bar
$= 658.33 \sqrt{\dfrac{2.179}{1.463}}$
= 803.43 l/min

Step 33
Use above method to balance range (20–16) [steps 30 and 31]
$= 297.3 \sqrt{\dfrac{2.179}{1.866}}$
= 321.27 l/min

Step 34
Total flow through mains from control valves to the entry to the area of operation (20) = 803.43 + 321.27 + 406.2
= 1530.9 l/min

Step 35
Pressure at junction (20) = 2.179 bar
Friction loss in 65 mm main (21–20) = 0.229 bar
Friction loss in 80 mm main (22–21) = 0.626 bar
Friction loss in 100 mm main (23–22) = 0.159 bar

Friction loss in 100 mm main (24–23)	= 0.162 bar
Friction loss in 100 mm main (25–24)	= 0.029 bar
Friction loss in 100 mm main (26–25)	= 0.019 bar
Static loss (5.0 m)	= 0.500
Total loss	= 3.903 bar

Step 36
The flow and pressure at the control valves to satisfy the area of operation F1 are 1530.9 l/min at 3.903 bar.

11.1.8 Installing pipework

The sprinkler rules[1,2] apertaining to pipe supports are shown in section 5, sub-section 22.

It is important to decide possible fixing points at an early stage in the design, as quite often they will determine the pipework routeing.

Sprinkler pipework must be securely supported directly from the building structure or from a primary support bracket designed specifically for the purpose. It must be borne in mind that the pipe supports must be capable of withstanding the forces exerted when pipework is filled, often with pumps (automatically started) and also, when water is discharging from open sprinklers.

The maximum spacing between support brackets are given in the sprinkler rules[1,2].

Pipes up to and, including 65 mm diameter	4.0 m apart
Over 65mm up to and, including 100 mm diameter	6.1 m apart
Over 100m up to and including 250 mm diameter	6.5 m apart

The above distances apply to either vertical or horizontal pipes and are measured along the length of the pipe.

Examples of the location of pipe fixing is shown in Figure 11.17. There are special requirements for mechanically jointed pipes.

All pipework must be drainable (with the exception of drops to single sprinklers below false ceilings). The rules[1,2] stipulate the drainage falls (clause 21.2.5, table 42) and size of drain valves clause 20.1.6 table 39).

Wet installations. Drainage slope 2 mm/m for all pipe diameters

Alternate wet/dry, dry and pre-action installations. Drainage slope 12 mm/m for all pipes up to and including 40 mm; drainage slope 4 mm/m for all pipes above 40 mm.

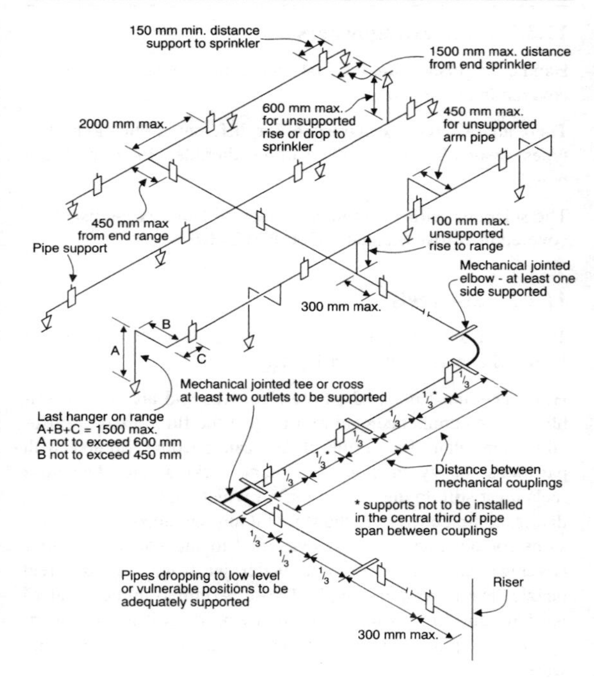

Figure 11.17 Pipe support locations

The minimum sizes of drain valves for trapped sections of pipework are:

Pipes above 80 mm diameter	50mm
Pipes 65m and 80 mm diameter	32mm
Pipes up to 50 mm diameter	20mm
Trapped range pipes above 50 mm diameter	25mm

11.2 HYDRANTS AND HOSE REELS

Manually operated fire-fighting equipment consists of hand extinguishers, hose reels, external hydrants, dry and wet risers.

Hand extinguishers and hose reels are for the use of relatively untrained personnel, whereas, hydrants, dry and wet risers are for trained Fire Brigade use only.

11.2.1 Hand extinguishers

Hand extinguishers are the first line of attack when a fire is discovered in its early stages.

There are different extinguishers for different applications. The types available are water, carbon dioxide, foam and dry powder.

The selection, siting and maintenance of hand extinguishers is covered in British Standard 5306:Part 3 [10].

11.2.2 Hose reels

To design a hose reel system it is necessary to refer to the British Standard 5306 : Part 1 [11].

Hose reels must be sited adjacent to exits and always accessible. At the commencement of a project the floor areas are usually open plan. Later, the floors can become riddled with partitions, heavy machinery, storage racking, etc. The hose reels must still do the correct job for which they were intended, that is, completely cover the floor taking into account all diversions for obstructions. It is permitted to include, in the floor coverage calculations, a 6 m water jet from the hose reel nozzle. The maximum length of hose permitted in the standard is 45 m, but 30 m is more commonly used, as it is more manageable to pull around obstacles when fully charged with water.

The flow through each hose reel should not be less than 24 l/min. when the two highest (remote) reels in the system are operating simultaneously. However, some Fire Authorities ask for the water supply to give a higher flow rate than the British Standard. For example in London the water supply required to the system is 136 l/min. Apart from the connection to the hose reel, the mains pipework is usually 50 mm diameter with a 25 mm diameter inlet pipe to each hose reel. As a guide, the friction loss in heavy quality galvanised steel pipe at a flow of 136 l/min is 11.8 mb/m for 40 mm and 3.6 mb/m for 50 mm pipe.

Usually, in a several-storey building, a towns main is not adequate to supply hose reels directly and a pumping system is required. The pumping system consists of a 1125 l water storage tank supplied from the towns main by a 50 mm connection and ball valve. Duplicate electric driven hose reel pumps draw water from the tank. The duty pump is automatically started on a drop in pressure in the system.

11.2.3 External hydrants

Although underground fire hydrants are installed on towns

mains in public highways, they may not be adequate to cover a new project. In these circumstances, the Authorities or Insurance Company may ask for a private hydrant system to be installed from a connection into the towns main.

British Standard 5306:Part 1[11] gives information on the siting of hydrants, details of underground hydrant pits, etc.

The maximum distance between hydrants	150 m
The maximum distance from building entry	70 m
The minimum distance to a building	6 m

Due to the higher risk in chemical plants the above figures are reduced, as shown in British Standard 5908:1980 [12].

The underground main should be designed as a ring suitably valved and fed (if possible) by more than once source. The main should be sized to feed the most remote hydrants by the longest route with one section of the main shut down.

11.2.4 Dry and wet riser

Building heights between 18 m and 60 m are installed with dry risers. Above 60 m a wet riser is installed complete with an automatic pumping system. The risers are usually located in ventilated lobbies or fireproof staircases, with landing valves on all floors except the ground floor. Where possible, a roof valve is installed for test purposes.

British Standard 5306: Part 1[11], also gives information on dry and wet risers.

There should be one riser to every 900 m^2 of floor area. If more than one riser is required then they should not be more than 60 m apart. No part of the floor should be more than 60 m from the landing valve.

The diameter of the riser, where there is one valve per floor is 100 mm. If two valves are permitted per floor from the same riser then the diameter is increased to 150 mm.

The wet riser pumping system consists of a water storage tank with a minimum capacity of 45 m^3 supplying duplicate fire pumps, each capable of 25 l/s (1500 l/min) giving a minimum valve pressure of 4 bar when up to three most remote landing valves are operating simultaneously. The capacity of the tank plus the flow from the towns mains must be capable of supplying sufficient water to enable the pump to operate at full duty for 45 minutes.

The pumping system is similar to the sprinkler pumps shown in Figure 11.8.

11.3 GAS EXTINGUISHING SYSTEMS

The design and installation of all types of gas extinguishing systems must take into account the precautions stated in the Health and Safety Executive Guidance Note GS16 [13] and the relevant Standards.

The majority of gas extinguishing systems in the past have been either Halon 1301, Halon 1211 or Carbon Dioxide.

11.3.1 Halons

Unfortunately, although Halons 1301 and 1211 were safer to use from a human point of view, they were considered to have an effect on the ozone layer, along with other CFCs. At a meeting in Montreal in 1987, the major countries signed a Protocol restricting their future use.

New extinguishing gases are being developed by the major manufacturers to be environmentally friendly, i.e. zero ozone depletion, as well as being safe and efficient.

For the standards applicable to Halon 1301 see references 14 and 15 and for Halon 1211 see references 16 and 17.

11.3.2 Carbon dioxide

Carbon dioxide is a natural gas and has been used as an extinguishing agent for many years. The method of extinguishment is to reduce the oxygen level to a point below which combustion cannot continue. The systems are designed either for total flooding of the protected area or local protection to a particular risk.

The standards to be used are either BS 5306:Part 4 [18] or NFPA 12 [19].

The systems can be automatically operated either by electrical, pneumatic or mechanical detection. If electrical detection is used, it must be designed in accordance with BS 5839 [8].

Due to the hazard to personnel from the carbon dioxide, it is essential that the system is locked off whenever the area is occupied. A system monitoring panel is located at each entrance to the protected area. The panel shall include the following indicators:

Red lamp	CO_2 discharged
Amber lamp	Automatic and manual control
Green lamp	Manual control

The protected area must be reasonably gas tight to maintain the concentration. A method of testing the integrity of the enclosure is to conduct a door fan test and monitor the leakage points.

11.4 DELUGE SYSTEMS

In Section 11.1 it was shown that the sprinklers in a sprinkler system operate independently of each other. This means that only sprinklers affected by the heat from the fire would operate. There are, however, situations which need complete simultaneous water spray to cover a whole area of a risk at the outbreak of a fire. This type of system is called a deluge system.

The deluge system consists of a network of pipes, complete with open spray nozzles surrounding the risk to be protected. The control of the system is through a set of control valves which can either be operated (tripped) by a detection system with an emergency manual override, or just manually operated.

11.4.1 Design standards

The two main standards used for the design of deluge systems are issued by the National Fire Protection Association (abb NFPA)[20] and the Loss Prevention Council (LPC)[21]. Reference to these standards will give the design density (water application rate) required for the protection.

11.4.2 Deluge control valves

There are several manufacturers of deluge control equipment on the market. The simplified schematic of one make is shown in Figure 11.18 to show the operating principles.

The deluge valve has three sections. An upper chamber (C) which when pressurized will close a clapper to seal the water inlet (A) from the outlet (B). Any method of releasing the pressure in the upper chamber will allow the clapper to lift and water to flow into the system and out of the open nozzles.

11.4.3 Nozzles

Nozzles should be carefully selected for their performance characteristics, taking into account the spray pattern, pressure required at the nozzle, distance of throw and orifice size.

Nozzles can be either medium or high velocity type. High velocity nozzles are used on fires involving flammable liquids with flash points of 66°C and above. Liquids with flash points below 66°C are not always possible to extinguish but medium

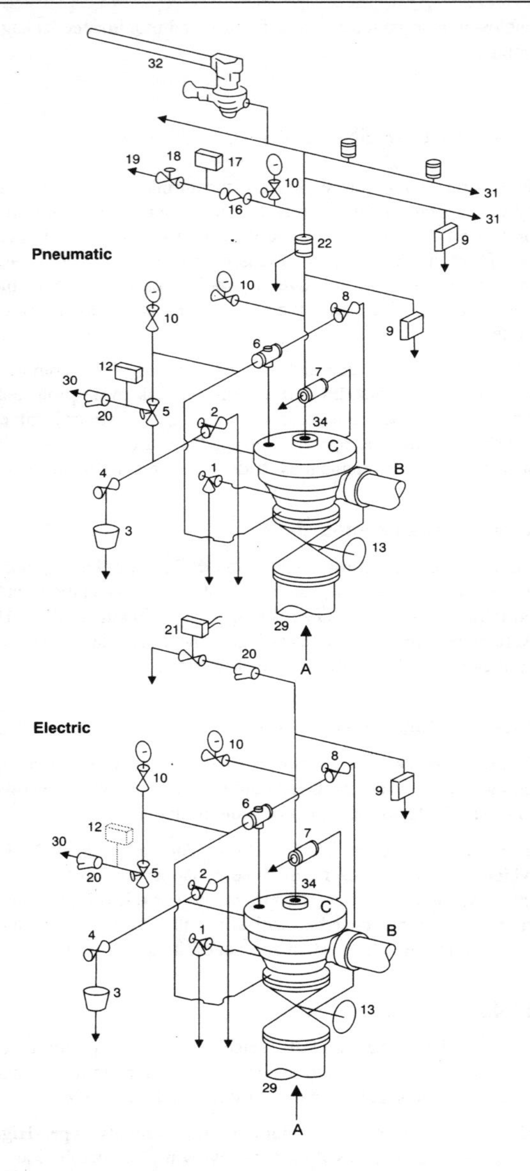

Figure 11.18 Typical deluge system control valves (courtesy of the Viking Corporation)

velocity nozzles can be used to control. They can also be used for exposure protection to structures and equipment, cooling to flammable liquid storage tanks, electrical cable trays, etc.

The nozzle manufacturer's data sheets must be followed as there is usually quoted a minimum and maximum nozzle pressure and K factor. Equation (11.1) in Section 11.1.3 can also be used to determine the flow from a nozzle.

11.4.4 Pipework

The pipework and fittings must be galvanized and in most cases is to British Standard 1387 heavy quality. The diameter of the pipework must be hydraulically calculated taking every open nozzle in the system into account.

11.4.5 Methods of operation

The method of operation can be by one, or a combination of the following, depending on the type of risk.

Pneumatic detection system

This type of detection system consists of air-filled pipework fitted with heat detectors, surrounding the risk. The heat detectors are either rate-of-rise (32) and/or fixed temperature similar to a glass bulb sprinkler without the deflector, although standard sprinklers (31) can be used instead. When any detector sprinkler operates, the sudden drop of air pressure in the pipework will operate the deluge control valve releasing water to the open spray nozzles.

Manual releases (lever operated valves) (9) can be placed at strategic points around the risk, connected to the pneumatic detection system, for use in emergencies.

Hydraulic detection system

This system is the same as the pneumatic system but uses water in the detector pipework instead of compressed air.

Electrical detection system

This system uses whichever detector is most suitable for the risk involved, i.e. smoke, heat, flame, etc. operating a solenoid valve release (21).

Manual electrical releases can also be incorporated.

A mechanical release (9) is always installed at the control valves, which can be used in an emergency should the automatic detection system fail for any reason.

11.5 SMOKE VENTILATION

11.5.1 The purpose of smoke ventilation

Most fatalities in a fire arise from the inhalation of smoke and associated toxic gases, rather than from the flames themselves. Since the presence of smoke inhibits the movement of people in a fire emergency it follows that smoke control is essential. Smoke ventilation is necessary for the following purposes:

(i) To protect the life of the occupants of the building by assisting them to escape to a safe place.
(ii) To help the fire brigade to fight the blaze by keeping smoke away from the approaches to the fire.
(iii) To protect the building structure by releasing heat from inside the building.

The methods most commonly adopted to achieve these ends are:

(a) *Containment.* The spread of smoke is minimized by providing lobbies that are naturally ventilated to outside and by using closed doors and barriers generally.

(b) *Dilution.* Visibility within the smoke is prevented from deteriorating for as long as possible, by the removal of smoke and the introduction of outside air, in order to give the occupants the opportunity of escaping.

(c) *Ventilation.* Fans or natural ventilation openings keep buoyant, hot smoke above the upper level of the escape route.

(d) *Pressurization.* Certain critical routes, such as escape staircases, are pressurized to discourage the entry of smoke and facilitate escape from the building.

(e) *De-pressurization.* Hot smoke is removed from escape routes.

(f) *Clearance.* Smoke is evacuated from the building after the fire has been quenched.

11.5.2 The development of a fire theory

The growth of a fire from its ignition has been studied and methods of prediction developed [22–26]. The amount of smoke produced by the primary combustion of a fire and the entrained airflow rate depend on the perimeter of the fire, the flame temperature and the effective height of the hot gases above the fire. A Large Fire Theory has evolved that predicts the rate of smoke production in kg/s and m^3/s for a given set of initial assumptions. From this, and a knowledge of typical burning rates for various materials stored in a building, smoke temperatures and the necessary ventilation rates can been calculated for effective smoke removal. A design fire is specified as having a plan area of 3m × 3m with a burning rate of 5 MW and this is taken as a safe basis on which to design fan-powered ventilation systems for both sprinklered and unsprinklered areas. Hence a specification has been developed [26] that gives the temperatures and associated durations that a fan handling hot smoke has to be able to withstand. These are given in Table 11.1.

Table 11.1 Temperature–time specification for fans handling hot smoke and gases.

Category code	Suitable for operation in the following conditions Temperature (°C)	Time (h)
650/1	560/600	1.0, 1.5, respectively
400.2	400	2.0
300/1	300	1.0
300/0.5	300/250	0.5/2.0, respectively
150/5	200/150	0.5,2.0, respectively

11.5.3 Fan types

Belt-driven centrifugal fans are unsuitable for extracting hot smoky gases because the belts are vulnerable at high temperature and cease to be effective at temperatures well below 150°C. It follows that axial flow fans are the usual choice for mechanical smoke extract systems.

Fans must be able to handle both hot smoke and cool smoke under emergency operating conditions. When high temperature smoke is handled the volume stays the same but the mass flow rate is reduced since the density of hot smoke is less than that of air at normal ambient conditions. The fan power is also reduced, according to the fan laws, but it is strongly recommended that the motor is rated for the full duty at normal conditions.

Driving motors are close-coupled to fans and may be in hot, smoky gases up to category 400/2. Beyond this, for category 650/1, the motor must be outside the hot airstream and a bifurcated version of the axial flow fan is used. The motor still needs cooling and a ducted supply is sometimes needed. Fans should be selected for a higher category than the calculated temperature predicts.

11.5.4 Installation and maintenance

Ducts used in mechanical smoke exhaust systems should be able to withstand anticipated temperatures for the required period of time and should have a 4 h fire resistance. In general terms this means that ducts should be made from double steel sheets separated by appropriate insulation. Jointing materials, gaskets and supports must be able to withstand the temperatures.

Although lubricants that operate at very high temperatures are in use they are unsuitable for re-lubrication. This must be considered when formulating maintenance programmes. Certification of performance is necessary.

When fans operate under emergency or normal conditions, they must not generate noise that drown the public address system, used for informing the building occupants during an emergency. Hence care must be taken in fan selection and the choice of silencers. Silencers must be made of materials that retain their integrity and acoustic performance under fire emergency conditions. Anti-vibration mountings must also be effective when the fan is handling hot smoky gases. This is possible if spring mountings are used. Flexible connections must be of suitable materials.

Electrical connections should be by copper covered, mineral insulated cables although other insulating materials, able to withstand high temperatures, are also available.

Standby fans may not be required but an independent electrical supply is usually needed. This can take the form of a standby emergency generator or an independent electrical supply from another substation.

In the case of a smoke system also used for general ventilation in the building under normal conditions, a problem is that a fan used in an emergency is unlikely to be suitable for general use afterwards. However, it is essential that good maintenance under normal conditions ensures the fan will not fail when required for an emergency. It is particularly important that a standby electrical generator receives good maintenance. These are much noisier than the generators used in CHP installations and there may be some reluctance to give them test runs at the necessary frequent intervals. Nevertheless, thorough regular test runs are essential.

11.5.5 The supply of inlet air

Air must be supplied to a building under emergency fire conditions to make good the air extracted. The supply is usually through inlet openings without the use of fans. Providing insufficient area of inlet openings prevents the smoke extract system from working properly. The air supplied must be beneath the level of the smoke layer and at a low velocity in order to avoid disturbing the smoke above. With natural exhaust ventilation the inlet areas must be from 1.5 to 2.0 times the exhaust areas but this is not so important with mechanical extract. However, with fan-powered extract, the face velocity over the inlets should be about 3 m/s and should never exceed 5 m/s. Inlets may take the form of doors that fail to the open position in an emergency, automatic louvres in the walls or the roof interlocked to open when the smoke exhaust fans are energized, roof vents in adjacent smoke reservoirs that are interlocked to stay shut if the reservoir is smoke logged, and inlet fans.

11.5.6 Smoke reservoirs

In the malls of shopping centres it is usual to provide reservoirs in the ceilings with openings that allow the smoke to escape and keep the mall a smoke-free escape route.

When a smoke plume rises from a fire it entrains air, its volume increases and its temperature and buoyancy reduce. The smoke escaping from the shop into the mall entrains more air which further reduces its buoyancy. For the smoke reservoirs to be effective the buoyancy of the smoke must be enough for it to rise and enter the reservoir. There is then a balance between air entering the mall to feed the plume and the buoyancy of the plume taking it out of the reservoir. The depth of smoke in the reservoir depends on the mass flow rate of entering smoke, its temperature and the width of the reservoir. It is independent of the method of extraction (natural or fan-powered). Reference 27 provides information on the minimum depth of a reservoir, with various widths and mass flow rates of smoke, to deal with a design fire (3m × 3m and 5 MW).

11.5.7 Pressurized staircases [28,29]

A system for pressurization should be readily available when a fire starts, to give a smoke-free protected route to safety, and be reliable, simple and economic. Air must be supplied uniformly throughout the height of the staircase and a single supply point is not acceptable unless it feeds three storeys or fewer. Hence the supply duct should run vertically with supply outlets not more than three storeys apart. Pressurization levels up to 60 Pa are recommended and methods are given for estimating the air leakage from the staircase through cracks around access doors and windows. The Code of Practice [28] does not apply to shopping centre malls.

11.5.8 Natural ventilation

To be effective this relies on the natural buoyancy of the hot smoke being enough to carry the smoke up and out of the openings, which must be properly sized by calculation[22–26]. Adequately sized inlets, at low level, must be provided to make good the air exhausted through the outlets. The system must comply with reference 30 and have a full maintenance schedule.

Several methods have been used for opening natural vents:

(a) Fusible links holding the vent shut against springs.
(b) Electric motors or magnets that open vents on receiving a signal from smoke detectors.
(c) A pneumatic device holding a vent shut against springs.

Fusible links are now forbidden for means of escape, electrical methods need wiring and are not desirable and hence pneumatic actuation seems to be the preferred solution.

References

1. British Standards Institution. British Standard 5306:Part 2:1990 - Fire extinguishing installations and equipment on premises. Part 2 Specification for sprinkler installations.
2. Loss Prevention Council. Rules for automatic sprinkler installations. Incorporating BS 5306:Part 2:1990 and LPC Technical Bulletins.
3. Factory Mutual Insurance, 1151 Boston – Providence Turnpike, P.O. Box 9102, Norwood, MA 02062
4. National Fire Protection Association, 1 Batterymarch Park, Quincy, Massachusetts, USA.
5. Loss Prevention Council, Fire Offices Committee 29th edition Rules for Automatic Sprinkler Installation. (incorporated in [2] above).
6. National Fire Protection Association. Standard for the Installation of Sprinkler Systems, NFPA 13 Chapter 9 Large-Drop and Early Supression Fast Response (ESFR) Sprinklers.
7. Factory Mutual Insurance. Loss Prevention Data Sheet 2–7; Installation Rules for Sprinkler Systems using Large-Drop Sprinklers.
8. British Standard 5839 – Fire detection and alarm systems for buildings Part 1 Code of practice for system design, installation and servicing.
9. B. E. Boulton, Examination of a USA idea, *Institution of Fire Engineers Journal*, June 1979.
10. British Standard 5306:Part 3 – Code of practice for selection, installation and maintenance of portable fire extinguishers.
11. British Standard 5306:Part 1 – Hydrant systems, hose reels and foam inlets.

12. British Standard 5908 – Code of practice for fire precautions in chemical plant.
13. Her Majesty's Stationery Office. Health and Safety Executive Guidance Note GS 16.
14. British Standard 5306:Parts 5.1 - Code of practice for Halon 1301 total flooding systems.
15. National Fire Protection Association. NFPA 12A Standard on Halon 1301 fire extinguishing systems.
16. British Standard 5306:Part 5.2 - Code of practice for Halon 1211 total flooding systems.
17. National Fire Protection Association. NFPA 12B Standard on Halon 1211 fire extinguishing systems.
18. British Standard 5306:Part 4 - Specification for carbon dioxide systems.
19. National Fire Protection Association. NFPA 12 Standard on Carbon Dioxide extinguishing systems.
20. National Fire Protection Association. NFPA 15 Standard for water spray fixed systems for fire protection.
21. Loss Prevention Council. Tentative Rules for Medium and High Velocity Water Spray Systems.
22. Thomas *et al*, Fire Research Technical Paper No 7, 1962, BRE.
23. J. P. Morgan, BRE Information Paper 34/79.
24. J. P. Morgan, BRE Information Paper 19/85.
25. E. C. Butcher, Fire Progression, Spread and Growth, BRE Information Paper, Oct 1987.
26. G. A. Courtier and J. A. Wild, The development of axial flow fans for the venting of hot fire smoke, C401/016, I Mech E, 1990.
27. H. P. Morgan, Smoke Control Methods in Enclosed Shopping Complexes of One of More Storeys. A design Summary. BRE 34 (K2.3) 1976.
28. BS 5588: 1978, Fire precautions in the design and construction of buildings Part 4, Code of Practice for smoke control in protected escape routes, using pressurisation.
29. Fire Research Note No. 968, HMSO.
30. BS 7346:Part 1, 1990. Specification for natural smoke and heat exhaust ventilators.

12 Coldwater services

12.1 DISTRIBUTION SYSTEMS

The usual arrangement for coldwater systems is for a water supply to be brought into the building from the Water Company mains and directly feed all drinking water draw-offs and storage tanks. Coldwater for all other purposes is then drawn from storage tanks. However, for small establishments or dwellings some water companies may not require the installation of storage tanks and will permit all draw-offs to be supplied directly from the mains water supply. Early contact with the water supplier is therefore desirable to establish the requirements that have to be complied with.

Wherever possible water for drinking or culinary purposes should be taken direct from a mains water supply but drinking water can be distributed from storage providing that the tank used is properly constructed and installed to store potable water (see Section 12.4).

Distribution of drinking water should be planned to avoid long runs of pipework to points with very little or infrequent usage. Ideally drinking water systems will be arranged so that the entire contents are drawn-off and replenished afresh every working day.

Coldwater distribution must be designed to prevent contamination occurring from back flow through draw-off taps and valves. This protection can take the form of separate down services, pipework arrangements or mechanical devices in the pipelines (see Section 12.3).

Pumped or boosted coldwater and drinking water distribution systems may be required where the available water pressure will not adequately serve the needs of all draw-off points (see Section 12.5).

Care needs to be exercised when planning routes for coldwater and drinking water distribution pipework to ensure that it is protected against both heat sources and freezing and is readily accessible.

12.2 WATER HYGIENE

Materials used for storage and distribution of water must not contain any toxic substances or impart odours or tastes. This includes solders, which must be lead free, and jointing compounds which must not support any form of bacteriological growth.

Pipework conveying drinking water or cold water must be kept out of direct contact with hot water or heating pipework. The whole pipework installation and the storage tanks should preferably be insulated but especially any runs of pipe which are within areas of high ambient temperature. The purpose of these precautions is to inhibit bacteria growth by keeping the water temperature below 20°C.

Newly completed installations must be sterilized by being treated with chlorine in solution before being put into service. Sterilization should be repeated at least annually and after any work has been done on the pipework or storage tanks.

12.3 BACKFLOW PROTECTION

Pipework systems must be protected against water flowing back into the system from storage tanks or draw-off taps.

Water by-laws require that the inlets to water tanks and draw-offs to fitments are provided with air gaps to prevent the stored water from coming into contact with the water flowing into them. Type 'A' air gaps (Figure 12.1) are required where the stored water is contaminated or where the fitment served contains substances which are harmful to health.

Type 'B' air gaps (Figure 12.2) are required where stored water or the fitments that are served may be contaminated by substances which are harmful to health.

Specific situations in which the type 'A' and type 'B' air gaps are to be provided are scheduled in British Standard 6700: 1987 Section Two.

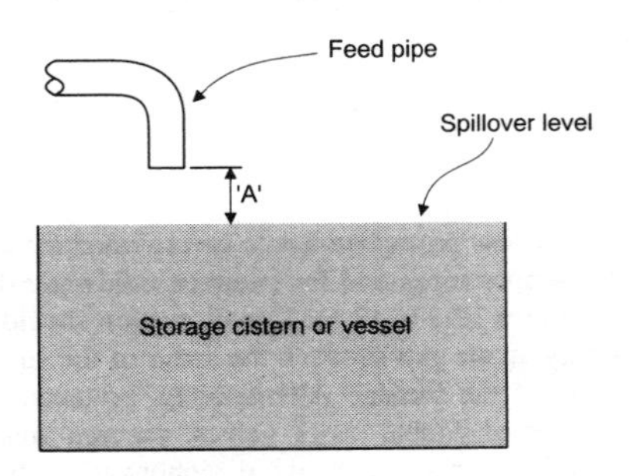

Figure 12.1 Type 'A' air gap

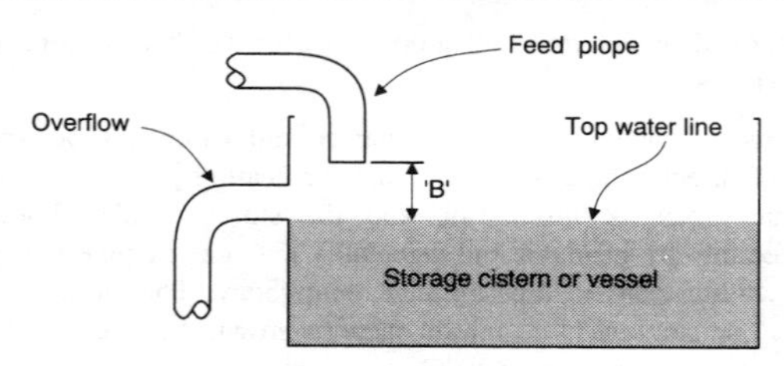

Figure 12.2 Type 'B' air gap

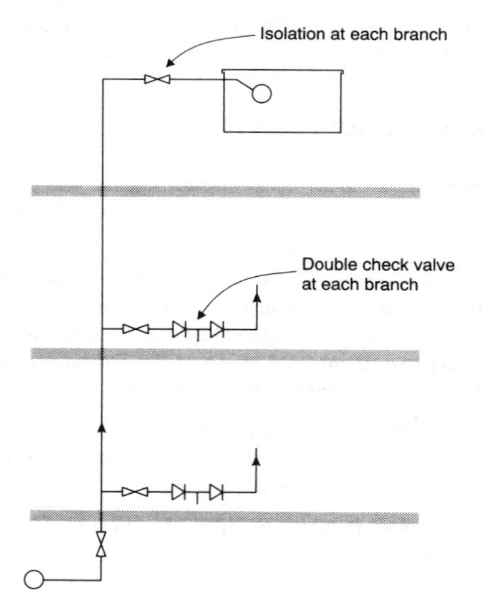

Figure 12.3 Secondary backflow protection on common mains water supply

Secondary backflow protection has to be provided for common mains water supply pipes and for common coldwater distribution pipes (Figures 12.3 to 12.5). This protection should ideally be provided by an air gap between the spout of the tap and the spillover level of the fitment. Alternatively, protection can be achieved by use of double check valves, vacuum breakers or pipework arrangements which avoid siphonage or backflow occurring.

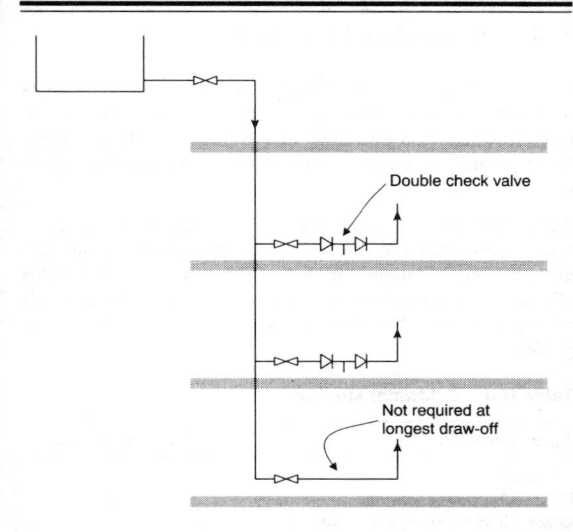

Figure 12.4 Secondary backflow protection on common distribution pipe where discharge on draw-off tao does not provide adequate air gap

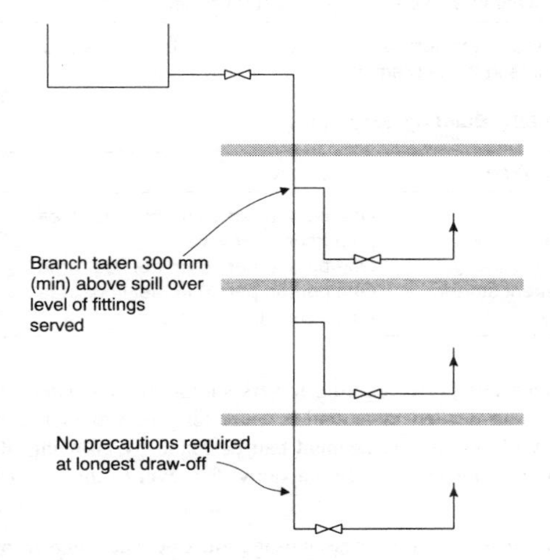

Figure 12.5 Secondary backflow protection alternative to mechanical protection

12.4 COLDWATER STORAGE

Coldwater storage is normally provided to cover for 24 hours interruption of mains water supply. Water by-laws give a minimum requirement for dwelling of 115 litres or 230 litres where the coldwater storage also provides a feed to hot water apparatus.

Tables 12.1 and 12.2 may be used to establish domestic coldwater requirements for various types of buildings. Table 12.1 gives recommendations for storage rates per head and Table 12.2 may be used to estimate building occupancy. For hospitals reference should be made to specializing Health Service guides.

Table 12.1 Coldwater storage

Type of building	Storage per head
*Hotels	135
Hostels/boarding schools	90
Offices and factories with canteens	45
Offices and factories without canteens	40
*Department stores with canteens	45
*Department stores without canteens	40
Day schools	30
Restaurants	7 per meal
Public toilets (meeting places and entertainment)	15

*Note: Additional storage will be needed for public toilets and restaurants in these premises.

Table 12.2 Building occupancy

Building Type	Occupancy
Offices	One person per 10 m^2 net floor area
Factories	30 persons per w.c.
Shops	One person per 10 m^2 net floor area
Department stores	One person per 30 m^2 net floor area
Schools	40 persons per classroom

Coldwater storage for cooling towers should be calculated on the make-up needed to replenish water lost from evaporation and blow-down. An assessment can be made by allowing 40 litres of coldwater storage capacity for every kilowatt of refrigeration.

Storage for industrial processes usage must be calculated from quantities to be advised by the user.

Tank hygiene

The contents of coldwater storage must be of potable quality

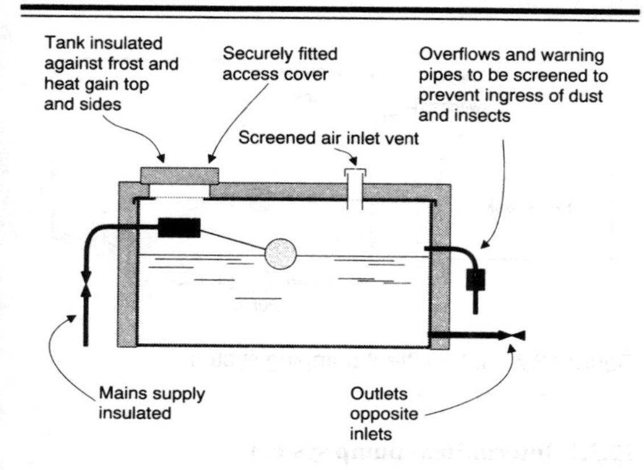

Figure 12.6 Storage tank requirements

and therefore protected against contamination and stagnation. To this end, all tanks must have close-fitting covers and the overflows, warning pipes and vents must be provided with screens. Inlet and outlet connections should be at opposite ends of tanks to prevent areas of stale water (Figure 12.6).

Tanks should be adequately insulated or otherwise protected against freezing conditions and insulation should be considered in conditions where heat gain by the stored water is a possibility.

Overflows

All tanks must have an adequate overflow pipe fitted, additionally tanks with capacities over 4500 litres must have a 25 mm diameter warning pipe arranged to indicate when the water is the tank reaches a level 50 mm below the overflow pipe.

Every inlet and outlet connection, apart from overflows and warning pipes, should be controlled by an isolating valve.

12.5 BOOSTED COLDWATER SERVICES

Boosted water services are necessary where mains water supplies do not have sufficient pressure to feed rooftop storage tanks or where water storage is all at ground or basement level and needs to be distributed in the upper floors of a building.

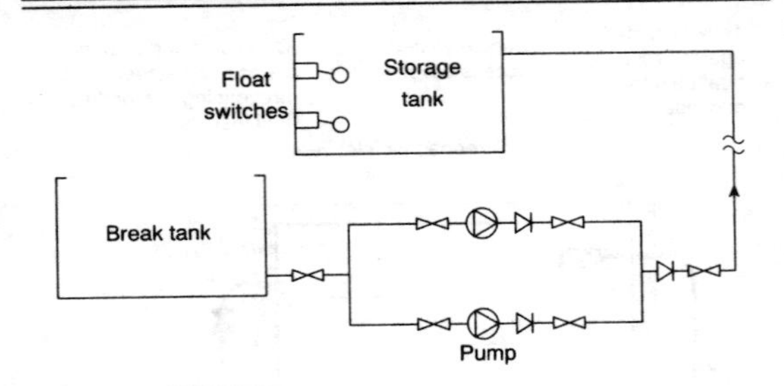

Figure 12.7 Intermittent pumping system

12.5.1 Intermittent pump system

Where the only requirement is to provide a feed for a single storage tank, then a system with float switch controlled pumps may be used. The pumps in this system are operated only when the water level is replenished (Figure 12.7).

12.5.2 Hydro-pneumatic system

In situation where there are a number of tanks to be fed, and the resultant float switch controls would therefore be complex, or where there is a requirement to provide a pressurized distribution system, then a hydro-pneumatic system should be used. In addition to duty and standby pumps this system utilizes an air and water filled vessel, or a hydraulic accumulator, to maintain a constant system pressure. When demand reduces the volume of water in the system then the duty pump starts and continues to run until demand ceases and water volume is replenished (Figure 12.8).

12.5.3 Constant pump system

An alternative to the hydro-pneumatic system is a constant running system where the duty pump runs continuing to maintain system pressure, and support pumps cut in and out to meet pressure fluctuations caused by rising and lowering of draw-off demand. To prevent overheating of the duty pump when operating against closed valve conditions, a small bore relief line must be taken from the pump discharge side and connected to the suction side or run back to a break tank (Figure 12.9).

Alternatively a variable speed pump giving a constant outlet pressure.

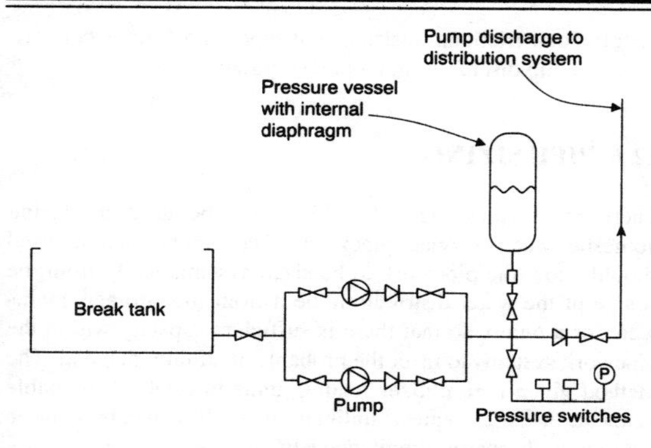

Figure 12.8 Hydro-pneumatic system

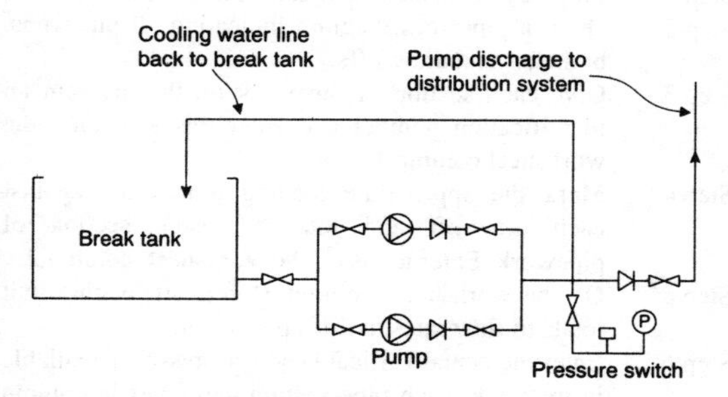

Figure 12.9 Constant running pumping system

12.5.4 Water demand

The pump capacity should be equal to the total peak demand of the system it is serving. For systems feeding domestic storage tanks, the peak demand should be calculated at 0.07 litres per second for every 1000 litres to be stored. Pressurized distribution systems should take the total simultaneous draw-off as the peak demand.

12.5.5 Pump pressure

The pump should be the sum of:

(a) Static pressure necessary to reach the topmost draw-off point.
(b) Residual pressure required at the topmost draw-off point.

(c) Total frictional resistance of pipes and fittings between pump discharge and topmost draw-off.

12.6 PIPE SIZING

There are various methods which may be used for sizing domestic water service pipework. The actual method used should allow the pipework to be sized systematically from the source of the water distribution, be it from roof storage, mains water or a pump, so that there is sufficient capacity within the pipework systems to meet the probable maximum demand. The method given here utilizes loading units to establish probable demand, it is extremely unlikely that all fitments will be required to discharge simultaneously.

Procedure

Step 1 Prepare a worksheet as illustrated in Figure 12.10.

Step 2 Draw a pipework diagram indicating all pipe runs, branches and draw-offs.

Step 3 Give each section of pipework on the diagram an identification number and enter this also onto the worksheet column 1.

Step 4* Mark the appropriate loading unit value against each draw-off and total for each section of pipework. Enter totals on the worksheet, column 2.

Step 5* On the worksheet (column 3) convert loading unit totals to flow rates in litres per second.

Step 6 Enter the actual vertical head (or pressure) available in metres to each pipe section and enter in column 4.

Step 7 Assume a pipe diameter which will provide the required flow rate at a velocity not greater than 2.0 metres per second. Enter in column 5.

Step 8 Using the assumed pipe diameter make preliminary calculation of head loss and compare it with the head available in column 4. Too great a head loss will require a large pipe diameter to be selected and an excessively lesser head loss will indicate that the assumed pipe diameter is too small.

Step 9 Make necessary adjustments as indicated by Step 8 and continue to next pipework section.

* Tables of loading units, conversion charts and pipe sizing graphs are published in British Standards 6700 and by professional institutions.

1	2	3	4	5	6	7	8	9	10	11
Pipe No.	Loading units	Flow l/s	Metres head available	Pipe dia.	Measured pipe run	Loss of head metre/metre run	Equivalent pipe lengths	Effective length	Head consumed	Total head consumed (4-10)

Figure 12.10 Example of pipe sizing worksheet

13 Hotwater Services

13.1 HOTWATER DISTRIBUTION

A conventional hotwater system is illustrated in Figure 13.1. Water is heated in the storage cylinder and is circulated, usually with the aid of a pump, to serve the various draw-off points. Circulation of the hot water restricts the temperature drop in the system so that there is hot water available quickly at all draw-off points, avoiding the need to first run-off relatively cold water to waste.

An economic alternative to the use of circulation pipework, as described above, is a 'dead leg' system. This system utilizes electronically heated tape to maintain the desired temperature range in the distribution pipework. To obviate the need for

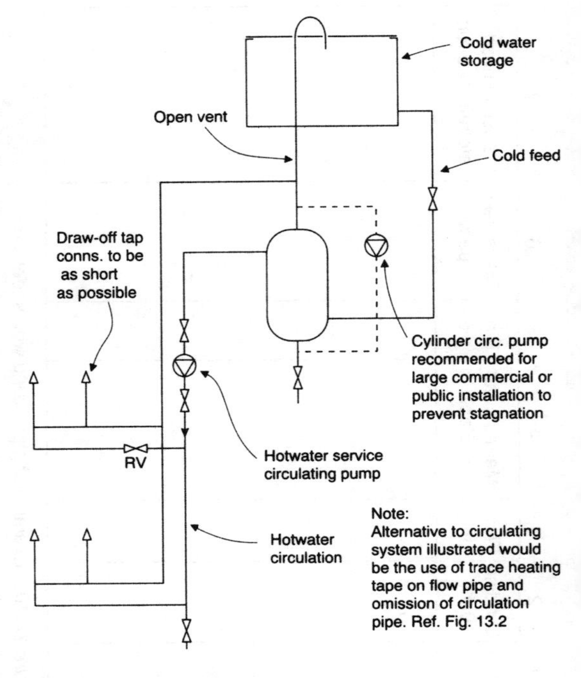

Figure 13.1 Vented hotwater system

complex controls, the tape used for this purpose should be of the self-regulatory type and should be applied directly to the outside of the pipe and covered by the normal pipe insulation material.

Expansion of the water as it is heated up is relieved by an open vent pipe which terminates above the water level of the feed tank.

The hotwater flow pipe should be sized to meet the probable maximum draw-off from the system as for coldwater pipe sizing. In a circulation system, the return pipe is sized so that the water is circulated through the pipework system with a temperature drop of between 10°C and 20°C. Likewise the flow rate for the circulatory pump should be sufficient to maintain the desired temperature drop and its head pressure should be equal to the frictional resistance of the circulation pipework.

13.2 UNVENTED SYSTEMS

Figure 13.2 is a schematic layout of an unvented hotwater system. Such systems must be designed so that any expansion water is either contained within the pipework system or by the provision of an expansion vessel.

Building regulations require that an unvented system must be either a packaged unit which is British Board of Agreement approved or constructed of individual components which are so approved. This to ensure that the following safety features are incorporated.

(a) The flow temperature is controlled between 60 and 65°C.
(b) High temperature cut-out to automatically shut off the primary heat source should the thermostat fail and a temperature of 95°C be exceeded.
(c) Temperature and pressure relief which must automatically operate at a temperature of 95°C to prevent the stored water from exceeding 100°C.

Pipework which conveys the discharge from relief valves must discharge in a safe and visible location preferably outside the building. It must fall continuously through its length and must not exceed 9 metres without an air break, or contain no more than three 90° bends.

13.3 HOTWATER STORAGE

Hot water for domestic purposes is normally stored at a temperature of not less than 60°C and not greater than 65°C.

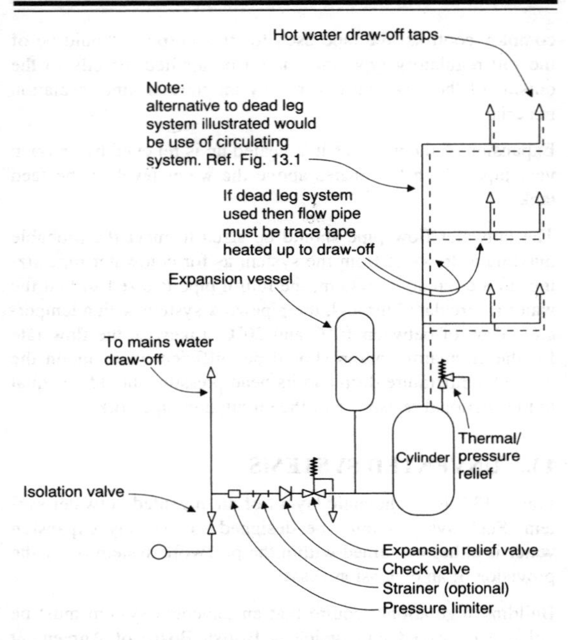

Figure 13.2 Unvented hotwater system

Table 13.1 Hotwater storage capacities

Building	Litres/person	KW/person	Recovery (hours)
Hostels	27	0.70	2.5
Hotels	36	1.00	2.0
Houses and flats:			
Single bathroom	33	0.75	2.5
Multi-bathroom	45	1.00	2.5
Offices	4.5	0.10	3.0
Restaurants*	6.0	0.20	2.0
Schools/colleges:			
Day	4.5	0.10	3.0
Boarding	25	0.70	2.5

*1 per meal

Adequate hotwater storage capacities must take into account rates of consumption and heat-up periods. Table 13.1 gives water consumption, energy consumption and heat-up periods for various types of building; usage of this data will provide a hotwater supply of reliable capacity for the majority of installation.

For buildings with large populations and variable periods of peak demand the figures given in Table 13.1 may give rise to overcapacity. This may be checked by plotting peak periods of use over the operating hours of the building and make any necessary adjustment to suit.

13.4 PRECAUTIONS AGAINST LEGIONNELLA

To prevent the growth of the bacteria which cause Legionella, hotwater systems should incorporate the following arrangements with regards to storage and pipework arrangements:-

(a) Storage temperature
Hotwater must be stored at 60–65°C and this temperature range must be uniform throughout the storage vessel. Ideally, the vessel should be capable of being isolated and heated to 70°C as a regular maintenance procedure.

(b) Storage vessels
Cylinders and calorifiers with concave bottoms should be avoided as these can provide stagnant zones with relative low water temperatures.
Pipework connections should be arranged to avoid temperature stratification within the storage vessel (see Figure 13.1).

(c) Pipework
Wherever possible, long runs of pipework with very little draw-off should be avoided so that 'dead legs'are at a minimum length. To achieve this, distribution pipework should be a circulatory system to within 300 mm of draw-offs or alternatively the distribution pipe should be provided with electric trace heating tape to maintain the draw-off temperature at not less than 50°C. Insulation of all distribution pipework is essential.

(d) Draw-off points
Infection is due to inhalation of bacteria in water droplets. The obvious sources are therefore shower and spray taps. Showers should incorporate a self-draining device and mixer taps must not have spray outlets and for added security should be dual flow type.

14 Building drainage

14.1 DRAINAGE SYSTEMS

The preferred practice in the designing of building drainage is to provide separate systems of pipework to carry the foul water, the surface water and car park or garage drainage. These separate systems may all combine at a single sewer outfall providing that proper disconnection traps or interceptors are installed.

Foul water drains collect the discharge from the buildings sanitary fittings, i.e. water closets, baths, basins and sinks, etc., and convey it away from the building for disposal. The method of disposal is normally by connection to a public sewer but there are still some rural areas where sewers are not available and on site sewage treatment or storage is required. The design of sewage treatment plant is a subject in its own right and where the need to have this form of foul drainage disposal arises, specialist advice should be sought from the responsible water authority.

Foul drain runs and principal branch drains must be ventilated to prevent the negative and positive pressures caused by discharges from blowing trap seals and allowing smells and gases to then escape into the building and its environs. Ventilation is needed at the head of the main drain run, at the point where the drain leaves the building and on any branch drain over 7 metres long. Most ventilation requirements are achieved by the discharge stacks but occasionally additional vent stacks may have to be provided where the arrangement of discharge stacks does not give ventilation to all of the points where ventilation is needed.

Surface water drainage collects the rainwater run-off from roofs and paved areas and takes it away from the building for disposal to a sewer, a soakaway or a water course. The actual disposal method is dependent upon the policy of the responsible water authority.

Where surface water drains combine with foul drains at a common sewer connection the surface water drain must pass through a disconnection trap immediately before combining with the foul drain.

Car park drains are a system of drain runs and trapped gullies which discharge through a petrol interceptor before connection to a foul sewer or water course.

14.2 DESIGN CONSIDERATIONS

14.2.1 Pipework

Ideally drain pipe runs should be external to buildings but obviously this is not possible in large buildings or buildings in city centres. Where it is not possible to locate drains externally, they should be routed through areas of the building where easy access can be obtained without a lot of disruption to the day-to-day activities. e.g. corridors, plant areas and car parks.

Pipe runs must always be laid in straight lines and to even gradients between points of access. Where changes of gradient or pipe size are needed this should only be made at manholes.

All connections between branch drains and main drain must be in the direction of flow and preferably at manholes, especially on foul drain systems.

Inlets to foul drains are always to be trapped except where connection is made to a foul discharge pipe or drain ventilating pipe.

14.2.2 Access

The usual method of providing access to drains is the construction of manholes but rodding eyes and gullies with access plugs are also acceptable in appropriate situations, such as surface water drains and car park drains.

The size of a manhole is dependent upon the number of branches that are to be connected to it, normally no more than five on each side, and the depth from ground level to invert. Figure 14.1 gives suitable sizes for various depths and construction of manholes.

Figures 14.2 to 14.4 illustrate the different types of manhole and construction forms. Pre-cast concrete manholes are normally the most economical to construct and are therefore the first choice for drains external to buildings. Brick manholes are useful for shallow manholes and manholes that have to be constructed within buildings.

14.2.3 Materials

Cast iron is the most frequently used material for drains which run under buildings. It is a very strong material, not prone to physical damage, and is available as proprietary systems with quickly formed mechanical joints.

Clayware drains may also be used under buildings providing that the correct strength of pipe is used, the correct joint is used

Number of branches and arrangement	Metres deep to invert	Sealed chamber manholes						Open channel manholes					
		Main drain diameter						Main drain diameter					
		100m		150m		225m		100m		150m		225m	
		Long	Wide	Long	Wide	Long	Wide	Long	Wide	Long	Wide	Long	Wide
No branches	1	900	450	900	450	1025	600	600	450	600	450	600	600
	1 - 2	1025	675	1025	675	1025	800	1025	675	1025	675	1025	800
	2 - 5	1240	1025	1240	1025	1240	1025	1240	1025	1240	1025	1240	1025
	1	900	800	900	800	1025	800	800	800	800	800	800	800
	1 - 2	1025	800	1025	850	1025	800	1025	800	1025	800	1025	800
	2 - 5	1240	1025	1240	1025	1240	1025	1240	1025	1240	1025	1240	1025
	1	1025	800	1025	800	1240	800	800	800	800	800	800	800
	1 - 2	1025	800	1025	800	1240	800	1025	800	1025	800	1025	800
	2 - 5	1240	1025	1240	1025	1240	1025	1240	1025	1240	1025	1240	1025
	1	1240	800	1240	800	1400	800	1025	800	1025	800	1240	800
	1 - 2	1240	800	1240	800	1400	800	1025	800	1025	800	1240	800
	2 - 5	1240	1025	1240	1025	1400	1025	1240	1025	1240	1025	1240	1025
	1	1540	800	1540	800	1750	800	1350	800	1350	800	1540	800
	1 - 2	1540	800	1540	800	1750	800	1350	800	1350	800	1540	800
	2 - 5	1540	1025	1540	1025	1750	1025	1350	1025	1350	1025	1540	1025

Figure 14.1 Manhole sizes

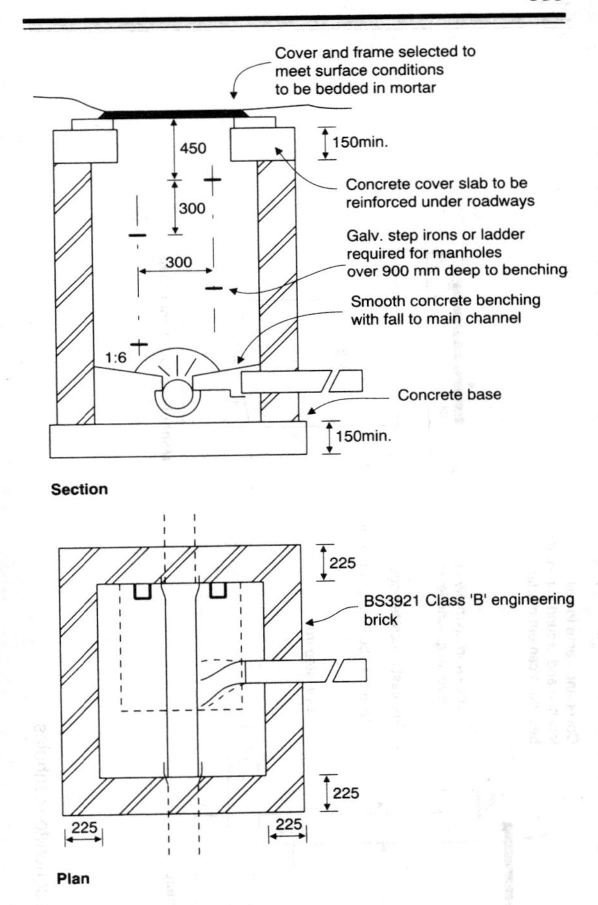

Figure 14.2 Typical brick manhole 1–2 m deep

and care is used in the selection of bed and surround materials in back filling.

Plastic pipes and fittings are available for both foul and surface water domestic drain installations. Care must be exercised in the laying of plastic drains to avoid damage from the backfill material and to prevent loads causing the cross-sectional area of the drains to be deformed.

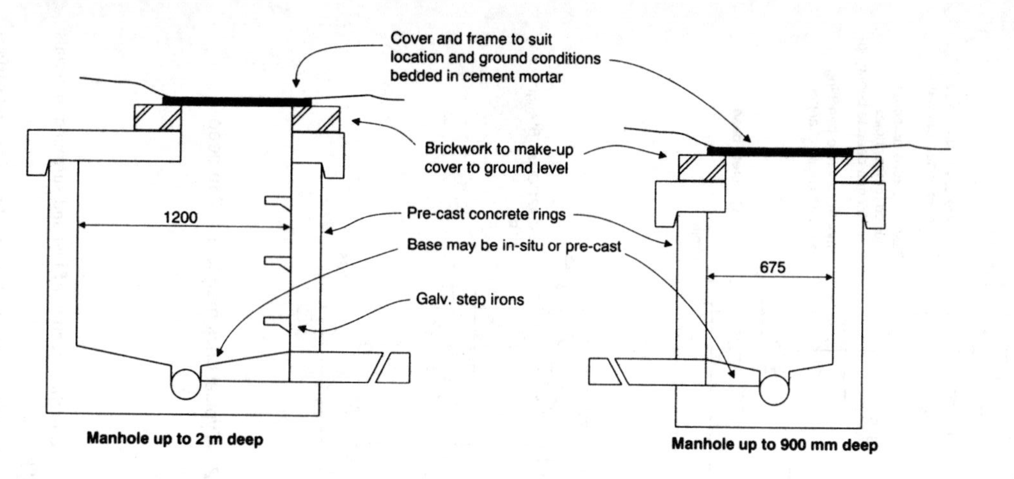

Figure 14.3 Pre-cast concrete manholes

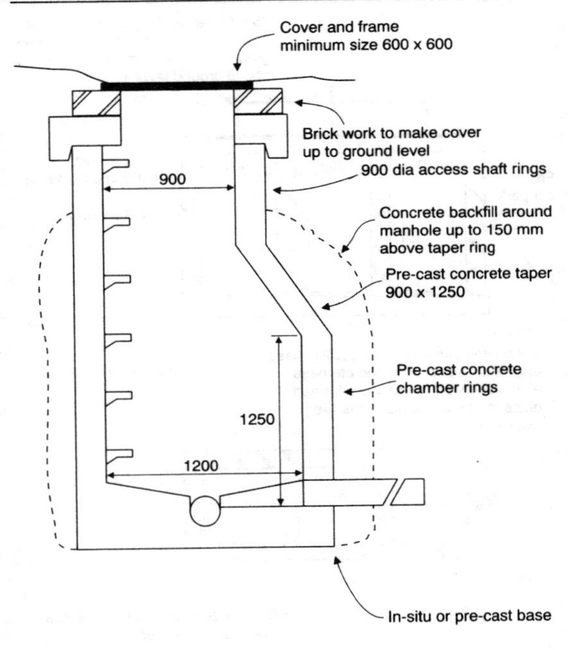

Figure 14.4 Pre-cast concrete manhole 2–5 m deep

14.2.4 Grease traps

Grease traps are designed to be installed where kitchen effluent is discharged into the drain system, the intention being to prevent grease coating the inside walls of the drain pipes and causing blockages. Grease traps of no matter what type become neglected and insanitary and their use should be avoided for all but large commercial kitchens where the drain system involves pumping or discharge to a private sewage disposal plant. Grease will not allow sewage pumps to operate properly and will have a deleterious effect on the bacteriological action of the sewage plant.

14.2.5 Petrol interceptors

It is illegal to discharge into a public sewer petrol or oil which gives off an explosive or flammable vapour. To prevent this happening it is necessary to install a petrol interceptor to collect the run-off from car parks or garage drains before it leaves the premises. A traditional interceptor is illustrated in Figure 14.5

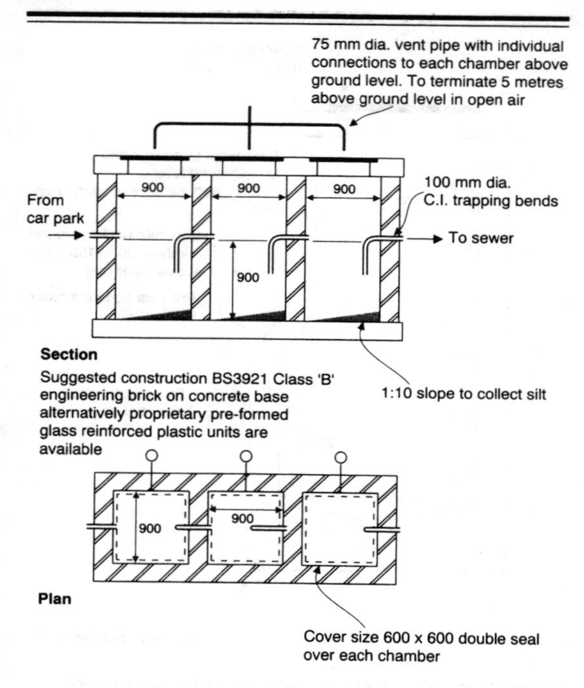

Figure 14.5 Typical petrol interceptor

but there are various pre-formed units available which are constructed from glass reinforced plastic materials.

14.3 PIPE SIZING

14.3.1 Foul drains

The minimum permitted size for foul underground drain pipes is 100 mm diameter. To achieve a relatively self-cleansing drain, the carrying capacity of foul drains should be restricted to two-thirds full flow with a velocity of 0.75 m/s.

Flow rates can be estimated from the discharge units used for stack loading (see Section 15.4) converted to litres per second.

14.3.2 Surface water drains

Surface water drains need not be restricted in the same way as foul drains and can therefore be sized to run full at velocities of between 0.75 m/s and 3.0 m/s.

Rates of flow are derived from the area to be drained multiplied by the rainfall intensity:

(a) For external paved areas where standing water is acceptable during and immediately after heavy rain then a rainfall intensity of 50 mm/h (0.014 l/s m^2) may be used.
(b) For paved areas where quick run-off is desirable and also roof areas where there is no danger of overflow into the building, a rainfall intensity of 75mm/h (0.021 l/s m^2) may be used.
(c) Areas of roofs or paving where there must be no risk of water ingress into the building then a rainfall intensity of 150 mm/hr should be selected (0.042 l/s m^2).

14.3.2 Sizing method

To a scale of at least 1:200, prepare a scheme drawing to show all drain rain at branch drains, all manholes and all discharge stacks and other drain inlets.

Starting at the manhole nearest to the outfall, give each one a reference number and indicate its finished cover level.

Mark the flow rates, in litres per second from each drain inlet. Total these flow rates for branch drains and main drains downstream to the outfall.

Using the criteria given earlier appropriate pipe sizes and gradients can be read from gravity drainage flow charts which are available from various sources.

14.4 SEWAGE PUMPING

The pumping of sewage effluent will utilize a free standing packaged pump set, a dry well arrangement or a wet well arrangement.

14.4.1 Packaged pumpsets

Packaged pumpsets incorporate a suction tank or cylinder with one or more sewage pumps mounted on the outside. The unit usually comes complete with integral controls and all interconnecting pipework. This is the most versatile method as it does not require the construction of any sumps or pump chambers and can therefore be installed within mechanical plant rooms. It is especially suitable for new installation into existing buildings.

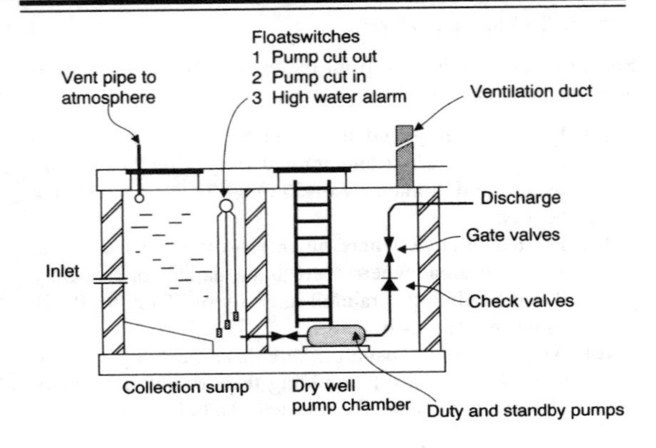

Figure 14.6 Dry well pump arrangement

14.4.2 Dry well systems

The dry well arrangement consists of two specially constructed chambers, one to form the suction sump and to receive the effluent and the other to house the pumps. The advantage of this arrangement is that the pumping plant can be easily isolated for maintenance or repair without having to get access to the wet sump or suspend usage of the drain system. It is by far the most costly sewage pumping arrangement and is only practical for new buildings (Figure 14.6).

14.4.3 Wet well systems

A wet well arrangement is a single sump containing 'submersible' sewage pumps. The effluent is discharged into this sump and is pumped directly by the sewage pumps. This is the most commonly used arrangement due to economy of construction over the dry well arrangement (Figure 14.7).

14.4.4 Storage and pumping capacities

The storage capacity of the wet well or suction sump is measured between the cut-in and cut-out levels of the pump control floats. The size must be a balance between the effective use of the pumps and the need to avoid septicity from the effluent being stored for too long a period. A good practical solution is to provide a capacity equal to four minutes peak effluent inflow and to size the pumps to run for two to three minutes.

Duty and standby pumps should be provided, with each pump capable of handing the full design. Two pumps is the normal

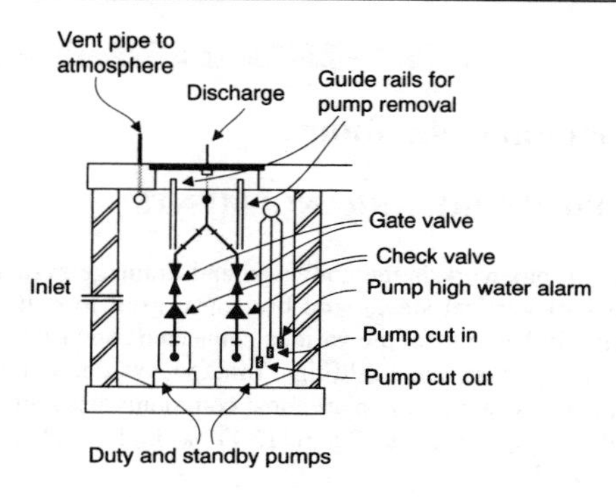

Figure 14.7 Wet well pump arrangement

requirement where foul drainage effluent only is to be pumped but for surface water drainage on combined foul and surface water drainage, three pumps should be considered. These would be arranged as duty and support with the third pump on automatic standby.

Discharge pipework and joints must be securely fixed and anchored to resist the hydraulic thrust pressure when the pump cut-in. Each pump must have an isolating valve and a non-return valve on its outlet side.

15 Sanitary plumbing

15.1 PIPEWORK ARRANGEMENTS

Sanitation pipework is the above-ground drainage system that consists of vertical stacks and branches, used to collect the effluent discharge from the various connected sanitary fitments and convey it to the building drainage system below the ground. The systems which are most commonly used are either vented one pipe systems (Figure 15.1) or single stack systems (Figure 15.2).

Vented systems have common discharge stacks into which all branch discharge pipes are connected. Additionally they have an interconnecting arrangement of vent pipes to stabilize any pressure fluctuations within the discharge stacks and to prevent siphonage from sanitary fitment traps. A vented system is the most flexible arrangement and can be adapted to most building types and sanitary fitment configurations.

Single stack systems consist of discharge stacks and branches without any ventilation pipes. Pressure fluctuations and the risk of siphonage are avoided by restrictions on the size, length and method of connection of the branch discharge pipes. These restrictions are illustrated in Figure 15.2 and must be strictly adhered to if trap seals are to be maintained. Single stack systems are the least flexible and are mainly adopted where there

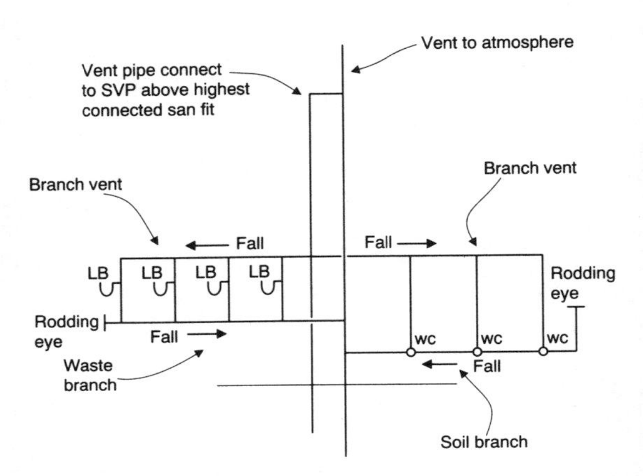

Figure 15.1 Typical fully vented system

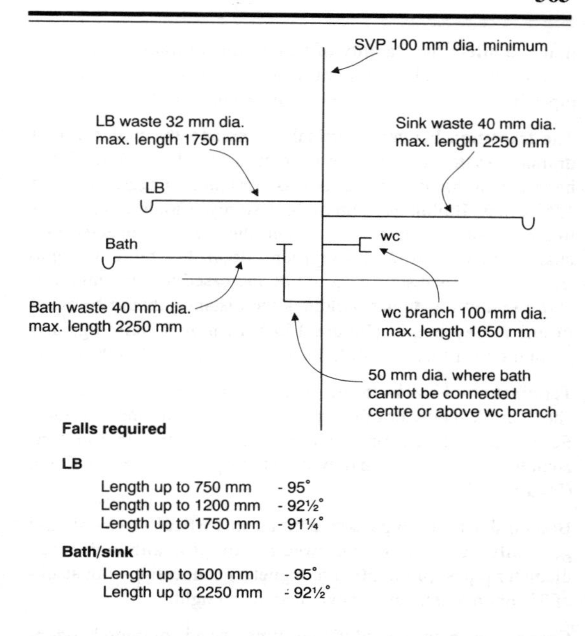

Figure 15.2 Single stack system

are small groups of similar bathroom layouts such as may be found in hotel buildings and housing developments.

15.2 DESIGN CONSIDERATIONS

15.2.1 Performance

Sanitation pipework should be designed to be as simple an arrangement as possible to quickly and quietly carry away the discharge from all connected sanitary fitments and appliances. The completed installation must be capable of retaining traps seals under all discharge conditions so that foul air is prevented from entering the building.

15.2.2 Discharge pipes

Discharge stacks should be, as far as is possible, without off-sets between the topmost branch connections and the foot of the stack where it connects to the below ground drain. Offsets in this 'wet' part of the stack may lead to compression or suction and affect the stability of water seals in traps. Where offsets are unavoidable they must be provided with relief vent pipes immediately above the offsets (Figure 15.3).

The transition between vertical stacks and the underground drainage systems should be by means of a long radius 92½° bend with a throat radius of at least 150 mm, or preferably two 135° bends. Buildings of up to three storeys should have a minimum distance of 450 mm between the lowest connected discharge branch and the underground drain. For buildings upto five storeys this distance should be increased to a minimum of 750 mm and for higher buildings the distance should be equal to one storey height (Figure 15.4). In higher buildings any ground floor fitments should be directly connected to the drain.

Termination of stacks in the open air must be at least 900 mm above any window or other opening into the building envelope. Furthermore, the termination must not be closer than 3 m horizontally to any such window or other opening into a building (Figure 15.5).

Branch discharge pipe connections to vertical stacks should generally be swept in the direction of flow although small diameter pipes, (up to 50 mm diameter) connecting with stacks of 75 mm diameter and above, may be straight.

Connections between individual fitments and common horizontal branch discharge pipes should always be swept in the direc-

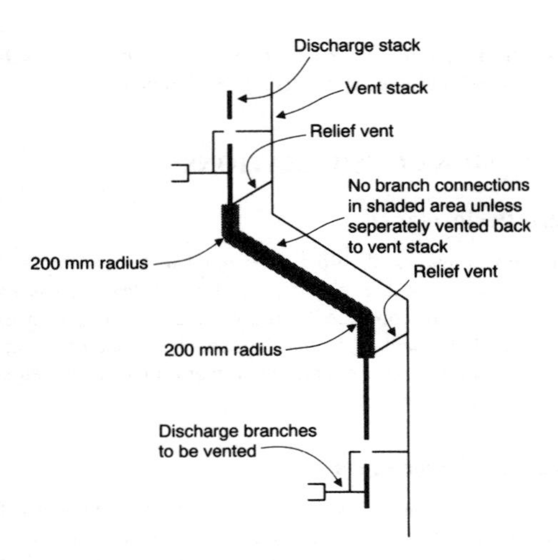

Figure 15.3 Relief vents to offsets in discharge stacks

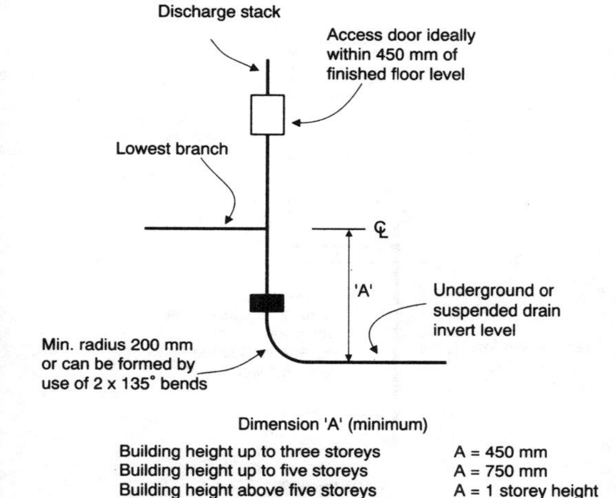

Figure 15.4 Discharge stacks termination at foot

tion of flow with a minimum throat radius of 50 mm. Alternatively they should enter at an angle of 135°. Opposed branches, especially in the horizontal plane should be arranged to avoid any possibility of cross flow occurring between one fitment and the other. This is particularly important where small diameter pipes oppose larger diameter pipes.

The gradients at which branch discharge pipes are installed is a factor in the avoidance of trap seal loss through self-siphonage. The fitments most at risk are those that have fast discharge rates, such as wash basins fitted with plugs and chains. Gradients should be uniform to within 250 mm of the stack connection. Where vent pipes are provided, the minimum gradient should not be less than 1¼°. Where no separate vent pipes are to be provided, e.g. single stack systems then the gradient should be relative to the length of branch discharge pipes as follows and illustrated in Figure 15.2.

Wash basin branch lengths
  Up to 750 mm long = 95° max
  Up to 1200 mm long = 92½° max
  Up to 1750 mm long = 91¼° max

Bath or sink branch lengths
  Up to 500 mm long = 95° max
  Over 500 mm long = 92½° max

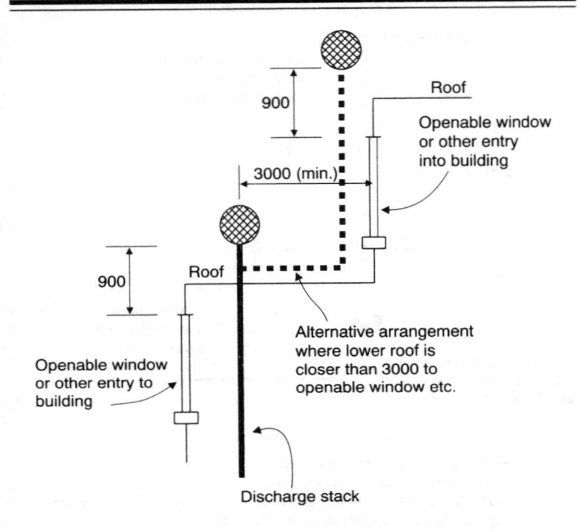

Figure 15.5 Discharge stack termination at roof

15.2.3 Ventilation pipes

Ventilation stacks should start with a connection at or below the lowest discharge branch into the discharge stack (Figure 15.6). The termination of ventilation stacks should be with a connection back into the discharge above the topmost discharge branch or, alternatively, taken separately through the roof to atmosphere.

Branch ventilation pipes should be installed at, or preferably above, the flood-over level of the sanitary fitments that they are serving and should have a continuous back fall into the discharge branch that they are connected to.

Ventilation pipes to individual fitments should be as close as possible to the crown of the trap that is attached to the fitment, never more than 300 mm away, and always connected to the top of the discharge branch pipe.

15.2.4 Access

Access to vertical stacks is needed at the foot and at every floor level where branches discharge into the stack.

Ventilation stacks should ideally have access at the top and at the bottom of the stack and at three-floor intervals in between.

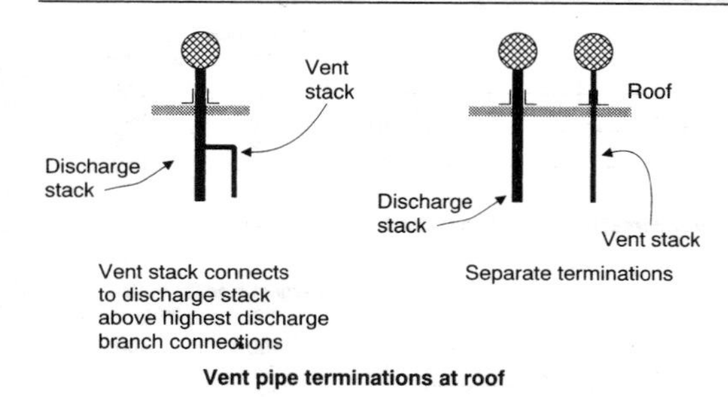

Vent pipe terminations at roof

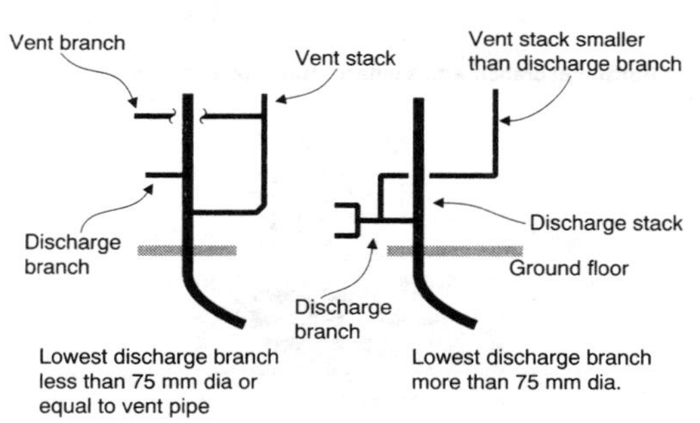

Vent pipe terminations at foot

Figure 15.6 Ventilation stack connections

The actual positioning of access doors and rodding eyes should be such that they are not obstructed by the structure of the building or by other services. Branches serving ranges of sanitary fitments should have rodding access from the end furthest away from the vertical stack. Wherever possible high level branch pipework which takes the discharge of sanitary fitments from the floor above should be accessible so as to be roddable from the floor on which the sanitary fitments are located.

Access to horizontally suspended pipe runs should be on top of the pipe and positioned to allow a minimum clearance of 225 mm between the pipe and any overhead obstruction.

15.2.5 Pipe spaces

Pipe spaces need to be sufficient to accommodate the pipework and also to provide room for access into pipework. Typical pipework situations are illustrated in Figure 15.7 and give dimensions from which suitably sized spaces can be planned.

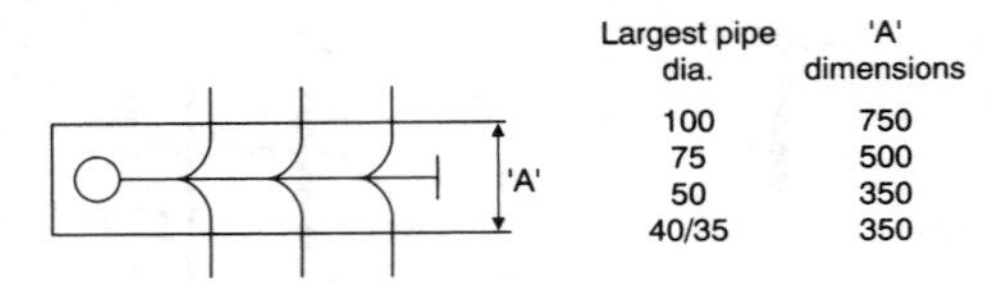

Largest pipe dia.	'A' dimensions
100	750
75	500
50	350
40/35	350

Horizontal branch with opposed sanitary fittings

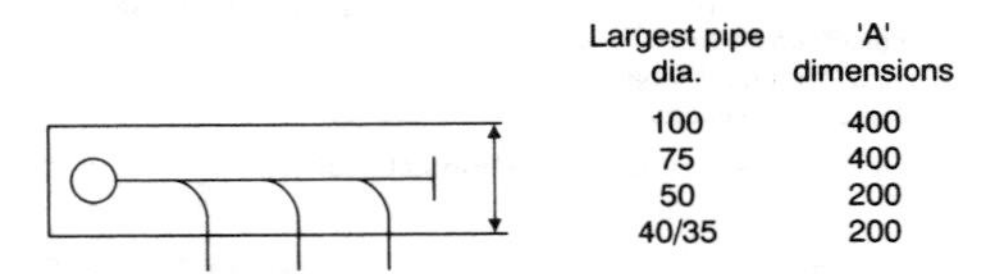

Largest pipe dia.	'A' dimensions
100	400
75	400
50	200
40/35	200

Horizontal branch with sanitary fittings on one side

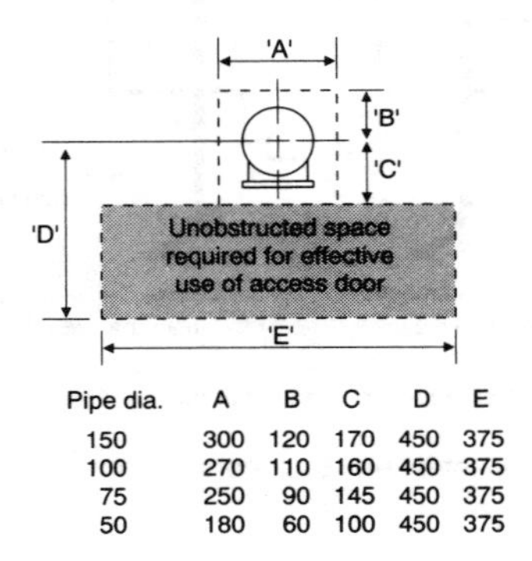

Pipe dia.	A	B	C	D	E
150	300	120	170	450	375
100	270	110	160	450	375
75	250	90	145	450	375
50	180	60	100	450	375

Vertical pipe space

Figure 15.7 Typical pipe spaces requirements

15.3 TESTING PROCEDURES

15.3.1 Air testing

The installation should be tested in sections for air tightness as the work proceeds and before pipes are encased in ducts and ceilings. Upon completion of the installation the work must be tested to the requirements of the local building control authority. Normally an air test pressure of 335 Pa, held constant for three minutes, with all water seals charged and openings plugged, will satisfy these requirements.

It is good practice to carry out a full test and do any remedial work before inviting the authorities to witness formally the final test. Prior to testing, all pipe joints, connections to fitments and access points should be inspected for soundness and where necessary re-fitted.

15.3.2 Performance testing

In order to prove performance during peak working conditions discharge test should be applied to the completed installation. Branch discharge pipes serving wash basins or sinks should be tested by filling the fitment to overflow level and abruptly discharging its contents. Ranges of fitments should be discharged in rapid succession starting at the fitment furthest from the stack. For discharge stacks, a selection of the fitments connected to each stack should be discharged simultaneously, as nearby as possible.

Table 15.1 gives the recommended selection of fitments according to the total number served by the stack. At least one W.C. should be discharged from the top floor and the remaining fitments should be discharged on adjacent floors.

Table 15.1 Simultaneous discharges

Usage category (see Table 15.2)	Number of fittings on stack	Number of fittings to be discharged	
		W.C.	Sink or basin
Domestic	1 – 9	1	2
	10 – 24	1	3
	25 – 35	1	5
	36 – 50	2	5
Public	1 – 9	1	1
	10 – 18	1	2
	19 – 26	2	2
	27 – 50	2	3
	51 – 78	3	4
	79 – 100	3	5
Peak	1 – 9	2	2
	10 – 18	2	4
	19 – 26	4	4
	27 – 50	4	6
	51 – 78	6	6
	79 – 100	6	10

15.4 SIZING OF SANITATION PIPEWORK

15.4.1 Minimum pipe sizes

A discharge stack or branch to which at least one W.C. is connected must have an internal diameter of at least 100 mm.

Outlets from wash basins must be provided with 32 mm minimum diameter branch discharge pipes and sinks and baths must have branch discharge pipes of 42 mm diameter.

15.4.2 Intervals of usage

The usage of sanitary fitments varies according to the type of building in which it is installed. The loading this usage imposes on the sanitation pipework system dictates the intensity of flow and hence the pipe sizes to be installed.

Table 15.2 gives appropriate intensity categories for various types of buildings and Table 15.3 gives loadings in discharge units for these three intensities of usage.

15.4.3 Sizing procedure

For large or complex installations a diagrammatic layout should be prepared to show all vertical stacks and branch pipes, indicating all connected fitments. Select discharge unit ratings for each appliance, mark these on the diagram and total for each branch and stack. The sizes of branches should be read from Table 15.4a and the sizes of stacks from Table 15.4b.

Main vent pipe sizes can be read from Table 15.4c.

EXAMPLE

A 10-storey office building has four WC pans, four wash basins and a sink on each floor. The storey height is 3 m.

(a) From Table 15.2 usage is public.
(b) From Table 15.3 each WC rates 14 discharge units, each wash basin rates 2 discharge units and each sink 14 discharge units.
(c) Each waste branch carries 22 discharge units (4 × 2 + 14) which from Table 15.4 indicates that a 50 mm diameter pipe will be required.
Each No. 1 branch carries 56 discharge units (4 × 14) for which a 100 mm diameter pipe size is required.
(d) Stack loading will be 780 discharge units (10 × 78 discharge units) and will require a pipe size of 100 mm diameter.

Table 15.2 Intensity categories

Domestic	Public	Peak
Houses	Offices	Cinemas
Flat	Factories	Theatres
	Hotels	Stadiums
	Schools	Sports Centre
	Hospitals	

Table 15.3 Discharge unit values

Appliance	Usage	Discharge units
Wash basin	Domestic	1
	Public	2
	Peak	4
Wash basin with spray tap, without plug and chain		0.25
Bath	Domestic	6
	Public	18
Shower (per head)	Domestic	1
	Public	2
	Peak	4
Urinal (per stall or bowl)		2
WC (9 litre)	Domestic	7
	Public	14
	Peak	28
Sink	Domestic	6
	Public	14
Bathroom Group: 9 litre WC, bath, sink, 1 or 2 wash basins		14
Domestic washing machine; freestanding waste disposer; or fitting with 40 mm outlet not listed above		7
Other fittings not listed, with trap or outlet size 50 mm dia		7
65		7
75		10
88		10
100		14

*If a shower is *en suite* with a bath, its discharge can be ignored in stack loading.

Table 15.4a Stack capacities

Stack diameter (mm)	Maximum loading (discharge units)
50	20
65	80
75	200
88	400
100	850
125	2700
150	6400

Note that (1) stacks serving one or more wc pans shall be not less than 100 mm
(2) separate stacks with sinks discharging into them shall, if over five storeys high (a) be not less than 100 mm
(b) be connected direct to drain. This avoids nuisance from detergent foam in domestic situations.

The maximum stack capacities shown correspond to discharge for vertical pipe flowing quarter full; this limited capacity will reduce the risk of discharges forming solid plugs which cause loss of trap seals.

Table 15.4b Branch capacities

Branch diameter (mm)	Maximum capacity of branch (discharge units)		
	Fall on branch		
	1:100	1:150	1:25
32	–	1 wash basin only	
40	–	2	8
50	–	10	26
65	–	35	95
75	40	100	230
88[1]	120	230	460
100[2]	230	430	1050
125	780	1500	3000
150	2000	3500	7500

[1]Minimum for 1 wc
[2] Minimum for 2 or more wcs

Table 15.4c Vent pipes

Discharge stack or branch diameter (mm)	Vent stack or banch diameter (mm)
32	32
40	32
50	40
75	50
100	50
150	75

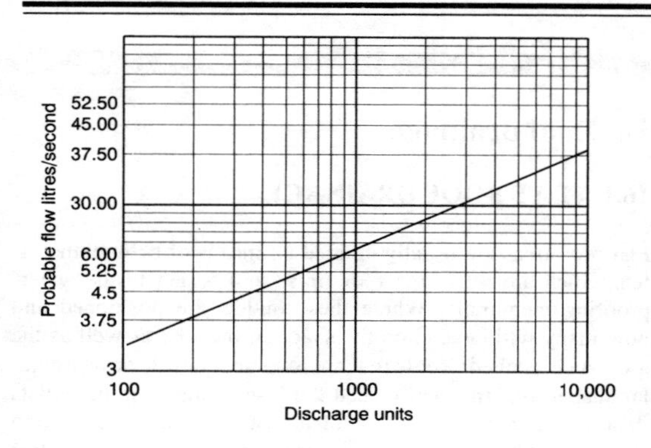

Figure 15.8 Conversion of discharge units to litres/second

16 Roof drainage

16.1 FLAT ROOF DRAINAGE

Flat roof areas are usually drained by means of bellmouth outlets which are set in the roof finish and sealed to the waterproofing membrane. Where these outlets are positioned and how many will depend on the shape of the roof as well as the area to be drained. Roofs that are long and narrow or of irregular shapes will frequently need the installation of more outlets than a relatively square roof of the same area. Consideration should also be given to any obstructions which might isolate areas of roof from outlets.

The choice of rainwater outlet layout should therefore be that which takes into account the foregoing and which then affords the most economical arrangement for both rainwater down pipes and ultimately the underground drainage system.

16.2 ROOF OUTLET CAPACITIES

Drainage capacities of outlets will vary and reference can be made to individual manufacturer's data. However, the following figures give estimated capacities for circular bellmouths with flat gratings and spigot outlets of 75 mm and 100 mm diameters. They are based on a rainfall intensity of 75 mm per hour and a depth of water above the outlet of upto 25 mm.

(a) Vertical spigot — 100m^2 (2.1 litres per second)
(b) 45° spigot outlets — 95m^2 (2.0 litres per second)
(c) 90° spigot outlets — 80m^2 (1.7 litres per second)

16.3 INTERNAL RAINWATER PIPES

Rainwater pipes that are to be installed inside buildings must have the same quality of materials as those required for sanitary pipework. This includes joints, means of access and final testing. Consideration should also be given to insulating and vapour sealing any horizontal pipework that runs through ceiling voids where condensation is likely to form on the outside of the pipes during periods of cold weather.

For the selection of vertical rainwater the capacities detailed in Table 16.1 can be used.

Table 16.1 Capacities of rainwater pipes

Diameter (mm)	Capacity (l/s)
75	1.8
100	3.4
150	7.2
225	18.0

Horizontal rainwater pipes which collect discharge from a number of roof outlets should be sized in the same manner as building drainage systems.

16.4 GUTTERS

Pitched roofs are drained to eaves gutters or valley gutters depending on the type of roof that is to be drained. Both types of gutter can be installed without falls providing that no ponding or backfalls are allowed to occur.

Flow capacities of half round or nominal half round eaves gutters:

(a) Straight runs of gutter with outlets one end.

Gutter size (mm)	100	150	200
Flow capacity (l/s)	0.8	2.5	6.0

(b) Straight runs of gutter with central outlets.

Gutter size (mm)	100	150	200
Flow capacity (l/s)	1.75	5.5	12.0

16.5 RAINFALL INTENSITIES

For the purpose of calculating rainwater drainage flow rates from roof areas a design rainfall intensity of 75 mm per hour is normally adequate. This rate is rarely exceeded and when it is it is for very short periods. Where any chance of ponding or overflowing would not be acceptable then a rainfall intensity of 150 mm per hour should be used for design calculating.

The flow rate for a rainfall intensity of 75mm per hour is 0.021 l/s m^2 and for 150 mm per hour 0.043 l/s m^2 of roof area.

16.6 DESIGN PROCEDURE

1. The first requirement is to establish the total area to be drained. Flat roofs are simply the plan area plus any sloping surfaces which drain onto it. Sloping roof areas

must take into account the pitch of the roof which will increase the actual plan area.

2. Calculate the rainwater run-off using the roof area and the selected rainfall intensity.
3. Decide the number of flat roof outlets required and layout to provide the most economical pipework collection system as discussed in Section 16.1. Alternatively, for pitched roofs, select the appropriate gutter sizes.
4. Position downpipes to achieve an economic layout for the below-ground surface water drain system.
5. The sizes of downpipes needed can be established from the total rainwater run-off from the area drained and the number of downpipes which serve the area.

INDEX